FIRMAMENT

BY ANGIRAS
PROPHET OF THE PAST

Originally published under the name *Dance of the Planets 1 - The Earth*

ISBN 978-0-7414-0070-3

Published by:

INFINITY PUBLISHING
1094 New DeHaven Street, Suite 100
West Conshohocken, PA 19428-2713
Toll-free (877) BUY BOOK
Local Phone (610) 941-9999
Fax (610) 941-9959
Info@buybooksontheweb.com
www.buybooksontheweb.com

PREFACE

In 1950 Immanuel Velikovsky, an unknown scholar of ancient myth, published a book titled *Worlds in Collision.* In it, the author advanced the amazing idea that the planet Venus was created within proto-historical times and made several very close passes to the Earth, devastating our world. He claimed that Venus subsequently ejected Mars from its inner orbit, causing it also to interact closely with the Earth, resulting in additional trauma for the human race.

The clues which led to his conclusions came from ancient myths, the scriptures, and the astronomical records of ancient cultures. Indeed, he concluded the primary purpose for the sacred myths of all cultures had been to record these cosmic events which were experienced by everyone living on the Earth in those times. Although cultural differences have caused the planets to be cast as different 'gods,' he was able to detect the common thread of events to which all these myths referred, that is, close interactions of the two planets, Venus and Mars, with the Earth.

Given the esoteric nature of the subject, it is surprising that the book immediately became a best seller. In fact, within the next fourteen years, twenty-nine reprintings of the hardback version were made in the U. S. and Great Britain. It is still possible to buy a paperback version of this book in the local book stores, forty-nine years after its first publication. It was particularly popular with college students of the time, because of the controversy that pitted it against one of the most doctrinaire branches of science, astronomy.

Based on a number of myths, Velikovsky concluded that close passages of the flaming, newly-born proto-Venus actually 'overturned' the Earth twice. In the interval of time between them, probably only a few years or tens of years, the Sun rose in the west and set in the east. These overturnings resulted in catastrophic flooding and scorching of the Earth. Establishment scientists claimed that Velikovsky's scenario violated the mathematical principles of astronomy, based on the foundations established by Isaac Newton and Albert Einstein - an assertion which was absolutely untrue. The term 'catastrophism,' refurbished from the previous century, was adopted to describe the radical planetary interactions of Venus and Mars with the Earth proposed by Velikovsky. It is surprising that the strongest reaction came from astronomers, because they had paid very little attention to the planets before 1950. In fact, the generally accepted picture of Venus was that its climate was similar to that of the Earth. Not only had no planet been visited, but the idea of space travel was not even planned at that date, indeed planetary astronomy was an insignificant 'department' of astronomy. One would expect that the strongest reaction would have come from geologists and archaeologists, who had infinitely more data, claiming that there were no signs of these traumatic encounters on the Earth. Velikovsky partially silenced this quarter by publishing a second book, *Earth in Upheaval*, in which he pointed out a number of geologic examples, which were difficult to explain without catastrophism.

Worlds in Collision triggered an unprecedented wave of activity in planetary science, but for the wrong reason, that is, to refute Velikovsky's premise. Unfortunately, the resulting hypotheses adopted to this end have become center pieces of that branch of the discipline, much to its detriment. The fact that Velikovsky's challenges derive primarily from his interpretations of ancient myth made it particularly unpalatable to the scientific community. This blasphemy, from an uncredentialed outsider was met with an unbelievably vicious counterattack from the scientific community. It was led by one of the best known astronomers of the day, Harold Shapely, who personally recruited ambitious younger scientists, such as Carl Sagan, to speak out, for the sole purpose of discrediting Velikovsky's claims. For the most part, these attacks were authored by people who had not even read Velikovsky's work, and who grossly misstated his claims. While the controversy undoubtedly helped the sale of the book, the powerful attack could not be ignored by the publisher, Macmillan, because a large fraction of their business was in publishing textbooks written by these same establishment scientists. As a result of the pressure applied by the scientists, a senior vice president of Macmillan was fired, and the publication rights were sold to a competitor, Doubleday, which did not publish textbooks and therefore was not vulnerable to the blackmail. To illustrate the power of this attack, the book was transferred from Macmillan to Doubleday while it was on the list of the top ten non-fiction books in the USA.

The following quote from the student paper, "The Daily Princetonian," (February, 1964) written in defense of Dr. Velikovsky, refers to this lack of ethical behavior on the part of the astronomical community:

> While it could have been assumed that anyone challenging the basic premises of Newton and Darwin might be laying himself open to a certain amount of argument, the personal vituperation, deliberate misrepresentation of facts, off-hand misquotations, efforts at suppression of the books containing the theories, and the denial of the right to rebut opponents in professional journals that Dr. Velikovsky encountered indicate that far more was going on than 'mere' challenge to established ideas. What the Velikovsky affair made crystal clear...is that the theories of science may be held not only for the truth they embody, but because of the vested interest they represent for those who hold them.

The original purpose of this book was to carry forward the work begun by Velikovsky, using data from space probes. However, as a result of studying the *Rig Veda* and Hindu myths, along with data from Earth and space exploration, we have discovered an amazing sequence of planetary interactions that go far beyond Velikovsky's ideas. The resulting Velikovsky/Angiras theory, presented here, transcends Velikovsky's theory to a greater degree than his transcended the accepted theories in 1950, when *Worlds in Collision* was first published. The implications of this 'new catastrophism' for the entire solar system are revolutionary.

In *Firmament* we interpret a number of ancient myths, not only from India, but also from Egypt and Greece, piece them together into a coherent scenario which covers some three millennia, and verify this using data from the Earth sciences, archaeology, anthropology and ancient history. A companion book, *Chaos*, presents the consequences of this recent planetary chaos for the other bodies in the solar system, both large and small. The present notions of the planets are more profoundly impacted by our scenario than anyone can imagine.

The many new concepts presented in both volumes are collected under the 'umbrella' of the Velikovsky/Angiras (V/A) scenario. Although Immanuel Velikovsky is no longer among us, and although the new theory differs in many details from his original ideas presented in *Worlds in Collision*, Angiras acknowledges his great spark of genius, which reestablished the long-lost link between past and present knowledge, by including his name in the first place of honor.

The process of determining past cosmological events from the myths, also reveals the meaning of the sacred myth of many cultures. Indeed, over the centuries more scholars may have devoted their lives to understanding the true nature of Indra, Vishnu, Shiva and Brahmā and their counterparts in the sacred myths of other cultures than have concerned themselves with the planets. Thus the revelations in this book constitute a two-edged sword.

But the sword is not the most appropriate analogy for our work. Alexander the Great, when confronted with the puzzle of the knot tied by King Gordius, proved he would be the future ruler of Asia, not by untiing the knot, but by cutting the rope with his sword. This was the appropriate solution for a conqueror. The eastern myths are the Gordian knot, and the deciphering of their true meaning finally unties it - solving the puzzle of the Orient - not in the way of a conqueror but in the way of a seeker of truth. Some may be interested in the planets, some in the ancient myths, and some in oriental religions. The V/A theory satisfies all these thirsts. It opens up fruitful pathways for future progress in all of these disciplines.

Firmament

Table of Contents

List of Figures

PROLOGUE

Ah Love! Could you and I with Him conspire
To grasp this sorry Scheme of Things entire,
Would not we shatter it to bits - and then
Re-mould it nearer to the Heart's desire!
The Rubaiyat
Omar Khayyam

This book is about mythology and the planets. In most minds this conjures up images of shepherds, familiar with the night sky, spinning tales in which the distant planets are the prime players. It is difficult for us, today, to imagine how these virtually stationary dots of light could be the source of myths involving the destruction of the world and battles between the gods. This requires even more imagination than picturing a particular group of stars as a Great Bear. The planets are more than sixty million of kilometers away and the astronomers tell us that the solar system has been the same for more than four billion years. Surely the planets could have had no immediate influence on the race of man! But the myths and their mysterious message cannot be dismissed. They wait silently, almost mockingly - a monument to our ignorance of the past in the age of supercomputers.

The definition of myth given in Webster's Dictionary is as follows:

> myth/'mith'/n [Gk mythos] 1: a usually traditional story of ostensibly historical events that serves to unfold part of the world view of a people or explain a practice, belief, or natural phenomenon. 2: parable, allegory. 3a: a person or thing having only an imaginary or unverifiable existence. 3b: an ill-founded belief held uncritically especially by an interested group. 4: the whole body of myths.

This pretty much expresses the view of western man in the twentieth century, although we might be surprised by the phrase 'of ostensibly historical events'. Originally the Greek word *mythos* referred to real events involving 'super-heroes' whose exploits occurred in a proto-historic time. The best known sources of these myths in western cultures today are found in the works of Homer. His works, the *Iliad* and the *Odyssey*, are far from a complete rendering of ancient Greek myths - leaning instead toward those portraying heroic deeds and ignoring those with darker implications. He avoided evoking religious or mythological themes that were not of interest to his essentially patriarchal and military auditors. It was this Homeric conception of the Gods that defines mythology in the western world. Hesiod on the other hand was a chronicler of Greek myth, and because his expositions were less political and poetic they are less well known. Hesiod narrates and systematizes a much wider range of Greek mythology, citing extremely archaic myths rooted deep in proto-history, but these are isolated myths, and provide no confidence that Greek mythology gives a complete, sequential record of cosmic events, as does the *Rig Veda*.

Although Homer is thought to have lived sometime between 1000 and 800 B. C. the Greek myths that he recorded come from a much earlier time. The Greek myths relate ancient events, such as Phaethon driving the chariot of the Sun too close to the Earth and scorching it. But by the time of the Classical period, the Greeks had completely lost the thread of the myths' true meaning, that is, the planetary encounters, which we claim, were the basis of these myths. As a result the rationalists fomented the erosion by which the ancient gods and myths were emptied of their original meanings. This ignorance of the meaning and value of myth has become consolidated in the modern era and particularly in the twentieth century, due to the arrogant, know-it-all, attitude of modern science.

A similar process resulted in the deterioration of the 'pure' myths in India, where the original 'knowledge' of the *Rig Veda* was gradually anthropomorphisized through the classics (Mahābhārata, Brahamana and Ramayana) and the Puranas, and lost altogether in the Upanishads. The stories in the latter books often distort the relationships of the Vedic deities, which were passed down for thousands of years by means of sung hymns. Fortunately, all of these books are still available to us today, including the most ancient *Rig Veda*, which although less popular, is still held in great reverence by the Hindus because of the purity of the 'knowledge' at contains.

There has been a surprising resurgence of interest in myth in recent decades within certain segments of western society. Unfortunately, much of this comes under the umbrella title 'New Age'. It is largely a reaction against the increasingly secular, technological, computerized nature of the world in which we are 'becoming.' Many have turned toward the orient because it seems to the westerner that these people have a stronger 'spiritual awareness.' We believe this trend is due in part to the less demanding nature of these religions.

There are serious questions as to the value of the present day teachings from the orient and their application in the western world. These teachings take the form of yoga, zen, the marshal arts, massage therapy, breathing therapy, crystal therapy, and dozens of other ways of supposedly attaining 'spiritual awareness.' The number of gurus with adopted Indian and Chinese 'stage names' almost outnumbers the 'seekers' in search of answers from them. Western men and women at their worst, want a quick shot of 'spiritual awareness' painlessly administered via audio CD while in rush hour traffic. At their best, they indulge in the well-intentioned practice of methods which are gross distortions of the ancient Indian forms of meditation. Westerners would be much better off studying the Bible, thereby getting in touch with their true spiritual roots. But the Christian God asks more than most of us are willing to give (or give up), and so many turn to the mushy thinking of the opportunistic 'guru,' whose teaching is empty, if not cynical.

The hackneyed eastern philosophy, that claims true 'spiritual awareness,' can only be gained by making ones mind a complete blank, is not only impossible for the modern western mind to comprehend, but it is also a gross distortion of meditation as advocated in the ancient texts. The Satapatha Brahmana (xi. 5, 6, 1) says the following about meditation:

> Study and teachings are loved. He (who practices them) becomes composed in mind. Independent of others he daily attains his objects, sleeps pleasantly, becomes his own best physician. Control of his senses, concentration of mind, increase of

> intelligence, renown, capacity to educate mankind [are the results of study] ... Of all the modes of exertion which are known between heaven and Earth, study of the Veda occupies the highest rank...The Veda is the eternal eye of the fathers, of gods, and of men; it is beyond human power and comprehension...As fire, when it has acquired force, burns up even green trees, so he who knows the Veda consumes the taint of his soul which has been contracted from works. He who comprehends the essential meaning of the Veda, in whatever order of life he may be, is prepared for absorption into Brahmā, even while abiding in this lower world. [1]

Thus the most ancient books, the Vedas, are to be studied diligently. This does not sound like making one's mind a blank.

In order to discover the knowledge of the cosmic encounters that is inherent in the ancient myths, they must be read with an open mind:

> ...the contents of these are not to be questioned by reason, since from them [a knowledge of] duty has shone forth. The Brahmin who, relying on rationalistic treatises, shall contemn these two primary sources of knowledge must be excommunicated by the virtuous as a sceptic and reviler of the Vedas.

As clarification, the following is offered:

> He and he only is acquainted with duty, who investigates the injunctions of the rishis...by reasonings which do not contradict the Veda. [2]

In other words, read the ancient hymns, not with skepticism or in an attempt to ridicule them, but *in the spirit of truly trying to understand the hidden knowledge they conceal.* This is the exact purpose of this book, and the hymns yield an amazing story of planetary interactions, which lasted almost three thousand years. Fortunately, there are some modern thinkers who seriously and studiously attempt to uncover the 'true' meanings of the most ancient myths, in hopes that they contain information on the origin of the world and the meaning of life. The most famous of these was Mircea Eliade. His definition of myth is much more in tune with injunctions of the Vedic priests (rishis):

> Myth narrates a sacred history; it relates an event that took place in primordial time, the fabled time of the 'beginnings'. In other words, myth tells how, through the deeds of Supernatural Beings, a reality came into existence, be it the whole of reality, the Cosmos, or only a fragment of reality...Myth tells only of that which *really* happened,

> which manifested itself completely...In short, myths describe the various and sometimes dramatic breakthroughs of the sacred (or the "supernatural") into the World. It is this sudden breakthrough of the sacred that really establishes the World and makes it what it is today. [3]

> myths are a constant reminder that grandiose events took place on Earth and that this 'glorious past' is partly recoverable. [4]

Eliade also points out that:

> in societies where myth is still alive the natives carefully distinguish myths - "true stories" from fables or tales, which they call "false stories." The Pawnee...include among the "true" stories in the first place all those which deal with the beginnings of the world; in these the actors are divine beings, supernatural, heavenly, or astral. [5]

Myths are thought to be purer in content in archaic societies in which they are repeated or reenacted on a regular basis, and in which they are passed from generation to generation by word of mouth. American Indians and Australian aborigines are examples of these peoples. In these societies men see themselves and their connections to the universe in terms of their myths, whereas modern men consider themselves as a result of history because they have lost the thread of meaning which links them back to their ancient past. The oldest, purest myths are the most logical source in which to search for our real place in the scheme of things, as opposed to those myths which have been modified by history and political expediency.

The far East, particularly India, represents an important exception to the idea that purest myth is found in purely archaic cultures. Although India certainly has been a part of modern history, there remains a strong spiritual aspect to the eastern psyche which is inaccessible to the western mind. Perhaps it is due to a collective subconcious 'memory' of the cosmic events on which the sacred myths are based. It is here that the most ancient myths have been preserved in pure form - passed from generation to generation for thousands of years - not 'written down' for many centuries after writing was introduced in India, because they were considered so sacred that only the Brahmin priests could have knowledge of them. Until relatively recently, a person of a lower caste could be put to death for simply laying his eyes on the Vedas. The *Rig Veda*, the primary repository of this knowledge, consists of over a thousand hymns, apparently praising various deities and expounding on their exploits. In fact, it is no more and no less than a complete, sequential record of compelling cosmic interactions between Venus, Mars and Earth, which occurred in the last five millennia. This knowledge is almost reportorial and was not formed into 'stories' until the classical period, at which time the planetary interactions had ceased.

There are three major themes that permeate the sacred myths of most cultures. These are (1) the myth of the great destructions, (2) the cosmogonic myth - stories of the origin of the world or of the people on it, and (3) the myth of the eternal return.

Prologue

The stories of great destructions are widespread. Most of them describe an *almost total annihilation* rather than an absolute one. Typically they describe destructions by earthquakes, fire, falling mountains, and tidal waves. All but a few people of the tribe survived, and most animals were also killed. The ancient Egyptians and Aztecs claimed that such great annihilations have occurred four times in the 'memory' of their people. The destructions are thought to be the result of the world 'growing old and wearing out.' Although most people are killed in these annihilations, a number of peoples believe that those killed will be reborn from their buried bones in the new world. The destruction is thought to be followed by a 'new world', which in many cultures is described as a 'new sky', implying a reorientation of the Earth relative to the stars.

The cosmogonic myths are those which narrate the beginning of the world. There are not a lot of myths which deal with this as a singular event in which the world or universe was created ab nihil (from nothingness). The creation of a 'new sky' following a great destruction, implies the same Earth, and therefore not a creation from the very beginning. Myths of the creation are sometimes referred to as being performed by a supernatural being in a dream time so ancient that men cannot remember. A surprising number of myths from archaic cultures claim that the Supreme Being began the work of creation of the world and man, and then abandoned his creations and withdrew into the sky, leaving the completion of a lesser deity, his son or representative. Others tell of a conflict between the original creating deity and a later-arriving one, in which the original creator is defeated and sent to oblivion, with the victor assuming the power of the formerly powerful deity. A specific example of such an abdication is given in the *Rig Veda*, where the powerful deity Varuna moves away from the Earth after the creation of a new sky, and is replaced by the less powerful deity, Indra, after a skirmish.

Myths of the creation of people, as opposed to the world, by a deity are much more common. These, however, usually assume an 'essential' framework that already exists; for example, man is created from the Earth mixed with the blood of a slain demon. The Hindu myth states that the deity Brahmā created the people of the various castes from the various parts of his body; the darkest of skin from his feet, up to the lightest of skin, the Brahmins, from his head. This was done by the sheer will of the deity. This same deity is said to have cast-off his body after each creation and assumed another for the purpose of each new creation.

The third predominant myth is that of the eternal return. This is related to the creation and destruction myth in the sense that the 'Earth' continues to exist. After the destruction the eternal return closes the cycle, leading to a new creation.

The Hindu myths describe these cyclic creations and destructions in terms of a day of Brahmā and a night of Brahmā. Each day begins with the gathering of the waters, the formation of the sacred egg within the waters, the birth of Brahmā from the sacred egg, and then a sequence of creations are performed. Over a period of time the created entities (gods, demons, fathers, and men) gradually deteriorate in vigor and finally, after an explosive catastrophe, Brahmā lies down on the coils of the mythical serpent Sesha and falls asleep, commencing the night of Brahmā. The lying down of Brahmā signals the destruction of all the created entities under the auspices of Siva (Shiva), the destroyer. The third god of the Hindu triad, Vishnu is the 'sustainer' who guarantees the eternal return leading to the next creation. The day and night of Brahmā are said to be of equal duration.

Firmament

The contents of the 'sacred' myths of the world, when interpreted properly, tell more about how religion developed among ancient peoples than it does about a 'true' god. In fact, the word religion comes from the Latin stems *re* and *ligio,* and therefore translates literally as 'linking back.' There is no direct reference to 'god.' Using this interpretation, this book should be classified under religion because our purpose is to rediscover, from the ancient myths, the cosmic events which occurred in proto-historical times. What we immediately find when we study these myths are words which are translated as 'gods,' and it is at this juncture where our interpretation of the events described come into direct conflict with the western concept of religion, that is, the relationship of man to a humanistic God. If our interpretation is correct, the gods of ancient myth hardly form the basis for a religion because we claim the 'sacred' myths are all expositions of causal, physical, cosmic events involving 'catastrophic' and 'chaotic' encounters of Venus and Mars with the Earth in proto-historical times.

The only logical relationship between these cosmic deities, which arose from these planetary encounters, and the 'true' God, lies in the belief that these events were set in motion by Him in order to create a new planet (Venus) and revitalize an aged or as yet incomplete Earth. This intervention involved the triggering of a long series of encounters between what we call priori-Mars and the Earth. It was priori-Mars in geosynchronous orbit around the Earth that was the firmament. Although this may seem implausible, even to the most religious person, the fact that the Bible states that God created the firmament and the prophets predicted the capture and release of priori-Mars 'must give us pause.' Moreover, the anger of Jehovah with the worship of 'pagan' idols such as Baal and Asteroth in the 'high places,' which were the very planets that he set in motion, indicated His displeasure with the Jews' failure to recognize His hand.

The Velikovsky/Angiras Theory of the Origin of Myths

The V/A hypothesis is that the sacred myths of all cultures are depictions of close approaches of Venus and Mars to the Earth within proto-historical time. They shook the Earth to its very foundations within the last 6000 years and were the common experience of all cultures. Velikovsky was certainly not the first one who had suggested a cosmic source for myths. Many myths clearly state that the deities were 'sky gods.' Up to the present time three completely different factors have combined to thwart the acceptance of Velikovsky's hypothesis.

The first of these comes from a fundamental aspect of myth which is brought out in our interpretations of the *Rig Veda*, the most ancient 'book' in the world. This is the tendency of the ancients to give names to the heavenly bodies based on their appearance - their apparent size, whether flaming or quiescent, whether they appeared in the day or night sky, and so on. As a result, a single body might be called by a number of names as these attributes changed, even though the enlightened of the age may have realized that it was one and the same physical entity.

The second is the anthromorphization of the physical forms and phenomena, that is, the tendency to attribute human-like characteristics to the planet as well as familiar actions, based on the phenomena that were observed. For example, the planet Mars was described as a 'green man with a red cloak', having 'two vicious dogs for guards', and it was said that it 'came down to Earth,' when it only approached the Earth. There was no ambiguity in the minds of the people that actually experienced these events, but even as soon as a few hundred years after the encounters ceased, the true idea of what the ancestors had seen was lost. All that remained was the myth itself and the human-like characteristics came to dominate the peoples collective 'memory' even though the driving events were purely physical in nature.

The ultimate obstacle to the acceptance of our astronomical explanation is the current uniformitarian stance of the scientific 'priesthood', which holds that the solar system has been the same for hundred of millions, indeed, billions of years! The perspective of modern man is completely different from the ancients. The sacredness of the myth has long been forgotten in the western world, indeed the meaning of the word 'myth' has become inverted to mean 'unreal' - the opposite of what it meant to the ancients. All the ancients had was the myth. They had no body of science such as we have today. We, on the other hand have science and myth - two bodies of knowledge, but the latter is discounted by all planetary scientists. The V/A hypothesis is that the knowledge bases in the two disciplines are complementary. We claim that myth deals with spectacular events involving the planets in past millennia and science deals with the them as they appear today. Because the scientists do not recognize the connection and therefore refuse to even consider the possibility that there are valuable data in the myths, they have adopted a number of seriously flawed assumptions, which will never be corrected by mathematics and physics.

Reading Suggestions

This book builds on the research of Immanuel Velikovsky, extending it based on new interpretations of Vedic, Hindu, Egyptian, Greek and other myths. Our determinations of *what* happened in proto-historical times come strictly from ancient myths. We then explain *how* these events happened using the laws of physics, and cite physical evidence to validate our hypotheses. *But science alone, as it is currently practiced, could never have uncovered what the ancient myths reveal.* The great majority of readers will find the first twelve chapters more difficult to read and assimilate, but these constitute the fountainhead from which everything else flows. Although it is like reading the last chapter of a mystery novel, we suggest that the reader who gets bogged down in the mythologic lore, skip ahead and read Chapter 13, which presents the entire scenario derived from the ancient myths. Also of interest are Chapters 14 thru 16, which give the physical evidence that validates our scenario. The last chapter should be of great interest to all, because it deals with the special implications of our scenario for the origin of modern man.

We do not recommend skipping any of our interpretations of the myths, only to take them at your own pace. After all, they provide the solutions of all the 'ancient mysteries' that so many authors have sought to plumb. The scenario which we develop was one of world shaking proportions and it took place within the last six millennia - not billions or even millions of years ago. We show that the myths of different cultures all describe the same events, even though they seem superficially different. Moreover, there are special 'gems' buried in these chapters, like the location of Atlantis, and the reason why the pyramids were built. Some, which will be expanded upon in our companion volume, *Chaos*, concern other bodies in the solar system. For example, the ancient peoples of Egypt actually saw the maria being formed on the Moon and their myths describe the nearby 'birth' of what we today call the planet Mercury.

Firmament

1
A SUMMARY OF *WORLDS IN COLLISION*

The most famous (or notorious) of the 'Biblical' catastrophists, again in America, was Velikovsky, the author of Worlds in Collision, who in 1950, from his studies of ancient tales, revived the old notion that a comet had shifted the Earth's axis, and even resurrected Radlof's notion that the planet Venus had hurtled past the Earth around 1500 B.C., flinging Mars around in the process...

Time, however, had marched on: what was perhaps seen as tenable in 1823 was now patently absurd in 1950. The decline of catastrophism, as a serious scientific proposition, was complete...

Velikovsky was a psychiatrist who had studied under Freud, but whose incursion into the domain of astronomers was to prove an intellectual disaster...

What is significant about the Velikovsky affair is not so much his scientific proposition, which is an easy target, as the vitriolic reaction of the American astronomical community to it.

Victor Clube and Bill Napier
The Cosmic Winter

This chapter is a straightforward but greatly abbreviated review of Immanuel Velikovsky's seminal book *Worlds in Collision.* Although there are a number of aspects of his planetary scenario with which we will take issue in this book, the details are not discussed in this chapter except to note that they are significant points of contention. These are declared at the end of this chapter but the details are left to the individual chapters in which they naturally fall.

The Prologue of *Worlds in Collision*

This consists of two chapters. The first states a broad brush view of the cosmology of the solar system, that, is, the origin, structure, and space-time relationships. Although cosmology is often applied to the universe as a whole, the focus of Velikovsky and the Velikovsky/Angiras (V/A) theory presented in this book, is the solar system, which, although only a tiny subset of the universe in space time, is of enormous importance because such systems are the sources of intelligent life in the universe. Currently popular concepts of the solar system are explained. Next, various questions of the origin of the planetary system are discussed along with their associated difficulties. In this connection the origin of the comets are specifically singled out, since the description of certain bodies such as comets in ancient texts are of particular significance in the development of Velikovsky's planetary scenario.

The second chapter of the prologue is devoted to Mother Earth, in particular to the many mysteries about her which have baffled science for years. These mysteries stem from findings that imply amazing global changes in the planet Earth, counter to the currently accepted notion of a *uniformitarian* solar system. They include apparent reversals of the magnetic field, the ice ages, and the mammoths found frozen in Siberia with tropical vegetation in their bellies, and so on.

These were the subject of an entirely separate book by Velikovsky, entitled *The Earth in Upheaval*. Repeated creations and destructions of the world - a common theme in the mythology of many ancient cultures, are described in terms of a 'new sun' or a 'new sky'.

Worlds in Collision Part I - Venus

With a prologue of only 31 pages as a backdrop, Velikovsky begins the first of two major parts of the book, entitled Venus and Mars, respectively. The part concerning Venus consists of 10 chapters occupying some 164 pages.

Strangely enough, Velikovsky starts his Venus narrative with a quote from Joshua 10:12-13 that implies that the sun stood still in the sky for most of a day. He urges the literal acceptance of this possibility, introducing the idea that this could be caused by a large body approaching the Earth. This idea is bolstered by another verse that says many large stones were cast down from the sky, killing many of the fleeing Canaanite kings and their armies. Velikovsky alleges the date of these occurrences to be the middle of the second millennium before the present era, that is, about 1500 B.C.

To prove the cosmic nature of this event he then attempts to establish a direct connection with an event described in the Mexican 'Annals of Cuauhtitlan' in which the night did not end for a long time. He reasons that an extension of the day at the longitude of the Middle East would have been experienced as an extension of the night on the other side of the planet, claiming that the myths describe the same global event.

Citing a similar occurrence 52 years earlier in the pre-Columbian writings as a clue, he then attempts to back up 52 years in the biblical record and look for a similar event. He concludes that the sought-for event was the Exodus of the Jews from Egypt. The claim that extraterrestrial influences played a crucial role in these events is one of the central themes in *Worlds in Collision*. Because Egyptian records make no specific note of the Exodus, it is not possible to date this event by the ancient and continuous Egyptian calendar. Nor is dating it from the Jewish point of view possible, because the story of the Exodus occurs in the early portion of the Bible, prior to the era when it can be relied upon as a historical document. Nonetheless, based on the biblical idea that it occurred during a period of extraordinary tribulations in Egypt, he suggests that the writings of a number of cultures (Mayan, Egyptian, Greek, and Babylonian) describing similar plagues, argues that these were the result of a single global event caused by an extraterrestrial agent. These plagues included a red dust that stains the rivers and spreads disease and hails of hot stones from the sky. He also introduces quotes from various ancient writings which mention a sticky fluid, bitumen, a rain of fire, and in the Jewish Midrashim, a stream of hot Naphtha that fell on Egypt. The Voguls of Siberia are quoted in this context as saying, "God sent a sea of fire upon the Earth...the cause of the fire they call the fire-water...the fires raged for seven years." Unfortunately, it is not possible for the reader to ascertain whether the quotes Velikovsky uses in this section come from the same time period.

Based on the descriptions and assumed global extent of these events, Velikovsky hypothesizes that a giant comet approached the Earth, at the time of the Exodus. At the point of the closest passage of the two heavenly bodies, the Earth passed through the tail of the comet, from which the stones, petroleum, and suffocating red dust fell. At the point of closest approach a darkness and a great storm enveloped the Earth for seven or nine days. The duration varies between the Egyptian and Jewish accounts. The Pharaoh pursued the fleeing Jews under these difficult conditions and although the Pharaoh and his troops were killed, many Jews also died because of the severe natural conditions.

Velikovsky then quotes additional writings of cultures around the world that describe similar catastrophes. He expands this event to include a deluge and a flood, due to the oceans breaking out of their boundaries after a great earthquake. The Jewish Midrashim states that the last night in Egypt was 'bright as the noon on the day of the summer solstice'. Velikovsky claims this was due to the proximity of the glowing body of the comet Venus. He claims that the Egyptians described this as the day of combat in which Horus waged war with Seth.

To emphasize the magnitude of the destruction on the Earth, he quotes a Buddhist text that states:

> when a world cycle is destroyed by wind, the wind turns the ground upside down and throws it into the sky and areas of hundreds of leagues in extent crack and are thrown upward by the force of the wind and do not fall again but are blown to powder in the sky and annihilated.

The theme of a hurricane is associated in the Babylonian Epic of Gilgamesh as being associated with the battle between Marduk and Tiamat. Myths from various cultures are quoted indicating enormous tidal effects which caused the oceans to pile up kilometers high and remain stationary like glass. He cites sources which claim that the ocean covered much of the arable land in China and it took 10 years to drain the fields so they could be farmed once more. These tidal forces were the same as those which allowed the Jews to cross the (Red) sea and escape the Pharaoh.

Ancient myths from all parts of the world are cited containing many common observations, related in terms which can be interpreted as the same phenomena, that is, red dust, darkness, storms, and floods. They all associate these events with mythical gods, for which each culture had unique names, therefore, they were supernatural (extraterrestrial?) in origin. If Velikovsky's hypothesis is correct, these supernatural events give the clues to allow us to relate the deity names in the different cultures. Even more important in Velikovsky's mind is to associate these events with one or two planets which approached very close to the Earth. His candidate for the time of the Exodus is the planet Venus.

His task is somewhat complicated by the fact that many of the myths describe a combat or battle between gods, implying the presence of more than one. Further traumas are described: great lightning bolts exchanged between deities and between them and the Earth; the boiling of the sea; theophony - a sound generated within the Earth due to continual, enormous tidal stresses. He claims that the name of the Chinese Emperor of the time, Yahou and the Jewish god Yahweh were derived from an alliteration of this sound made by the convulsed Earth.

Firmament

In Chapter 5 Velikovsky quotes sources claiming that the quarters of the Earth were displaced, that East and West were reversed along with the seasons on several occasions. He claims that these reversals were due to the passage of Venus very close to the Earth, causing the Earth to flip over in space reversing the orientation of the North and South poles. Because the descriptions which apparently relate the close interactions of two planets often include the exchange of enormous thunderbolts between the bodies, Velikovsky deduced that the planets had electric as well as magnetic fields and that these should be considered when calculating the forces and torques between bodies during close encounters.

In Chapter 6 he relates that thick clouds, which extended close to the surface, covered the sky for years, as the Jewish people wandered in the desert. Miraculously, an exotic food came from this thick sky, that is, it condensed like dew on the Earth during the night. This food allowed men to survive the times during which the dark clouds prevented agriculture. This food was called 'manna' by the Jews, 'soma' by the Vedic people of northern India, and 'ambrosia' by the Greeks. Although it could be argued that many of the events in the myths seem to be merely grander versions of existing phenomena, and are therefore easy to explain as exaggerations, this cannot be said of this food from heaven. Being so anti-experience makes it more believable, since numerous cultures vast distances apart describe this same phenomenon. This constitutes a strong argument in favor of the reality of this heavenly stuff, which not only nourished the bodies and souls of the people but also made the animals both healthy and strong. Velikovsky hypothesizes that this was 'manufactured' from petroleum in the atmosphere of Venus. Here he is obviously stretching credibility perilously thin, because the encounters with Venus were apparently transient in nature, whereas the heavenly food was regularly bestowed upon the Earth with the morning dew.

In this same chapter, Velikovsky also proposes that the walls of Jericho did not fall as a result of the blowing of trumpets by Joshua's army, but they fell during a planetary interaction in which the convulsion of the Earth due to tidal forces caused a trumpet-like sound from its interior. Associated with this sound were continuous earthquakes, which destroyed the walls of the city.

Chapter 7 continues the discussion of Joshua's battles with the Canaanites. Babylonian and Midrashic sources claim that a portion of the stones which fell to Earth at the time of the Exodus, had remained 'suspended in the air' until 50 years later, at which time they rained down on the Canaanites. This battle was accompanied by fierce earthquakes, but not of the magnitude claimed at the time of the Exodus.

In an important section entitled Phaëthon, Velikovsky reviews the legend of this Greek god as given by Ovid. In it, Phaëthon, driving the chariot of the sun, fails to control the horses and the sun comes too close to the Earth and sets it afire. Green vegetation is turned to white ash and volcanos erupt, Libya becomes a desert and the seven mouths of the Nile lie empty, filled with dust. The golden sands of Tagus melt in the intense heat. The rivers Don, Euphrates, Ganges, Phasis, Danube and Alpheus boil, and Spercheos' banks are aflame. The Thracian rivers, the Rhine, Po and Tiber are dried up. Great cracks yawn everywhere. Even the sea shrinks up with dry sand in its place.

Velikovsky attempts to relate these events to those in the book of Joshua, but the magnitude of these effects seem too great to have merely knocked down the walls of Jericho. He relates Plato's version of how Solon the Greek was told by Egyptian priests of ancient destructions of the Earth. The Greeks had no recollection of this. He also narrates the story of the lost continent of Atlantis. Velikovsky, in an uncharacteristic move, arbitrarily changes the date of these events, related as 9000 years before the time of Plato, to 900 years before that time, in order that it fit into his scenario.

Velikovsky discusses the two great floods noted in Greek proto-history and argues that the flood of Deucalion, the most destructive, preceded that of Ogyges. He asserts that Ogyges was a Hyksos King (Agog I or Apop I) who ruled Egypt and founded Thebes. As a result of some rearranging of the accepted ancient dates, he identifies this King as a contemporary of Moses, and thereby claims the flood of Ogyges was the Greek experience at the time of the Exodus. No clear statement of the time between the greater flood (of Deucalion) and that of Ogyges is attempted; however, at lease one scholar claims that the conflagration of Phaëthon corresponded to the flood of Deucalion. It was that earlier event which supposedly killed almost all human beings, thereby erasing all memory of the world before that time. Velikovsky believed that both of these floods occurred in the middle of the second millennium before the present era, but downplays the level of destruction associated with the Deucalion flood, perhaps because it did not fit into his scenario of the Exodus since both the Egyptian and the Jewish people survived this event.

Chapter 8 opens with a discussion of an ancient Mexican ritual based on the expectation of destruction every fifty-two years due to Venus' approach. Velikovsky attempts to establish that this was also the interval between the Exodus and Joshua's attack on Jericho, based on the idea that these were caused by the same extraterrestrial encounter. He then relates the Jewish law concerning the Jubilee year which has long since been abandoned. Every seventh year was a sabbatical year during which the land was to be left fallow and Jewish slaves set free. The year following the seventh sabbatical year was the Jubilee year when the land not only had to be left fallow, but had to be returned to its original owner. That is, land could not be permanently conveyed - a sale being understood as a lease good only until the next Jubilee year. The reason behind this custom has been lost in antiquity. Velikovsky claims it derived from the period of time between close encounters of Venus with the Earth.

Velikovsky claims that the 50 years of Mexican ritual and the 52 years of Jewish ritual stem from a common astronomical origin. He launches into a long series of arguments occupying chapters 8, 9 and 10 designed to prove that the active agent which caused the extraordinary events accompanying the Exodus and the battle of Jericho was *the newly born planet Venus*.

The Mexican description of the deity Quetzal-Cohautl (feathered serpent) was that of a serpent attacking the sun. This came closer until the Earth was shrouded in darkness. Later the snakelike body transformed itself into a great star. Today Quetzal-Cohautl is the name of Venus.

Velikovsky claims this was the same event that caused the plagues at the time of the Exodus and the darkness in the subsequent years of wandering in the desert. By this association, he attempts to identify the planet Venus, which is never specifically mentioned in the Jewish books, with the events of the Exodus.

He quotes a Roman historian, Marcus Varro who states that Venus suddenly changed its color, size, form and course, and cites other scholars who dated these portents in the reign of Ogyges - coincident, Velikovsky claims, with the battle of Jericho.

Samaritan and Chinese chronicles state that a brilliant new star was observed in the East. Velikovsky then quotes a number of myths that imply that Phaëthon and Quetzal-Cohautl, both Venus deities, became the morning star.

In the section entitled "The Four Planet System," he attempts to corroborate the idea that Venus was born around 1500 B. C. by citing even more ancient astronomical listings of the planets and showing that these did not include Venus. A Hindu table of the planets attributed to 3102 B. C. lists only Mars, Jupiter, Saturn and Mercury. Of course these planets were not called by these names in that ancient culture and so this interpretation must have *assumed* that these were the only four 'travelers' that were visible to the naked eye at that time - which may not be the case. Velikovsky's argument is weakened by the fact that he uses the opposite explanation in the case of Babylonian records from around 1000 B. C., citing in a footnote that a German scholar, Weidner claims that Venus was not included in the list of planets because it was part of a more important triad of the Sun, Moon and Venus.

Democritus (460-370 B.C.) is quoted as saying 'stars have been seen when comets dissolve.' The Pythagoreans were a group of ancient Greeks that kept secret their extraordinary knowledge of the past and the planets. Aristotle wrote of their interpretation of comets: '...the comet is one of the planets...' Velikovsky marshals an impressive array of quotations from ancient myths and writings that describe a deity as a very bright star that rivaled the sun, as a 'bearded' star, and 'a comet.' It is also described as having a bulls horned head. He claims that when Venus approached one of the planets, its horns grew longer.

In Chapter 9 evidence is given that the Greek goddess Pallas-Athene is Venus. In one legend 'she sprang from the head of Zeus-Jupiter.' In another 'she was the daughter of a monster Pallas-Typhon, who attacked her and whom she battled and killed.' Apollodorus (the Library) says that Athene slew Pallas and used his skin' which appears to refer to the envelope of Venus that previously formed the tail of the comet. In a footnote Velikovsky argues that Venus is also Isis of the Egyptians, Minerva of the Romans, Vishnu of the Hindus, and Astarte (Ishtar) of the Babylonians. We will take issue with several of these associations in subsequent chapters.

Velikovsky admitted struggling to comprehend whether it was the planet Jupiter or Venus which caused the catastrophe at the time of the Exodus - because the myths name both. He concludes however, that Venus was born out of Jupiter and interacted with the Earth while still a 'comet.' The Greek myth claims she sprang from his head, and this birth was accompanied by great disturbances in the celestial and terrestrial spheres. If this is true, it constitutes an important clue to the veracity of his claim. It would have taken a minimum of several years after this disturbance before Venus could have come close to the Earth and cause some real damage.

He indicates the many ancient cultures that worshipper the morning star. The worship of the bull and cow was also attributed to Venus because it appeared to have horns. The fall of ambrosia or manna on the Earth was associated with the cow aspect of Venus, nurturing the Earth.

Proto-Venus is also related to a number of names thought to be equivalent to Satan, such as Ahriman, Seth (Set), Lucifer and Beelzebub (Baal Zevuv). The Iranian Bundahis says that at the close of one world age:

> the evil spirit (Ahriman) went toward the luminaries...He stood upon one third of the inside of the sky, and he sprang like a snake, out of the sky down to the Earth...He rushed in at noon...the sky was shattered and frightened...Like a fly, he rushed out upon the whole creation, and he injured the world and made it dark...And noxious creatures were diffused by him over the Earth, biting and venomous, such as the snake, scorpion, frog, lizard, so that not so much as the point of a needle remained free of noxious creatures.

This is similar to the plagues in Egypt at the time of the Exodus.

Many cultures carried out sacrifices to the morning star up to relatively recent times. The Pawnee attached twelve eagle feathers to the victim's head, arranged like a fan, and painted her body half red and half black.

In Chapter 10 Velikovsky explains the synodical year of Venus and how it had been used in the Orient and Occident for thousands of years. The Venus tables ascribed to the first Babylonian dynasty give a sequence of dates of appearance and disappearance of Venus due to conjunctions. These are highly irregular. Velikovsky attributes this irregularity to the rapidly changing eccentricity of its orbit, which did not settle into its current circular form until about 0 B.C. Thus Velikovsky proposed that Venus' orbit changed from highly elliptical, that is, crossing the orbit of the Earth and reaching as far out as Jupiter in about 1500 B.C., to its current circular shape in only some 1500 years. This claim was the focus of a major attack from astronomers who said such a change was impossible in such a short time. He quotes Isaiah 14:12-13:

> How art thou fallen from heaven O Lucifer, son of the morning! how art thou cut down to the ground which didst weaken the nations. For thou hast said in thine heart, I will ascend into heaven, I will exalt my throne above the stars of God.

Velikovsky noted that more than a hundred generations of commentators have failed to find its true meaning. He interprets it to mean that Venus, which once wreaked destruction on Earth, is now fallen from heaven and became the morning star, which is only seen close to the horizon.

Worlds in Collision Part II - Mars

The second part of *Worlds in Collision* is concerned with events which Velikovsky attributed to encounters with Mars. He claims that about 750 years after the catastrophe at the time of the Exodus, the planet Mars began a sequence of interactions with the Earth, which although not as severe as those caused by Venus, were nonetheless traumatic. Detailed events and dates are much easier to come by in this period of time due to many ancient writings which have been recovered. The Bible stands out as the preeminent historical document of the Middle East during the Iron Age. The events of earlier times are more shrouded in antiquity, and absolute dates are difficult or impossible to determine.

He points out in the scriptures evidence that a catastrophe called a 'raash' (commotion) occurred in the days of Uzziah, King of Jerusalem. This was foretold by the prophet Amos. Using biblical passages Velikovsky attempts to establish a series of planetary encounters, which he claims are separated by periods of about fifteen years. These are mentioned in the Bible, not to record their historical occurrence, but because they were prophesied by Amos, Isaiah, Joel, Hosea, and Micah through the divine inspiration of Jehovah. They were manifested primarily via floods and earthquakes but were sometimes accompanied by anomalous motions of the sun and stars, resulting in coincident shortening or lengthening of the day on which they occurred, and in new orientations of the heavenly bodies in the sky.

Chapters 1 and 2 of Part II of *Worlds in Collision* bring together a number of detailed descriptions from the bible and other sources. Because of this richness of detail and the more accurate dating, we rely on the descriptions of these events and their timing, in developing the Velikovsky/Angiras theory. The most spectacular story concerns the destruction of the 185,000 man army of the Assyrian Sennacherib by a 'blast' from heaven, which Velikovsky dates at 687 B.C.

Velikovsky almost cries out in the section entitled 'Maimonides and Spinosa, the Exegetes' asking how it is possible that these writers do not comprehend the physical reality of the destructions described in the biblical passages. He chides these (long dead) philosophers and all those who have since assigned metaphors to the physical events, apparently because they could not believe they were real.

Velikovsky had found sufficient evidence in writings of approximately 800 B.C. which indicated that the orbit of Venus had evolved to the point that it no longer approached the Earth. At the same time he had compiled a number of biblical descriptions of destructive events of this era which, although they are tantalizingly detailed, are attributed to no physical causative agent. If the sun and stars are to move in the sky and the oceans to move out of their bounds and earthquakes to become the order of each day, he insists that there must be a large heavenly body causing these events. He found the Bible silent regarding Mars as the causative agent. It was obviously a problem for Velikovsky, who itemized these 'commotions' and pointed out the periodicity of their occurrence, but found no unequivocal cause in the Jewish literature.

To determine the source of these destructive events, he seeks out new deity names which appeared in this approximate time frame. Turning to the founding of Rome he points out the key role which Mars played in the history of Rome. He identifies the founder of Rome, Romulus, with Mars, based on a quotation from Plutarch. He also quotes sources claiming that during the time of Romulus, there was a rain of blood and there were almost continuous earthquakes. The death of Romulus was associated with darkness and the awful peals of thunder and furious blasts, and, amidst this storm, Romulus disappeared. Ovid says that both poles shook and Atlas shifted the burden of the sky. The Sun vanished and rising clouds obscured the heaven ... the sky was riven by shooting flames. The people fled and king Romulus, upon his father's steed (Mars), soared to the stars.

Velikovsky associates Mars with the deity name Nergal of the Assyrians and Babylonians. Nergal was a very important deity in these cultures. The rising, setting and periods of apparent retrograde motion of Mars were of extreme interest in Babylonian astronomy. Its 'disappearance and return', its position in relation to the equator, the change in its brightness and its relation to Venus were all watched closely. Mars was described as the raging fire-god and the lord of destruction. Sargon (724 B.C. to 705 B.C.), father of Sennacherib wrote on one occasion:

> In the month of Abu, the month of descent of the fire-god.

In Chapter 3 Velikovsky reiterates his interpretation of ancient myth - that interest in Mars waxed and interest in Venus waned around 700 B.C. He claims that Nergal was hardly ever mentioned in writings from before the ninth century B.C. He also claims that on the astronomical ceiling of Senmut (architect of Queen Hatshepsut of the Egyptian Middle Kingdom), Mars does not appear among the planets. He interprets this as meaning it did not play a conspicuous part in the early mythology of the celestial gods. Here he is inconsistent in his interpretation because it favors his planetary scenario. In an earlier section he interprets the absence of Venus as meaning it was in a state of flux, its orbit changing rapidly, and so was not yet considered a planet, but was *more important* by virtue of this unpredictable motion.

The Iliad and Odyssey describe a combat between Ares (Mars) and Athene (Venus). The descriptions of the battles are rich in detail and although cast in the mythical language of gods in conflict, it is relatively easy to explain the actions in more physical terms. Velikovsky reasons that if these books could be dated, it might make possible the dating of the planetary encounter. He further reasons that the transfer of orbital energy between the two is a mechanism whereby Venus' orbit became so rapidly circularized.

One difficulty with Velikovsky's approach is that the material incorporated into the Iliad and Odyssey may have been experienced much earlier than Homer's time. Velikovsky argues that the conflict of the gods was mirrored on Earth by the conflict of the Trojans and Acheans according to the story, therefore, the events must have occurred within Greek historical times. Since Greek writing only developed about the time of Homer, he may have been the first great writer to record the ancient myths that had been passed down by word of mouth through many generations.

Similarly in ancient Mexico, the Toltec patron god was Quetzal-Cohautl (Venus), whereas the later Aztecs venerated Huitzilopochtli, 'a great destroyer of towns.' The characteristics of this god are: 'a bow in his right hand and in the left, sometimes a bundle of arrows, sometimes a round white shield.' These images are similar to those describing Indra in the *Rig Veda*. In sections on Tao, Buddha, The Bundahis, the conflict of Mars, sometimes called the Wolf-Star, and the (Venus) deity are hypothesized to be part of Chinese, Hindu and Iranian myths.

When the encounters between Mars and Venus occurred close to the Earth, the tidal forces on the Earth suddenly increased because the vector sum of their tidal forces combined. At those times, there was great turmoil on Earth, including floods, earthquakes, insects and vermin, storms and darkness. He quotes the following passage from the Taoist text of Wen-Tze giving the flavor of these times.

> When the sky, hostile to living beings, wishes to destroy them, it burns them; the sun and moon lose their form and are eclipsed; the

> five planets leave their paths; the four seasons encroach one upon another; daylight is obscured; glowing mountains collapse; rivers are dried up; it thunders then in winter, hoarfrost falls in summer; the atmosphere is thick and human beings are choked; the state perishes; the aspect and order of the sky are altered; the customs of the age are disturbed [thrown into disorder]...all living beings harass one another.

It is also of interest that Mars is described in the Indo-Iranian Bundahis as 'the special disturber of the Moon,' implying perhaps that Mars had the effect of modifying the orbit of the Moon.

In the section entitled 'Lucifer Cut Down,' Velikovsky suggests that the interaction of Mars with Venus may have saved the Earth from the destruction it had experienced earlier due to close interactions with Venus. In this Section there is a statement that is important in the relationship of Velikovsky's theory to later ideas and subsequently to the V/A theory.

> Venus, which collided with the Earth in the fifteenth century before the present era, collided with Mars in the eighth century. At that time Venus was moving at a lower elliptical velocity than when it first encountered the Earth; but Mars, being only about one eighth the mass of Venus, was no match for her. It was therefore a notable achievement that Mars, *though thrown out of the ring,* nevertheless was instrumental in bringing Venus from an elliptical to a nearly circular orbit.

We have italicized the above phrase because we wish to discuss it later in the chapter.

In chapter 4 of Part II (Mars) Velikovsky suggests that the term 'sword-god' refers to the atmosphere of Mars which was drawn toward larger planets like Venus and the Earth when they approached. The water in this atmosphere froze in the coldness of space and formed ice crystals which made the sword bright white.

He explains the words of Isaiah[6] when prophesying a repetition of the catastrophe in which Sennacherib's army was destroyed, as referring to the sword of Mars.

> Then shall the Assyrian fall with the sword, not of a mighty man; and the sword, not of a mean man, shall devour him...and his princes shall be afraid of the ensign...And all the host of heaven shall be dissolved...for my sword shall be bathed in heaven.

He claims that the term 'wolf-star' was due to the tidal distortion of Mars caused by the gravitational field of the more massive Venus or Earth during close encounters with these bodies.

The following quote referring to the Chaldeo-Hellenistic historian Berosus is a small indication of the knowledge that the ancient sages possessed, primarily as a result of actual observations. Seneca wrote:

> Berosus, the translator of Bel, attributed to the planets the cause of these perturbations...His certainty in this matter was so great as to fix the dates of the universal conflagration and deluge. Everything terrestrial will be burned, when the stars which now follow different orbits will reunite in the sign of Cancer and will place themselves in one line, so that a straight line would pass through the centers of all these globes. The deluge will come when the same planets will have conjunction in Capricorn.

By 'stars' he obviously meant the planets. Note, he understood that these were 'globes', implying that the ancients had closer vantage points than we have today. Most importantly, he realized, undoubtedly from experience, that when the planets lined up, the Earth experienced great trauma. Today, this sort of alignment produces no noticeable effect because the planets are farther away. However, three thousand years ago, the planets Mars and Venus were both near the Earth.

The Roman Pliny, who deduced that the Earth was also a globe, described the thunderbolts that accompanied the close interaction of two planets 'by which the threefold mass of mountains fell.' He states that this happened when ' heavenly fire is spit forth by the planet as crackling charcoal flies from a burning log.' The thunderbolt occurred when the heavenly fire came down on the Earth. He claims that Lake Bolsena in Tuscany, which is nine miles by seven miles and 285 feet deep, resulted from a thunderbolt striking the Earth.

Because of the earthquakes caused by Mars, it was also referred to as 'the stormer of the walls', and 'the sacker of towns.' A Babylonian text is quoted which says 'When the Earth quakes through the whole day, there will be a destruction of the land. When it quakes continuously, there will be an invasion of the enemy.' This gives a small idea of the terror of those days.

Velikovsky claims that the frequent earthquakes of those times were due to the tidal forces exerted by Mars when it approached the Earth. In Chapter 5 of Part II Velikovsky discusses evidence that ancient people knew of the two tiny satellites of Mars, Phobos and Diemos. These are so small that they cannot be seen, even with most telescopes today. The notion that they knew of these moons implies that Mars was much closer to the Earth in what we call Vedic times, since these satellites must have been visible to the naked eye. Actually, the ancient myths imply that there were a multitude of Martian satellites that flowed, and traveled in 'packs.' Velikovsky claims the 'maruts,' described in the Vedic hymns were these same demonic companions of Mars that accompanied him into battle. He also claims that a number of stones, now held sacred in different cultures, are so held because they fell from the sky. The most famous of these is the black stone of Kaaba, the most holy thing in Mecca.

Velikovsky suggests that the biblical term 'archangels' refers to planets. Gabriel was credited with the destruction of Sennacherib's army by some. Another archangel, Michael, was the pillar of fire that guided the Israelites and protected them from the Pharaoh at the Sea of Passage. He claims therefore, that Michael is the planet Venus.

Firmament

There was considerable planet worship in the Middle East around the seventh century B.C. This is recorded in the Bible (II Kings 21:5, II Chronicles 33:3). The symbol of calf was used for Baal. The Lord was very upset by this worship of the 'host of heaven' (Jeremiah 19:13). Apparently the worship of Yahou or Yahweh was also planet worship, probably of Mars.

Chapter 6 deals with the tendency of peoples to gradually forget or misinterpret the physical catastrophes of the more ancient times in terms of the current day. The accounts of real events became more and more just 'stories' once hundreds of years passed, and the original stimulus became more and more remote in time. Also, as the natural catastrophes ceased, it became difficult for people to believe they really occurred because they had no such experience in their lifetimes. In spite of the loss of true understanding, powerful 'images' remain in the mythology the world over, such as the dragon, the shape of an evil witch riding a broom and the scorpion, which are probably analogies for comets.

Velikovsky notes that the decimation of Sennacherib's army is associated with a mouse with by the Egyptians. He also relates a myth told by the Menomini tribe of Algonquin Indians in which the sun was snared by 'ropes' and retarded in its movement across the sky until it was freed by a mouse which gnawed through the ropes. The similarities of two such accounts from cultures half a world apart offer strong corroboration of the notion that all people on Earth observed cosmic events in those days.

In Chapter 7 a variety of quotations and physically measurable artifacts are cited as providing evidence of different orientations of the poles of the Earth in the period from 1100 to 750 B.C. These range from descriptions of the pole star as lying in the great bear, Chinese and Babylonian records of the shortest and longest days at specific places, the orientations of ancient temples and obelisks, and the longest and shortest days given by an ancient shadow clock and water clock in Egypt, none of which are correct in the current scheme of things.

Velikovsky also mentions that the location of ice during the last ice age is skewed relative to the present poles, with the apparent center over Baffin Land in the North American hemisphere. Brahmin charts of the sky show a large difference from modern astronomy. Calcutta being removed 180° longitude from Baffin Land these charts indicate the North pole was at Baffin Land, close to the present magnetic pole. Finally, he cites mastodons, frozen alive in Siberia and large settlements in Alaska as evidence that these lay in warmer zones in the past.

In Chapter 8 he discusses evidence from a number of cultures that indicates that in past millennia the year consisted of exactly 360 days. This was not due to inexact reckoning or ignorance, because an error of five and half days per year would result in a shift of the seasons *of more than 200 days in a period of forty years - a single persons lifetime.* A number of ancient cultures are known to have had quite sophisticated methods of calculation. The cultures named are Hindu, Persian, Babylonian, Assyrian, Greek, Roman and Jewish. Many ancient cultures reckoned a year of 360 days plus 5 additional days which were kept separate. The Aztec calendar contains 360 days around its circular periphery and the five 'bad days' are inscribed separately. In the year 238 B.C. a reform of the Egyptian calendar at Canopus' was formally decreed in order to harmonize the calendar with 'the present arrangement of the world'. This decree ordered one day to be added to the year (of 360 days plus 5 days) every fourth year.

Summary of *Worlds in Collision*

As a result of exhaustive study of the writings from Hindu, Persian, Babylonian, Egyptian and Chinese cultures, Velikovsky discovered that a sub-period of nine days was associated with the moon for a certain period of time. He concludes from this that the lunar period became thirty-six days and from this value changed to its current period of 29½ days. This is consistent with a year of only ten months which was the case in the time of Romulus. A number of other cultures had years consisting of ten or eleven months. The Hebrews based the year on twelve lunar months of thirty days duration. The use of the moon's phases to determine the length of the year is an important point which we will refer to in Chapter 10 of this book.

In 'Reforming the Calendar,' Velikovsky concludes from detailed Chaldean astronomical records discovered at Nineveh that the world order changed repeatedly in the course of a single century. This caused them to repeatedly revise the calendar. Other analysts have considered these inconsistencies as errors. Daniel, the Jewish sage at the court of Nebuchadnezzar during the Babylonian captivity, said to the King:

> He changeth the times and the seasons. [7]

It is not known whether he was referring to the God the Jews currently worship or Yahweh.

Considerable evidence is presented that up to about 600 B.C. the length of the month and year changed so frequently that calendar reform was almost impossible. The correct reckoning of the solstice could only be done by observing the direction of the sun via specially constructed structures built for this purpose alone.

In Chapter 9 he begins by claiming that the interactions of Venus and Mars with the Earth also traumatized the Moon causing the 'plains of lava' or maria. He then turns to Mars as observed at the time that *Worlds in Collision* was first published. Although this book has been in print for some forty years, Velikovsky has declined to change it in light of new information gathered by space probes during that period because he wanted his theory at that time to be on record for all to read. As a result some of the material on Mars is outdated. He did note in 1950 that the interval between times that Mars and the Earth came closest to one another is about 15 years, the same interval that he surmised separated the biblical devastations at the time of Amos and Isaiah. He also pointed out the remarkable coincidence between the tilt of the axes of the Earth and Mars relative to their orbital planes and the rates of spin of the two planets. He claims that Mars exhibits an excess of heat and suggests it was caused by close interactions with Venus and to a lesser degree with the Earth.

He claims that petroleum was and is a significant atmospheric component of Venus. He predicts that Venus is quite hot as evidenced by the relatively large amount of heat radiated from the dark side, combined with its very slow rotation (not known at the time of publication), and reiterates that this is consistent with his theory of its recent birth and close interactions with Mars and Earth.

In 'The End' he restates the idea that catastrophes have repeatedly troubled the civilizations of the Earth. He mentions that some asteroids have been found which cross the orbit of the Earth so that, although the planets Venus and Mars appear to no longer threaten the Earth as they did in the last four thousand years, the danger is not completely over. He cites planets and satellites which cross the orbits of others. He warns that a large interstellar 'comet' could 'run into one of the planets' and 'push it from its orbit,' starting the chaos again.

In his epilogue he summarizes his theory, referring to his subject as 'Historical Cosmology,' and gives a glimpse of his next book, *Ages in Chaos*. He suggests that these catastrophic planetary events are fundamental to the understanding of religions and psychological thought. And that the 'collective unconscious mind' may contain more memories of real physically experienced events than Freud or Jung ever imagined. He also indicates the areas of study suggested by his catastrophism.

Finally Velikovsky addresses some of the criticisms of his ideas leveled by the scientists, such as the rapid circularization of Venus' orbit, the mechanisms by which the axis of rotation of the Earth could be changed, and the physical repercussions for life on the planet. He emphasizes the idea that electric and magnetic fields may modify the purely Newtonian nature that is currently assumed in celestial mechanics. He appeals for a serious multi-disciplinary effort to really understand the 'historical facts' that he has uncovered.

Unfortunately, no such effort has been undertaken by the scientific community. After a vicious negative knee-jerk reaction, the planetary astronomers returned to their neo-Aristotlean 'clockwork' solar system which holds that any catastrophes, if they occurred at all, happened many millions of years ago. Velikovsky's ideas have had an effect. The idea of a very ancient 'watered-down' catastrophism, disguised in the term 'chaos theory,' has become the vogue in the popular literature of science, but no mention is ever made of Velikovsky.

Although the old guard astronomers think Velikovsky's ideas are dead, and the younger ones may never have heard of him, it is our purpose here to show his basic ideas have considerable merit, and, as a result of the corrections and additions in the V/A scenario, they will have a most profound effect on our understanding of the solar system in which we live.

V/A Revisions of Velikovsky's Original Theory

The Velikovsky/Angiras theory has its basis in the planetary catastrophism advanced by Immanuel Velikovsky in *Worlds in Collision*. The Velikovsky/Angiras scenario, developed in this book, brings Velikovsky's ideas to fruition, using the same source that he advocated - ancient myth. In addition we bring to bear data from a number of disciplines such as archaeology, anthropology, physics, aerospace science, and planetary science. This leads us to a number of major revisions and additions to his original hypotheses.

The V/A theory accepts the following tenets of Velikovsky's catastrophism:

(a) Venus was 'born' a new planet, out of the Jupiter system within proto-historical times.

(b) Soon after its creation, Venus twice approached so close to the Earth that the spin axis of the lithosphere, the part of the planet on which we live, was rotated approximately 180° relative to the stars so that 'east' and 'west' were reversed. These two events probably occurred within a period of less than fifty years, and on those occasions most of the animals and whatever people populated the globe, were killed.

(c) Venus also interacted with Mars on numerous occasions, with the net result that the red planet's orbit was changed, bringing it close to the Earth. Close encounters between Mars and Earth continued well into the first millennium B.C.

As a result of a complex chain of reasoning which will become clear as the reader progresses through this book, we take issue with Velikovsky's dating of the initial catastrophic encounters with Venus at the time of the Jewish Exodus. *We propose the creation of Venus and its subsequent devastating encounters with Earth occurred more than two millennia earlier. We further propose that a series of Earth-Mars encounters began soon afterward and continued for some three thousand years. Finally we claim that these Earth-Mars encounters were of such a unique and bizarre nature that no one in modern times has understood them - even Velikovsky himself.*

As a result of the much earlier advent of Venus, we believe that the cosmic events which formed the backdrop of the Jewish Exodus and subsequent 'commotions' mentioned in the bible, were interactions between Mars and the Earth, and that Mars had already been interacting with the Earth periodically for some 1800 years before that time. As a corollary, we identify the primary deities in eastern cultures of that period with Mars, while Venus, still a brilliant comet-like body, ran a close second. Venus continued to play an important role in the dynamics of the Earth-Mars interactions for thousands of years.

In the V/A theory, we incorporate the important idea, never specifically stated by Velikovsky, but proposed by Lynn E. Rose, an advocate of his, that *prior to the disturbance of the solar system by the birth of the proto-planet Venus, Mars was in an orbit inside that of the Earth where Venus now resides.*[8] The V/A theory goes one step further, based on clues in a number of myths, and proposes that *at that time Mars was a living planet covered with green vegetation and probably animal life.* How advanced these species were, may be proven by future astronaut-anthropologists.

The V/A theory discounts Velikovsky's notion that the out gassed material from the newly formed proto-Venus, which engulfed the Earth at the time of the two close encounters, added significantly to the petroleum supply on Earth. However, *it is highly likely that the billions of animals drowned by tidal waves and immediately buried in silt or clay at those times, did contribute to this supply as their bodies decomposed in the absence of oxygen, and the oils seeped into geolithic traps in the crust of the Earth.*

We do not believe that electromagnetic forces are significant enough to affect the interaction of two planets, except at very close quarters. The 'overturning' of the Earth's lithosphere, due to the close passage of Venus, can be explained strictly in terms of the tidal force of Venus as it passed close by.

Although, in light of years of study and much new data, our conclusions differ somewhat from Velikovsky's original scenario, the stroke of genius which enabled him to recognize real cosmic events in the myths of ancient cultures is a heritage of inestimable value. The most profound new revelations in this book come primarily from the application of his approach - the study of ancient myth, more than the application of modern science.

From the perspective of the nineteen-nineties, a reader of *Worlds in Collision* might question why Velikovsky, after a tour de force advocating recent catastrophism in the solar system, did not carry his theory to its logical conclusions, and predict the characteristics of the planets that should result. He did predict that Venus would be hot; that strong radio emissions would be emanating from Jupiter; and that the planets would be found to have electric fields associated with them. All of these were subsequently confirmed, but were explained by scientists as being due to other mechanisms. Velikovsky, although highly intelligent, was not an astronomer or physicist, he was capable of understanding what the planetary astronomers knew about the planets at that time. The problem was that *almost nothing was known.* When *Worlds in Collision* was published, no satellites orbited the Earth and few people anticipated that vehicles would travel to the planets within the next three decades. As a result there was very little hard evidence from the planets to corroborate or disprove Velikovsky's claims. Unfortunately, a number of the theories created strictly to counter Velikovsky's hypothesis, have become integral parts of planetary science, and like a 'pied piper,' have misled the discipline for the last fifty years. Although much good data has been collected, it is always interpreted in ways which justify these erroneous theories.

For example, new observations of Venus were made in the microwave spectrum which confirmed the high surface temperature of Venus as predicted by Velikovsky, due to its recent 'birth.' Based on these observations, Carl Sagan initiated a theoretical investigation of what he called 'The Runaway Greenhouse Effect,' in order to counter Velikovsky's explanation. Also radio telescopes were trained on Jupiter and confirmed another prediction of Velikovsky - that Jupiter was the source of considerable radio frequency noise. All of this activity was undertaken primarily to refute Velikovsky's hypotheses, not to confirm them.

As the data have trickled in over the last 30 years, the scientific community has studiously avoided interpreting the data in any way which might bring his catastrophism back into the limelight, but the data are there and the secrets that it contains lay patiently awaiting the true interpretation which ties them all together into a single coherent weltanschauung or world view.

2

ADITI AND AGNI

No idea is so antiquated that it was not once modern. No idea is so modern that it will not some day be antiquated...
To seize the flying thought before it escapes us is our only touch with reality.

Ellen Glasgow

Immanuel Velikovsky did the pioneering work in recovering the fragile thread of an almost lost record of recent chaotic events in the solar system. His legacy was twofold: he advanced a scenario of planetary-scale chaos in our solar system, not within hundreds of millions, or even millions of years, but within proto-historical times; at the same time he showed that an imposing body of knowledge concerning these events exists in the ancient myths of many cultures around the world.

The details of the planetary interactions he pieced together in *Worlds in Collision* were incomplete, particularly in regard to the planet Mars. This was due, in part, to the fact that his scenario was derived from the myths of a number of cultures, written in different languages, telling different stories. Also important was the fact that innumerable planetary interactions occurred over a period of several thousand years. It would be almost unthinkable that any of these primitive cultures could systematically record and preserve the record of these events to the degree that we would like to have it today. However, one culture did indeed accomplish this feat. Strange to say, but we know very little about the people who initiated this body of knowledge. We refer to them as the Vedic people or more specifically the rishis, the priestly class to which it was entrusted. But, as will be discussed in Chapter 7, they are most likely to be the mysterious *Aryans*.

Unfortunately, Velikovsky failed to recognize the full depth of knowledge in the myths of this culture. We refer to the oldest books in the world, the Indian *Vedas*, and also the Hindu great epics and puranas that derived from them. Although he cites these books in perhaps half a dozen instances, Velikovsky overlooked an enormous treasure-trove of information. It is difficult to say why. Perhaps it was because he favored the myths of the middle east, with which he was more familiar. Perhaps he, like many other scholars, both Eastern and Western, was put off by the sheer numbers of deity names in the Vedic hymns and Hindu myths which made them appear less likely to relate to events involving only four planets. However, the Master having taught, the apt student learns.

After considerable study of (translations of) the sacred 'book' called the *Rig Veda*, and certain later selections from the Great Epics and the Puranas, we have come to the conclusion that *these books contain a detailed, sequential record of observed cosmic events in the solar system in proto-historical times. Moreover, that this was the sole purpose for their compilation, as implied by their very name, 'veda' which means 'knowledge'.* At the risk of offending the many sages, living and dead, who have studied and agonized about the symbolic and psychologic meanings of these cryptic hymns and myths, we advance a completely physical interpretation of these 'myths', based on the general scenario suggested by Immanuel Velikovsky in *Worlds in Collision*. When interpreted with his hypothesis as a backdrop, and in the light of modern science applied to data from recent probes of the terrestrial planets, these myths unfold a sequence of absolutely amazing events which no one else has even imagined in modern times.

Our sources for the Vedas and Hindu books are translations/interpretations by Oriental and European scholars. These are only a tiny, but representative fraction of the total volume of the books from which they are quoted. Nevertheless they supply a richness of observational detail which has made it possible to interpret them in terms of planetary interactions. It is of interest to note that the translators were scholars of oriental languages, endeavoring to understand the Sanskrit in light of, and in the context of nineteenth century knowledge. Most of the translations and interpretations referenced in this book are at least one hundred years old. Therefore these scholars had few preconceived notions of planetary interactions, and lacked the modern scientific vocabulary which might have biased their translations in favor of such events. Generally speaking this provides a 'purer' source as a starting point for our interpretations.

The notion that the origin of some mythologic gods may have been physical in nature was espoused by a few, as evidenced by the following quote, published in the last century:

> Again, it is still as true as ever that a large number of the gods, whether early or late, are nature-gods whose origin, we regret to say, has been somewhat obscured by later processes. They again make up for the student of the Veda a class, the most important as well as the most difficult theme of investigation. Every nation's mythology must contain gods of this class. They bring with them problems that will never be dismissed until they are finally answered - and that, paradoxically, may never be. We have in mind gods like Varuna, Indra, and the Acvins. To some Vedic scholars it seems without a doubt begging the question to speak of nature-gods in cases when we do not know for certain what was the natural object that was personified. No one can say at this time that the origin of either Varuna, Indra, or the Acvins has been definitely settled. Yet, for my part, I confess that faith, because I remembered that such uncertainty represents in truth the normal result of mythologic development. As a rule, a nature-god does not remain transparent forever: the opposite happens far more frequently, as may be seen, again and again, in Hellenic or Teutonic mythology. Really durable myths are, as a rule, mixed myths, and, therefore, more or less obscure myths. A certain amount of the complications and entanglements of human life must be imported into mythology before it becomes mythology. Otherwise it remains philosophy, primitive cosmic philosophy, or primitive empirical natural science.

The Proliferation of Deity Names in the *Rig Veda*

Our approach in determining the meaning of the myths is a combination of inductive and deductive reasoning. We begin with the assumption of a scenario of the type suggested by Velikovsky as a given. We also have come to assume, after much study, that the subject matter of the Vedas is no more and no less than a detailed record, although in non-technical language, of these cosmic events. We assume the 'gods' or 'deities,' given human or super-human characteristics in the Vedas are the planets, or their adjuncts. We then proceed, much like a detective to try and interpret the meaning of the 'word pictures,' which are the clues.

The interpretation would not be difficult if each mythological god could be identified as a unique planet or phenomenon. The confusing characteristic of the Vedic hymns, and the one that has prevented the revealing of their knowledge to date, lies in the proliferation of deity names. Dozens of gods are named and described, whereas the scenario of Velikovsky involves only Venus, Earth, Mars, and to a lesser degree, Jupiter. To the casual reader, therefore, any correspondence with Velikovsky's theory seems unthinkable. However, the *Rig Veda* itself provides clues to this seeming inconsistency. For instance:

> What is but one, the wise call (by) manifolds (names). [10]

> ... grant it to us, Thy friends, to know Thy countless manifestations - the highest, the lowest, and those between.[11]

In these passages, the priestly class, responsible for the Vedas, is letting us know that different names were used in reference to the same physical bodies (planets). In addition to these general statements to this effect, there are also numerous passages relating two, three, or more specific deities, that is, implying that they refer to one and the same entity. These will be cited herein at the appropriate place. Still further, the association of a particular deity with another can be reinforced by similarities in their descriptions in the Vedas. By linking these associations together it is possible to show the association of a number of them, and thus make the connection with the planets in the V/A scenario.

If, as mentioned above, the Vedas are meant to convey the general scenario proposed by Velikovsky, the number of physical entities involved must be only a few. Yet, the earliest deity, chronologically speaking, Aditi, the mother of the Adityas, the glowing ones, is said to have seven or eight children. Thus the reader is immediately confronted with more gods than are required for the entire scenario. In addition, many more are introduced in subsequent hymns.

One rather sinister explanation why a planet would be given multiple names is that the rishis, the priestly class who compiled, maintained and passed these hymns from generation to generation, deliberately 'encoded' the 'knowledge' in such a way that the average people would not be able to understand it. One fact reinforcing this explanation is that down to relatively recent times, the Brahmins were the only caste whose eyes were ever allowed to see the Vedas. For a person of lower caste to glance at one line of the Vedas was punishable by death.

More plausible reasons for multiple names being associated with one planet were the long times (decades) between major cosmic events and the actual changes in the appearance of these bodies that occurred during these intervals. We claim that the interactions of Venus and Mars with the Earth occurred at intervals for a period of about 3000 years. Therefore different generations witnessed different events and may have called old bodies by new names, particularly considering that both proto-Venus and Mars underwent major changes in appearance during this 'Vedic period'.

Clues to the Age of the Vedas

The Vedas are records of the most ancient knowledge of mankind. The name signifies 'knowledge' that was passed on by voice (*sruti*), because these hymns were accumulated and passed down from generation to generation by chanting, for over a thousand years before they were first written in Sanskrit, perhaps as early as 2000 to1500 B.C. The passing down of the knowledge is described in the following quote from the *Rig Veda*:

> When one of these (the disciple) repeats verbatim the words of the competent teacher - that indeed is considered as a high festival for both. [12]

The Vedas are four in number. The *Rig Veda* is the oldest, followed by the *Yajur Veda*, *Sama Veda*, and the *Atharva Veda*. The relative age can only be estimated based on the content and styles of the Vedas themselves. The *Rig Veda* is the oldest and most important, and is the primary source from which many later period books draw their basic material.

The Vedas were originated by a mysterious people from whose language, proto-Indo-European, are derived practically every modern European language. They called themselves Aryans, a term that came to mean 'noblemen' when they moved into India and apparently assumed stewardship over the indigenous people. The date of this movement is currently believed to have been between 4000 and 1000 B.C. This estimate is based partially on an ancient treaty uncovered in Turkey (Boghaz Keu) between a Hittite king Shippiluliumash and the king Mitanni Mattiwaza. The latter is an Indo-Iranian, who worshiped some of the gods of the *Rig Veda*, and invoked their blessing in the document. The treaty is dated about 1400 B. C. and the named deities do not include the earliest of the Vedic period. This only indicates that the *Rig Veda* covers a significant period prior to this, but how long a time is difficult to surmise.

> (The hymns) have come to us from the midst of a darkness which is difficult to penetrate. The obscurity which surrounds the origin and early development of the religion... has been increased in recent years by what the excavations at Harappa and Mohenjo Daro have revealed. Its beginnings seem to recede farther and farther into the dark backward of time.[13]

We claim that the planetary events described in the Vedas took place over three thousand years, a period longer than that from the birth of Christ to the present day. The following quote from the *Rig Veda*, which the reader will notice does not always refer to the gods in a reverent manner, is of interest in this context:

> The (this) gang of gamesters, pretending to be god-like and true as the sun, ventures the game time and again (literally fifty-three times); but the dice will not truckle to the rage of a contestant, however truculent; even the king must eat humble pie before them.[14]

This quotation, coming from the middle of the Vedic period, makes the analogy between close encounters of proto-Venus and Mars with the Earth, and a die being tossed onto a board, causing good fortune and bad for the hapless human beings, who, whether they are poor men or kings have no control over the cosmic scale events unfolding in the heavens. Since the number fifty-three is specifically stated, the implication is that close encounters had occurred on that number occasions, at the time when the hymn was written. *However, it is later stated in the Rig Veda that there were a hundred Indras, indicating that there were a hundred encounters with Mars. This number is corroborated in the later Hindu myths.*

In a few cases geologic clues are present, such as the mention of a sea which covered all of northern India in proto-historical times, but they only lead to conflicts with the currently accepted geological dogma of uniformitarianism, examples of which are (1) the notion that mountain-building (orogeny) takes millions of years, and is the result of thousands of small earthquakes, each of which raises the land only a few inches; (2) that the total amount of water on the Earth has remained the same for many millions of years: and (3) that the Ice Ages were the only periods in the past during which the sea was lower than its current level. Catastrophism, the term which best describes the events we interpret from the myths, violates all of these tenets, currently considered 'laws' in Earth science. It is wise to reserve judgement on the matter of the seas mentioned in the Vedas at this juncture for reasons that will be revealed later. Also, care must be taken when applying the dating methods used currently by geologists and archaeologists, because they are also based on certain assumptions concerning the uniformity of the Earth's environment which the ancient myths indicate are invalid.

The Beginning - The Birth of Aditi

The first cosmic event described in the *Rig Veda* was the 'birth' of a deity called Aditi. It is extremely important this is the very first event described because it confirms what Velikovsky proposed as the trigger of the recent catastrophism, that the *Rig Veda* portrays the cosmic events in the order of their occurrence, and that the recording of these events was its sole purpose.

> Probably the term Aditi, 'the boundless,' was originally employed as an epithet of Dyauspitar, the Heaven Father...

> Aditi has the honour of being almost the only goddess mentioned by name in the Rig-Veda as the mother of any of the gods. [15]

Aditi was born from Dyauspitar (Jupiter), the father of the 'high heaven'. That is, the planet that dominated the night sky, as opposed to a morning or evening star (planet). As we now understand, this implies an orbit completely outside the orbit of the Earth. No mother is mentioned, rather Aditi is said to have been born from the combining of Jupiter and his 'maya,' or spiritual power. Aditi was said to have seven or eight sons, called Adityas, but all their names are not given, and some of the names given in different passages are not consistent:

> Judging from the number of hymns addressed to them, some of the deities occupied a conspicuous position in the Pantheon whilst others are named once or twice only, and then in connection with their more illustrious brethren.

As discussed above, the Vedas represent the same physical entities by 'manifold names' due to the fact that their appearance changed with time. Each of these names has associated with it important characteristics of the body as it evolved, which when interpreted correctly, add to the information conveyed, even though this is confusing to the casual reader.

Aditi was the original manifestation of the proto-planet Venus which was 'born' out of Jupiter as the result of an enormous impact. Thus the 'sons' of Aditi represent a series of aspects of the planet Venus as it evolved before the eyes of the Vedic people. During the decades after this impact the appearance of proto-Venus changed very rapidly, due to interactions with the Sun and the great variations in its distance from the Earth. The hymns reflect amazement at the many different aspects this deity takes on.

Just after the birth of Aditi, a strange thing is observed:

> Of the eight sons of Aditi, who were born from her body, she approached the gods with seven, and cast out Marttanda (the eighth) ... The eighth son was deformed. His brothers, seeing his deformity, improved his appearance. He was afterwards known as Vivasvat. From the superfluous flesh cut off his body, an elephant was formed, hence the proverb, 'Let no man catch an elephant, for the elephant partakes of the nature of man'.[16]

> The [Sanskrit] word Mrttnda [Marttanda] is compounded of two words, "mrtta", apparently derived from "mrityu", death, and "anda," an egg or place of production. [17]

Our interpretation of this story is as follows: a large hypervelocity 'traveler,' probably not visible from the Earth, struck Jupiter. As a result, a white hot cloud of plasma was ejected from the planet out into space. It must have been many times the size of Jupiter itself, otherwise the Vedic people could not have distinguished its elephantine shape. This event was actually witnessed by the Vedic people. The incandescent plasma cloud initially was irregular in shape, like the white of an egg, and then the main portion, having sufficient velocity to escape Jupiter's gravity, coalesced into a spherical shape (the yoke of the egg) under the influence of its internal gravitational forces. The resulting incandescent spherical body was known as Vivasvat. This was the third name given to the planet Venus; Aditi being the first and Marttanda the second. At this time proto-Venus was at its hottest, and glowed with a white light that was still impressive even though it was quite far from the Earth. Circumstantial evidence of the birth of Aditi exists to this day in the form of the Great Red Spot of Jupiter, which is the manifestation of material still jetting out of the crater formed by the impact.

Marttanda was a very transient aspect of the planet, and was only mentioned at the beginning of the *Rig Veda*. However, the name Vivasvat was invoked whenever proto-Venus shined with a white light, a great distances from the Earth. The earliest aspects of proto-Venus were striking, but were not seen as a direct threat, because it was still a great distance away. However, the impact did greatly perturb the Earth, a fact that portended subsequent doom.

The initial velocity of the coalescing incandescent proto-Venus was such that it went into a highly elliptical orbit about the Sun. Its closest distance from the sun, perihelion, was probably less than the radius of its current orbit. If it were a rigid body, its orbit would have continued to swing back out to the orbit of Jupiter where it originated. The *Rig Veda* indicates that it did swing out as far as the Earth's orbit but only for an extremely short time astronomically speaking. Assuming aphelion at Jupiter's orbit and perihelion at Venus' current orbit, its initial period was approximately five Earth years, and it crossed the orbital path of the Earth twice during each period. Of course, close encounters would only have occurred when the Earth happened to be passing the same point where and when proto-Venus crossed its orbital path. As its eccentricity decreased and period decreased, the probability of such encounters would have increased.

The Birth of Agni

The first major aspect of proto-Venus, that is, as a god to which numerous hymns were completely devoted in the *Rig Veda*, came several years after the birth of Aditi, when its orbit first brought proto-Venus close to the Earth. This was the great god Agni (fire). The flaming, proto-planet was literally a seething ball of magma as large as the Earth itself, on fire, at temperatures over 10,000 degrees, spewing enormous amounts of smoke out into space which formed two great tails similar to a comet, but much longer and black. Explosions in its interior repeatedly blasted huge clouds of flaming plasma into space, most of which fell back onto the surface. It raged into the inner portion of the solar system and, after relatively few orbits, came close enough to the Earth to cause devastation. The first devastating approach was described in the *Rig Veda* as the 'birth' of Agni, with no specific recognition that this was the same physical body which was born years before and called Aditi at that time. Actually the recognition *is* present in the Vedic hymns, because Agni is referred to as an Aditya, a son of Aditi. This was the only deity to be 'born in heaven'. Others, such as Indra (Mars), are described as changing their aspect, or being 'released' from their bonds, that is, having their orbit *changed*, but Agni was unique and awesome to behold. The birth of Agni is described in the *Rig Veda* in quite cryptic terms:

> In these old hymns Agni is spoken of as dwelling in the two pieces of wood, which being rubbed together, produce fire; and it is noticed as a remarkable thing that a living object should spring out of dry (dead) wood.

A slightly different translation of this comes from the following verses by Dr. John Muir, author of the preeminent English translation of the *Rig Veda, Original Sanskrit Texts*.

1 Great Agni, though this thine essence be but one,
Thy forms are three; as fire thou blazest here,
As lightning flashest in the atmosphere,
In heaven thou flamest as the golden sun.

2 It was in heaven thou hadst thy primal birth;
By art of sages skilled in sacred lore.
Thou wast drawn down to human hearths of yore,
And thou abid'st a denizen of Earth.

3 Sprung from the mystic pair, by priestly hands
In wedlock joined, forth flashes Agni bright;
But oh! ye heavens and Earth, I tell you right,
The unnatural child devours the parent brands.

4 But Agni is a god; we must not deem
That he can err, or dare to comprehend
His acts, which far our reason's grasp transcend;
He best can judge what deeds a god beseem.

5 And yet this orphaned god himself survives:
Although his hapless mother soon expires,
And cannot nurse the babe as babe requires,
Great Agni, wondrous infant, grows and thrives.

6 Smoke bannered Agni, god with crackling voice
And flaming hair, when thou dost pierce the gloom
At early dawn, and all the world illume,
Both heaven and Earth and gods and men rejoice.

7 In every home thou art a welcome guest,

The household tutelary lord, a son,
A father, mother, brother, all in one,
A friend by whom thy faithful friends are blest.

8 A swift winged messenger, thou callest down
From heaven to crowd our hearths the race divine,
To taste our food, our hymns to hear, benign,
And all our fondest aspirations crown.

9 Thou, Agni, art our priest: divinely wise,
In holy science versed, thy skill detects
The faults that mar our rites, mistakes corrects,
And all our acts completes and sanctifies.

10 Thou art the cord that stretches to the skies,
The bridge that spans the chasm, profound and vast,
Dividing Earth from heaven, o'er which at last
The good shalt safely pass to Paradise.

11 But when, great god, thine awful anger glows,
And thou revealest thy destroying force,
All creatures flee before thy furious course,
As hosts are chased by overpowering foes.

12 Thou levellest all thou touchest; forests vast
Thou shear'st, like beards which barber's razor shaves.
Thy wind-driven flames roar loud as ocean waves,
And all thy track is black when thou hast past.

13 But thou, great Agni, dost not always wear
That direful form; thou rather lov'st to shine
Upon our hearths, with milder flame benign,
And cheer the homes where thou art nursed with care.

14 Yes! thou delightest all those men to bless
Who toil unwearied to supply the food
Which thou so lovest - logs of well dried wood,
And heaps of butter bring, thy favorite mess.

15 Though I no cow possess, and have no store
Of butter, nor an axe fresh wood to cleave,
Thou, gracious god, wilt my poor gift receive:
These few dry sticks I bring - I have no more.

16 Preserve us, lord; thy faithful servants save
From all the ills by which our bliss is marred;
Tower like an iron wall our homes to guard,
And all the boons bestow our hearts can crave.

17 And when away our brief existence wanes,
When at length our earthly homes must quit,
And our freed souls to worlds unknown shall flit,
Do thou deal gently with our cold remains.

18 And then, thy gracious form assuming, guide
Our unborn part across the dark abyss
Aloft to realms serene of light and bliss
Where righteous men among the gods abide.

This lyrically translated hymn is rife with images, which when we are armed with the general scenario uncovered by Velikovsky and a knowledge of physics, give us a picture of those ancient events almost as clear as a television documentary. This was the intent of the rishis. It also gives profound insights into the thinking of the people, and shows the very origin of religious thought arising out of awe and fear - fear of immense and uncompromising power.

Verses 2, 3, 4 and 5 detail Agni's primal birth, that is, Agni's birth as opposed to Aditi, termed his 'mother' in the style of the *Rig Veda*. It was witnessed by everyone on Earth. The descriptions used here are a bit cryptic, but they yield their information, if begrudgingly. As proto-Venus, recently blasted out of Jupiter, still incandescent and out gassing enormous amounts of smoke and gases, approached the Earth from the sun side, it was largely hidden by the two smoke tails which stream ahead of it toward the Earth. Just as the tails of comets, the smoke streamed in the direction away from the sun, regardless of the direction of motion, due to the pressure of solar radiation. The two smoke trails are described here as '...the mystic pair...,' ' ... priestly hands ... joined...,' and '...the parent brands...'.

We know from comets and man-made experiments that at least two streams of particles result when an object exudes gas or smoke in space. This is due to the fact that some of the particles are ionized (have an electric charge), while others are electrically neutral. Due to the very high temperature of proto-Venus at this time a larger fraction of the smoke or gas particles were ionized than would be true with ordinary comets, which are primarily ice; therefore the second tail was as prominent as the first. The interplanetary magnetic field caused the (singly) ionized particles to move in a different direction than the neutral particles. Doubly ionized particles explain the presence of a third leg, which is sometimes used in descriptions of Agni:

> In pictures he is represented as a red man, having three legs, seven arms, dark eyes, eyebrows and hair. [18]

As the flaming proto-Venus (Agni) approached the Earth from the Sun side it was barely visible since these streams of smoke were dark and extended in the general direction of the Earth. They appeared as dark sticks, as opposed to the tails of comets, which are made up of ice particles which, when viewed at night, appear light. Thus the 'mystic pair' are the two 'sticks' which converge to a common point at which the fire (Agni) is found. These same dark forms are also imagined as two irons of the type used in branding cattle, with their ends resting in the fire. As proto-Venus closed with the Earth, and perspective increased, it was pictured as two priestly hands joined in prayer. The very idea of placing the hands together in prayer most likely came from this sky-image.

Then verse 3 states 'the child devours the parent brands,' characterizing the smoke columns as Agni's parents, and imagining that, as the brightness of the planet itself begins to dominate due to its increasing proximity to Earth and the smoke tails become disbursed due to the magnetic field of the Earth, the planet is consuming its own parents.

Note in verses 1 and 2 how Agni is first in heaven then close to Earth. This intercession between the Earth and the heavens suggests the 'religious' metaphor for the observed action of Agni, of delivering the souls of departed humans to the gods in verses 10 and 18. The obvious interpretation of this is that the orbit of the newborn proto-Venus brings it closer and closer to Earth, for example, in verse 5 ' Great Agni, wondrous infant grows and thrives'. Then after a close encounter with the Earth it gradually moves back to a more star-like aspect 'with the gods'. Only when it was close enough to appear bright and glorious, was it praised, worshiped, and offered sacrifices as in verses 7 and 13.

However, the raw destructive power of proto-Venus was unleashed when it approached closer to the Earth. This is expressed very graphically in verses 11 and 12. The translation can hardly be misinterpreted. According to our theory, the entire proto-planet Venus was actually a ball of plasma or molten lava. It was cooling by radiation, and emitting prodigious amounts of ultra-violet, visible and infrared radiation. 'Thou levellest all thou touchest, the forests vast thou shearest, like beards which barber's razor shaves... and all thy track is black when thou hast passed'.

Mixed in the remaining verses are worshipful expressions which attempt to appease this menacing 'god,' through praise and sacrifices. The sacrifices were of burning wood, since when Agni came really close to the Earth, he consumed even the green forests. The other offering, the *hom*, was butter, because when the proto-planet Venus approached the Earth, glowing masses of hot plasma were shot out in space with explosive force, only to be drawn back again, by the planet's gravity. These 'oblations' appeared golden to observers on Earth, and as they fell back to the surface, it appeared that melted butter was being consumed by the god.

This was a 'god' no one ignored, because he approached the Earth menacingly, becoming larger and larger, causing *continuous* creaking and quaking of the Earth, to the point that, according to verse 6, 'with crackling voice' he slaughtered every one and every thing in his path. It was only by an apparent miracle that he did not collide with the Earth. Although a collision did not occur the Earth was changed in an irreversible way.

The fact that sound was actually heard from Agni during one or more of its encounters with the Earth, implies that an 'atmospheric bridge' was formed temporarily between the two planets, since sound waves cannot propagate across the vacuum of space. This is consistent with passages from other cultures cited by Velikovsky indicating that the crackling of flames could be heard on Earth. Professor Williams gives the following account of Agni:

> Bright, seven-rayed god, how manifold thy shapes
> Revealed to us thy votaries: now we see thee
> With body all of gold; and radiant hair
> Flaming from three terrific heads, and mouths,
> Whose burning jaws and teeth devour all things.
> Now with a thousand glowing horns, and now
> Flashing thy lustre from a thousand eyes,
> Thou'rt borne towards us in a golden chariot,
> Impelled by winds, and drawn by ruddy steeds,
> Marking thy car's destructive course with blackness.

The 'seven' rays, 'three heads' and 'glowing horns' are the same promenances that inspired the Mexican name Quetzal-Cohautl, the feathered serpent. These were plumes of hot gases from explosions on the seething proto-Venus.

The Aryans who composed the Vedic hymns probably lived in what is now the Punjab, in northern India or Pakistan, in the foothills of the Himalayas, although they may have migrated to this area after the devastating encounters with proto-Venus. The reason for their movement toward the Himalayas will become clearer as we discuss subsequent events of the Vedic period. They did not actually go up into the Himalayas because they were highly unstable at that time. The reason for this is implied by the following quote:

> The mountains are the eldest children of Prajapati (the Creator). They were winged (birds). They kept flying forth and settling wherever they liked. At that time the Earth was unstable. (God) Indra cut off their wings. By means of the mountains he made firm the Earth. The wings became clouds. Therefore these clouds ever hover about the mountains. For this is their place of origin. [19]

Our interpretation is that proto-Venus, because of its great mass, exerted tremendous tidal forces on the Earth during the close encounters, and even during not-so-close encounters, because of its great mass. Early in the Vedic period entire mountain ranges were uplifted and the crust of the Earth was unstable.

Thus, during the early Vedic period when the massive proto-Venus still came fairly close to the Earth the lithosphere was in chaos. The mention of Indra's effect refers to later interactions at which time Mars became the dominant god.

The orbit of proto-Venus evolved rapidly due to a combination of factors which will be discussed below. After two destructive encounters in which it incinerated a significant portion of the Earth's surface and caused tidal waves that swept across entire continents, its eccentricity decreased rapidly to the point that its orbit no longer crossed that of the Earth. Mythology implies that this took the surprisingly short time of less than fifty years. Although astronomers and astrophysicists will undoubtedly dispute that such a reduction in eccentricity could occur is this short a time, they have no concept of the extraordinary forces exerted on proto-Venus by the Sun, due to its high temperature and ionized state. After that date it no longer came close enough to cause the complete devastation of the Earth that it had on the two earliest encounters. However, its influence on the Earth and Mars was still great at times when its aphelion corresponded to its inferior conjunctions with the Earth. This took the form of almost continuous earthquakes and effected interactions between Mars and the Earth. The Vedic people called the mysterious power of the Agni deity his 'maya'. They recognized that the gravitational force of proto-Venus was much greater than that of Mars because of the earthquakes that it caused even when at considerable distance from the Earth. Mars had to come much closer to the Earth to cause a comparable level of disturbance.

Agni was the first major (life threatening) deity. The hymns devoted to him are not as numerous as are those devoted to some of the later 'gods'. Nevertheless he was the most destructive of the deities. He destroyed most of the people and animals on the Earth, and therefore much of the memory of those devastations was lost. The Aryans were one of the few cultures which survived and so were able to relate the story of Aditi and Agni. The legends of the proto-Venus destructions were undoubtedly overshadowed by the many cosmic events of similar nature which occurred later due to Indra (Mars), to which many millions of people in the eastern hemisphere were witnesses. The following verses are from a *Rig Veda* hymn entitled 'To the Fathers (Funeral Hymn)":

> Now let us pay homage to the Fathers, to those
> who passed of old and those who followed,
> Those who have rested in the earthly region, and those
> who dwell among the mighty races...
>
> Fathers whom Agni's flames have tasted, come ye nigh:
> ye kindly leaders, take ye each your proper place.
> Eat sacrificial food presented on the grass: grant riches
> with a multitude of hero sons

Epithets for Agni

There is a subtle difference in Sanskrit between the actual name given a deity and a so-called epithet. The latter are descriptive 'pronouns' which refer to a deity by means of one of his physical characteristics, or some action he is known to be involved in. For example, in English, we sometimes refer to Mars as the 'red planet'. In Sanskrit translations these are combined into single words and capitalized in translation, thus appearing exactly as names in the English versions. The more important a deity is, the more epithets he accrues throughout his 'lifetime'. Some scholars compile lists of the epithets associated with the popular deities, and these are helpful as sort of a summary, or an additional corroboration of the physical theory in the case at hand. For example the following list of epithets for Agni appears in one reference[20]:

Vahni,	'He who receives the hom [offering of melted butter] or burnt sacrifice'
Vitihotra,	'He who sanctifies the worshipper'
Dhananjaya,	'He who conquers (destroys) riches"
Jivalana,	'He who burns'
Dhmketu,	'He whose sign is smoke'
Chhgaratha,	'He who rides on a ram'
Saptajihva,	'He who has seven tongues'

In several cases deities in the Vedic period were given epithets which at the time were merely descriptive terms. However during the Puranic or Hindu period which followed, in which an entire pantheon of new god names was adopted, one epithet was elevated to the highest level of a deity name, and the major deity name from the Vedic period was downgraded or completely eliminated. The best example was Siva, which meant 'the auspicious' in the Vedas, who became one of the three major deities of the Hindu period, that is, Brahmā, Vishnu, and Siva.

3
VARUNA

To me the converging objects of the universe perpetually flow,
All are written to me, and I must get what the writing

Walt Whitman
Song of Myself

Agni (proto-Venus, literally 'fire') faded from dominance (but not prominence) but as it continued to pass in the vicinity of the Earth, where it could clearly be seen as a 'world,' it came to be called by a new name, Varuna. This reflected both the change in its destructive nature as well as its appearance. To the best of our knowledge, no other scholar of the *Rig Veda* has recognized that these are one and the same physical body. It is not as though the common identity of Agni and Varuna is without hints in the *Rig Veda*:

> And now, as I am come before his presence, I take the face of Varuna for Agni's.[21]
>
> When God Agni struggles toward the *rta* he is said in a remarkable passage to become for the time being God Varuna.."[22]

In the *Rig Veda*, the name *Varuna* was applied to proto-Venus when it approached close to the Earth after its eccentricity had been reduced to the point that it no longer crossed the orbit of the Earth. It still loomed very large at certain inferior conjunctions but had evolved to a more benign state. A number of other names are associated with proto-Venus during the times between close approaches, when it was farther from the Earth. These will be discussed in the following chapter.

Varuna is often referred to using a pair of names, *Mitra-Varuna*:

> In the hymns addressed to Mitra and Varuna together, almost the same terms are employed as when Varuna is addressed alone. Both are spoken of as righteous, and as promoters of religion. They are said to avenge sin and falsehood. [23]

We propose that this is indeed a very intimate relationship, but one resulting from a unique physical relationship between these two 'gods.' To corroborate the idea that Mitra and Varuna constitute a unique pair in the scheme of the Vedas, note the following quotation:

> But Varuna is a close partner in a partnership which is expressed in the dual number. It consists of himself and the god Mitra, who is, however, little more than a silent partner in the combination. Such partnerships are frequent in the Veda, but exceedingly rare in the Persian Avesta. Yet the Avesta, in a matter-of-fact manner, joins Ahura and Mithra in the same dual partnership as the Veda does Varuna and Mitra.[24]

The literal translation of Mitra is 'compact,' and of Varuna is 'envelop' or 'encompass'. Keeping this in mind, consider what was happening to proto-Venus at this time. As an isolated body in space, the most likely physical change that would have happened to Agni (proto-Venus) early in its life would be to form a 'crust,' just as glowing lava does today on the Earth. Even though no cooling by contact with an atmosphere could have occurred early in its 'life,' there would have been rapid cooling at the surface due to radiation into space, which is essentially at absolute zero temperature. Thus its blazing brightness would have diminished and been replaced by a deeper red glow and eventually a dark hot crust. At this stage the crust was very thin and fragile. This led to verses like:

> ...he watches with a thousand eyes...

The crust was easily ruptured by any tidal distortion due to another planet because it was still a molten ball of boiling rock, with no rigidity. Thus when proto-Venus approached the Earth its entire shape was distorted by the tidal force of the nearby Earth. The magma interior radiated through the myriad of cracks, which were seen as 'a thousand eyes'. This very theme was applied in Egyptian myth, where some hieroglyphs for Osiris are translated 'many eyed'.

Upon closer approach, the newly formed crust was completely torn, causing the proto-planet to again blaze with fire, and Agni was said to be 'born again'. This happened three times according to the *Rig Veda*.

> Three several places of his birth they honor, in mid-air, in heaven, and in the waters. [25]

The distorted shape of the still liquid proto-Venus during these encounters is the explanation for epithets such as:

> ...the horse faced god...

In order to understand the pairing of Mitra-Varuna, it is necessary to introduce a second major Vedic deity, Indra. In addition to the formation of a crust, there was another major factor which influenced the evolution of the newly formed proto-Venus at this time. That is, interactions between proto-Venus and Mars. These occurred 'innumerable' times according to the Vedas, and the notion is reinforced by the continuing theme of interactions between Varuna and Indra in many of the hymns. Amazingly, the *Rig Veda* reveals that, early in the Vedic period both proto-Venus and Mars could be seen quite clearly as 'worlds,' since they were relatively close to the Earth at that time.

Early in the Vedic period, Agni had a close encounter with Indra which apparently deflected it from a close encounter with the Earth. This conclusion is based on a story from the Great Epic, the Mahabarata, in which Agni was said to have exhausted his vigor by consuming too many oblations of butter, and desired to regain his strength by consuming the entire Khandava forest, but:

> He was (at first) prevented from doing this by Indra.

Interactions with Indra (Mars) were the cause for the impression that Agni (proto-Venus) had lost some of his vigor. The *Rig Veda* tells us that Indra loved 'Soma' and took great draughts of this heavenly liquid from Varuna as often as he could. This means that when interactions occurred between proto-Venus and Mars, the Vedic people saw a 'bridge' of light material extend between these two deities. Velikovsky cited myths of other cultures in which this 'bridge' between the two planets is interpreted as a 'white sword'.

We suggest that what they observed was a portion of the atmosphere of Mars being drawn off the smaller planet by the tidal force of proto-Venus. The water vapor in this atmospheric 'bridge' froze into ice crystals in the near-absolute zero environment in space, making it appear white. It was not Soma going from Agni to Indra, but there was probably Soma passing from Mars to proto-Venus, along with the atmospheric gases. It was impossible for the Vedic people to tell the direction of the transfer, and made a good story to characterize what the Vedic people observed in the heavens, agreeing with their own experience, with the Soma (Chapter 6) that fell to the Earth. The acquisition of an atmosphere partially shielded the radiation emitted from proto-Venus and thereby gave the impression that its strength had been diminished. Also the cooler atmospheric gases captured from priori-Mars helped retain the gases being vented from the interior of proto-Venus, such as sulfur.

In some hymns the intoxicating nature of the Soma was also pictured as energizing Indra, causing him to become larger. This notion was due to the fact that the interactions resulted in the transferring of orbital energy from proto-Venus to Mars, further reducing the eccentricity of proto-Venus' orbit and increasing the eccentricity of Mars' orbit. Conservation of orbital energy and angular momentum demand that if one orbit is reduced, the other must be extended. *The implications of this scenario are profound. It means that, at the beginning of the Vedic period, Mars was in an orbit inside that of the Earth - much like the current orbit of Venus. It was ejected from this orbit by proto-Venus, and as a result moved outward toward the Earth.* This is clear from the fact that its apparent size increased after a number of encounters with proto-Venus.

Mitra-Varuna

Having explained how proto-Venus had begun to acquire an atmosphere, we are now prepared to explain the physical nature of the dual deity Mitra-Varuna:

> Varuna, wearing golden mail, hath clad him in a shining robe.[26]

In this quotation lies the explanation for the special pairing of proto-Venus deity names. Mitra literally meant 'compact'. This was a reference to the solid body of proto-Venus, and Varuna, meaning 'to envelop' referred to proto-Venus when its 'shining robe' or atmosphere was visible. As Mitra-Varuna approached the Earth, both could be clearly seen.

> Mitra became the guardian of the day, while Varuna's part was reduced to the guardianship of the night. [27]

This statement is particularly significant, and is absolutely consistent with our hypothesis. The dully glowing body of proto-Venus could be seen even during the day, but the atmosphere of proto-Venus could only be seen during the night, because the daylight sky of the Earth was too bright during the day. The following quote suggests the existence of two 'firmaments':

> All this king Varuna does behold - what is between the two firmaments, what beyond. [28]

This is a mysterious term when applied to the sky, where there is nothing 'firm'. We believe that it refers to other planets that were close enough to the Earth to be recognized as 'worlds.' Thus this may refer to proto-Venus and Mars, which, as discussed in subsequent chapters, was also close to the Earth during the Vedic period.

Although not many hymns were composed to Varuna, some of the most beautiful hymns in the *Rig Veda* are devoted to him. In the following excerpt, the word '*kine*' is used. This is an English word that has passed out of general usage since the time of the translation, somewhat more than a hundred years ago. It means cattle, which were the primary measure of wealth to the Vedic people. Similarly, the word '*meath*' was a drink made from Soma, an ambrosia that fell from the sky.

> He knows the pathway of the wind, the spreading, high and mighty wind;
> He knows the gods that dwell above.
>
> Varuna, true to holy law, sits down among his people; he
> Most wise, sits there to govern all.
>
> From thence perceiving, he beholds all wondrous things, both what hath been
> And what hereafter will be done.
>
> May that Aditya, very wise, make fair paths for us all our days;
> May he prolong our lives for us.
>
> Varuna, wearing golden mail, hath clad him in a shining

robe;
His spies are seated round about.

The god whom enemies threaten not, nor those who tyrannize o'er men,
Nor those whose minds are bent on wrong.

He who gives glory to mankind, not glory that is incomplete,
To our own bodies giving it.

Yearning for the wide-seeing one, my thoughts move onward unto him
As kine unto their pastures move.

Once more together let us speak, because my meath is
brought priest-like
Thou eatest what is dear to thee.

Now saw I him whom all may see; I saw his car above the
Earth;
He hath accepted these my songs.

Varuna, hear this call of mine: be gracious unto us
this day,
Longing for help I cry to thee.

Thou, O wise god, art lord of all, thou art king of Earth
and heaven;
Hear, as thou goest on thy way.

Release us from the upper bond, untie the bond between, and loose the bonds below, that I may live...

Strike us not, Varuna, with those dread weapons which,
Asura, at thy bidding wound the sinner,

Let us not pass away from light to exile. Scatter, that
we may live, the men who hate us.

As with the hymns to Agni, there is the idea that 'Varuna knows the gods' on the one hand and 'sits down' with people on the other. This reflects the fact that during this period proto-Venus' orbit takes it far away, and then brings it close to the Earth. Varuna's appearance is graphically described as wearing golden mail, covered with a shining robe. This is the still glowing proto-Venus surrounded by a transparent atmosphere. The phrase 'once more together' refers to a close approach of proto-Venus (inferior conjunction) that the hymnist is observing. The meath is the Soma being brought to the worshipper 'priest like'. It is clear that he sees Varuna's 'car above the Earth' and as 'thou goest thy way' implies proto-Venus is now moving away from the Earth, but the supplicant asks at this juncture for Varuna to 'release the bonds'. This describes a gravitational 'locking up' of the two planets at the time of proto-Venus' inferior conjunction, which must be broken off as they disengage. This was obviously a dangerous time at which great stresses were exerted on the crust of the Earth, causing continuous earthquakes, very high tides and even changes in the orientation of the spin axis of the Earth.

The following quotation summarizes the attributes of the god Varuna.

> There are relatively few of the hymns addressed to Varuna, though he is the guardian of the cosmic law (rta) and the universal monarch, the object of greater veneration than the more obviously Aryan deities. Varuna is the Prime mover of the universe. He is pictured as creating the universe, standing in the air, and by exercise of his creative will, or maya, and using the sun as an instrument, measuring out the three worlds and forming them; the heavens, the Earth, and the air between them...he is thought to be seated majestically in his thousand-columned, thousand-doored gold palace in the sky surrounded by his ever watchful spies. [29]

As mentioned above, Mitra, the solid planet proto-Venus, golden in color, was observed through the newly captured atmosphere. The meaning of 'his thousand-columned palace' is that vertically developed clouds, related to convection cells, appeared as columns. Such uniformly spaced column-like clouds are found in moist tropical regions of the Earth, such as Amazonia. However the column-like clouds on proto-Venus did not have their bases near the surface of the planet because the great heat there never would have allowed water to condense. Therefore the clouds must have extended upwards from altitudes in excess of 50 kilometers. Since the driving source of heat on proto-Venus is internal, that is, not linked to the incidence of solar radiation, these clouds could have been present over the entire planet. The fact that these clouds on proto-Venus could be seen from the Earth implies that proto-Venus came as close as 100,000 kilometers at that time.

The phrase 'thousand-doored,' refers to the many cracks and vents in the thin crust of proto-Venus, which were aggravated by the tidal force of the Earth, during close encounters. The freshly exposed lava glowed extremely brightly through these cracks and appeared as light streaming out through many doors. The following quote is another summary of the characteristics of Varuna:

> (Varuna) is the conserver of rta, cosmic law and truth. *Rta* is the equivalent of the later dharma and is equally difficult to define in any precise terms. It is the law that governs the universe... The purely

> physical traits attributed to Varuna appear to be secondary, and it is not all clear why, even in the Rig-Veda, he is associated with the waters and with rain. Like Indra, Varuna causes the rivers to flow ... but there is a marked contrast in the manner in which these deities perform the same function. Indra has to wage a terrific battle in order to set the waters free. Varuna does it effortlessly by his 'uncanny power' (maya)...Like the Indian monarchs of later days, Varuna has a host of spies 'undeceived and wise' whose sole function is to report on the misdoings of men...Varuna is the guardian of the law he has himself established; he brings order into a disorderly cosmos. 'The all-knowing Asura' propped up the heavens and measured out the expanse of Earth. All worlds did the great king take into his possession' (*Rig Veda*, 8.42.1). Once established the worlds are kept apart and in position by his law, and the sun, moon, and stars pursue their regular courses in accordance with his ordinances. [30]

Our interpretation of this passage is as follows: The mysterious law, or 'rta,' is the gravitational law which governs the orbits of the planets and their interactions. Although the priests could not understand the source of gravity as being due to the mass of the planets, they witnessed the results of the gravitational law in a much more impressive way than we ever can. They certainly recognized that some law or 'right order' was present, and further recognized that Varuna exercised this law with great power. We believe 'Propping up the heavens' refers to the uplifting of the Tibetan plateau and the Himalayas by proto-Venus, starting with Agni and continuing with Varuna, due to the strong upward gravitational attraction on the crust of the Earth at the point closest to proto-Venus during close encounters. The association with the waters could refer to the melting of mountain glaciers in the Himalayas or to ocean tides orders of magnitude greater than what we experience in modern times. Proto-Venus caused great tectonic activity within the Earth, producing almost continuous earthquakes and associated movement of magma toward the surface of the Earth. This was particularly true in the Himalayas, since they were freshly formed and still unstable, with innumerable fissures and vents. The resulting heat was strong enough to partially melt the mountain glaciers, and thereby releasing the water that had accumulated there. Varuna appeared to perform this effortlessly, that is, without the raining down of hot bodies and thunderbolts, that characterized later encounters with Indra (Mars), because its gravitational field was much stronger for its apparent size. The 'host of spies' were probably chunks of glowing rock ejected from proto-Venus at great pressure, which under the influence of the Earth's gravitational field, flew across the space between the planets when close, went into low Earth orbits, and subsequently burned up in the atmosphere. These circling bodies are anthropomorphised as observers sent out by Varuna to observe all around the Earth, and report any misdoing. They are not described as threatening entities, which caused damage on Earth. This is probably because they did not impact the Earth in the vicinity of the Vedic people. This is in marked contrast to the Maruts, which will be discussed in the section on Indra.

Even more profound is the clear implication, reinforced in more than one passage, that Varuna, as Agni in earlier times, exerted torques on the Earth, which changed its spin axis orientation. The following quote refers to the two destructive close passes of proto-Venus as Agni, which completely overturned the Earth:

> Governing in the east of earthly regions, the seasons hath he established in their order.[31]

The following quote reinforces our hypothesis, that some sort of 'locking up' took place between the Earth and proto-Venus during inferior conjunctions. Although the commentator tries to interpret this as having to do with moral laws, we believe a physical description was what the original author intended.

> Sin is the fetter with which Varuna binds men, and as he binds in his wrath, so can he loose in his loving kindness. 'O Varuna, untie the upper fetter, untie the middle, untie the lower' (RV,1.24.15), the sinner cries, 'for we would be sinless in [carrying out thine] ordinances.' This binding and loosing of the fetters of sin is the prerogative of Varuna and Mitra alone in the Rigveda; and the awe and Angst and God-fearing humility that these fetters inspire is unique in the whole corpus of literature. [32]

Physically, the 'binding and untying' is easily interpreted. The tidal force of proto-Venus on the Tibetan-Himalayan complex and the equatorial bulge, produced torques on the lithosphere, or rigid outer shell of the Earth, which, because it is a spinning 'top' in space, could reorient its spin axis relative to the stars. The Vedic people experienced these changes, which were accompanied with great groaning and howling sounds from the interior of the Earth.

The people saw this as a threatening aspect of Varuna, meant to punish men for sins, but since good men as well as bad were killed when he was angered, they felt that the 'law' by which Varuna judged, was unknowable to man. Thus the rishis often asked forgiveness for unwitting sins.

It does not seem logical that there are so few hymns devoted to Agni and Varuna, because these were the most powerful and destructive of all the deities. There are two possible reasons for this. First, the population in the early stages of the Vedic period was much less because most of the people had been killed in the first two destructive encounters with proto-Venus. As a result few people were alive to remember these devastations. In some cultures the lettered people were all killed and those that survived were incapable of passing down their experiences to subsequent generations. Although the population had begun to recover in the centuries which followed, while proto-Venus, as Varuna, still appeared large in the sky at certain conjunctions, there were still not nearly the numbers that existed in the later portion of the Vedic period. Secondly, the actions of Agni and Varuna were overshadowed by the many explosive onslaughts of Indra, which continued for almost three millennia.

4
THE MINOR ADITYAS

The human compulsion to find absolute truths, however noble, too often culminates in tyranny of the mind, or worse. Only an extreme skepticism toward science - and an open mindedness toward other modes of knowledge and ways of life, however alien - can help us avoid this danger.

John Horgan , paraphrasing
Paul Feyerabend's Farewell to Reason

A number of other gods were worshipper during the same period as Agni and Varuna, but were exalted to a lesser degree. Most have attributes that allow us to associate them with proto-Venus. They were also called Adityas, implying the rishis understood they were aspects of the same physical body as Agni and Varuna. Although the names quoted in different hymns are not always consistent, Aditi was said to have seven offspring. We have mentioned Agni, Mitra, Varuna, Marttanda and Vivasvat, so it is reasonable to expect that some of the other lesser deities were proto-Venus, or effects associated with it. Several of the names stem from visible effects associated with the approach of proto-Venus rather than the body of the planet itself. There are occasional hymns in the *Rig Veda* which appear to convey the common identity of the various deities as follows:

> May Pushan, Pavamana, Mitra, Varuna, Brihaspati, Maruts,
> Vayu, Acvins, Tvashtri, Savitri, and Saraswati come to
> us, glad and joyous, (not singly, but) in a body,- all of
> them, bounteous, with ascesis and deep piety (to God) ... [33]

> Agni awoke upon the Earth, and Surya riseth;
> Broad gleaming Dawn (Ushas) hath shown in brilliance.
> The Acvins twain have yoked their car to travel.
> God Savitar hath roused the world.[34]

> Cheerful in spirit, evermore, and keen of sight, with
> store of children, free from sickness and from sin.
> Long-living, may we look, O Surya, upon thee uprising
> day by day, thou great as Mitra is. [35]

The reason for the large number of minor deities, is that the very hot and glowing proto-Venus was visible, for considerable periods of time, at distances intermediate between its present orbit and the very close proximity, at which it was called Agni and Varuna. Its appearance changed rapidly and it was many orders of magnitude brighter than today. Thus it commanded attention.

O Aditya, self-effulgent and all-illuminating God, hasten to the succor of us, jeevas, before we are utterly lost. [36]

Surya, Savitar and Pushan

The common characteristic of all the *suras* was that they glowed. In fact the word for *god* was the word for *glowing* in many ancient languages. This is also closely related to the 'glory' commonly used in the Christian Bible, for example, '...the glory of the Lord shown round about them...'. Thus Agni was not visible by the reflection of sunlight, as the planets currently are, but due to its own bright glow. The brightness of proto-Venus perceived on Earth, depended on how far away it was. In addition, its temperature, and therefore its inherent brightness varied considerably. The primary determinant of its temperature was its relationship to the Sun. This was not because the radiation from the Sun 'shined' on it. When it passed closest to the Sun (perihelion), it was subjected to great tidal and electromagnetic forces, which reheated it to over ten thousand degrees. As a result, the brilliance of proto-Venus as it swung outward after passing though perihelion, was much greater than when it was moving inward in its orbit after having time to cool off, far from the Sun.

Another aspect of proto-Venus which caused it to be called by different names was its position in the heavens, relative to the Sun. Early in its 'lifetime,' proto-Venus' orbit swung out farther from the Sun than the Earth. During this period it could sometimes appear in the night sky, as does Jupiter. When this occurred, it was considered the deity that protected travelers at night, *Pushan*, since it banished darkness. When it appeared as a morning star, just prior to dawn and was brilliant, it was called *Savitar*, and when it appeared in the daytime, bright enough to be seen along side the Sun, it was given the name *Surya,* translated '*sun maiden*'. The following quote reiterates the common belief among scholars that Surya and Savitar were names for the Sun:

Surya and Savitri are two names by which the Sun is commonly addressed in the hymns.[37]

The following quote from the *Rig Veda* is probably the source of the idea that Surya is the Sun.

O Surya, Hari's staunch devotee, with that light of Thine
whereby all these worlds are clothed with radiance by day,
and shrouded in darkness by night,
lead us on day after day to higher and higher levels
of well-being and blemishless character.

However, we believe this is a misconception. Consider the following:

> A maiden by the name of Surya, that is Sun-Maiden, or daughter of Surya, that is Daughter of the Sun. [38]

Therefore, our interpretation of 'sun maiden,' is that proto-Venus glowed so brightly that it was visible in the sky during the day, alongside the Sun, like a daughter.

Savitar is another deity name which is thought to refer to the Sun. But if he were the sun, he would not be described as approaching on the dark blue sky as in the *Rig Veda*:

> God Savitar, approaching on the dark blue sky, sustaining mortals and immortals, comes on his golden chariot, beholding all the worlds.

Thus Savitar was the name for proto-Venus when it appeared in the sky just before dawn.

> He rides in a chariot drawn by radiant, white-footed steeds....He illumines the Earth...his golden arms stretch out to bless...the other gods follow him. [39]

The implication here is that Savitar is proto-Venus in the early stages of approaching the Earth. Thus 'the other gods follow him'. The steeds pulling his chariot, implies that there were already visible prominences, caused by eruptions on Proto-Venus due to its coming into the gravitational influence of the Earth. The 'arms stretched out to bless' are precursors of the characteristics of other deities which follow. Although proto-Venus did not have a magnetic field, it was moving into the magnetic field of the Earth as it approached. This field caused its ionized gaseous corona to divide into two promenances which arced toward the magnetic poles of the Earth. Such features would only be visible when the sky in the background was relatively dark.

> Weaving Night hath folded up her woof,
> In the midst of her performance wise Savitar suspends her work.
> He riseth from his couch and sets the seasons,
> With fitting plan God Savitar hath come hither.
> The scattered homes and all life
> The mighty flame of household fire pervadeth.
> The largest share the Mother has decreed unto her son;
> To do his own desire god Savitar hath sped hither. [40]

The statement that he 'sped hither' implies an entity which is growing brighter each day. This describes the events leading up to a conjunction of proto-Venus. Thus Surya and Savitar were precursors of each conjunction. However, it was only during those inferior conjunctions which corresponded to the aphelion of proto-Venus' orbit, when the planet and its surrounding atmosphere could be distinguished, that it was called Mitra-Varuna.

The Ushas and Asvins

The interpretation of Ushas as the dawn, that has been adopted by most scholars of the *Rig Veda*, is difficult to understand when the obvious meaning lies in its epithet, Aurora. Just as the people did not worship constant phenomena like the Sun, they also did not worship the dawn. The gods worshipper in the *Rig Veda* were part of a long series of extraordinary events. That is why the Vedas were written - to record these events for posterity, for us, if you will. As will be seen from the following quotations, the deity being worshiped is a phenomenon associated with the time of day that dawn occurred, but it was not dawn itself.

We have crossed to the other side of darkness,
Gleaming Aurora hath prepared the way.
Delightful as the rhythm of poem, she smiles and shines,
To happiness her beauteous face aroused us[41]

Her bright sheen hath shown itself to us;
She spreads, and strikes the black dire gloom.
As one paints the sacrificial post at sacrifice,
So hath Heaven's daughter put on her brilliance

4. Well-robed, enveloped, he is come, the youthful;
Springing to life his glory waxeth greater.
Contemplative in mind and god-adoring,
Sages of wise intellect upraise him.

9. Like swans that fly in ordered line,
Have come the pillars gay in brilliant colors.
They lifted up on high by sages, eastward,
Go forth as gods to the gods dwelling-places.

10. These posts upon the Earth, with ornate knobs,
Seem to the eye like horns of horned cattle.
Upraised by priests with rival invocations,
Let them assist us in the rush of battle!

The epithet Aurora is the best clue to the true nature of the phenomena being described, and the descriptions of the sacrificial posts with knobs at the tops, or the other analogy, that of a bull's horns, seems so explicit as to need little clarification. In verse (4) there is a very clear description of what we believe is the causative agent of the Ushas. Based on the previous descriptions, the terms used here, that is, 'well-robed' and 'enveloped' strongly suggest Mitra-Varuna. And since he is 'come' and is 'springing to life,' it is obvious that he was approaching the Earth. From these clues, the physical phenomenon that suggests itself as the Ushas is that as proto-Venus approached the Earth from the dawn side, the gravitational and electromagnetic fields of the Earth formed two 'bridges' between them, as suggested previously, and ionized particles from the very hot and out gassing proto-Venus were streaming across it to the Earth. Because the Earth possesses a dipole magnetic field, with a north and a south pole, the charged particles are forced to spiral around the magnetic field lines which converge toward the poles. This is why two posts are involved. When the streams of charged particles first hit the upper atmosphere they excited the air molecules producing a glow. These produced the 'knobs' at the top of the horns. The remaining particles spiraled downward toward the two poles as do the charged particles from the sun, which cause the auroras famaliar today. The source of light which constitutes these posts is the same as in the aurora borealis, the glow of air molecules which are excited by collisions with the incoming charged particles. Because of the proximity of proto-Venus and its high temperature, the number of particles was probably billions of times the present amount received from the Sun, and was more concentrated in the direction facing it - thus the horn analogy.

Fair Ushas, though through years untold
Thou hast lived on, yet thou art born
Anew on each succeeding morn,
And so thou art both young and old.

They round our generations run:
The old depart, and in their place
Springs ever up a younger race,
Whilst thou, immortal, lookest on.

All those who watched for thee of old
Are gone, and now 'tis we who gaze
On thy approach; In future days
Shall other men thy beams behold.

But some to holier thoughts aspire,
In hymns the race celestial praise,
And light, on human hearths to blaze,
The heaven-borne sacrificial fire.

And not alone do bard and priest
Awake-the gods thy power confess
By starting into consciousness
When thy first rays suffuse the east;

And hasting downward from the sky,
They visit men devout and good,
Consume their consecrated food,
And all their longings satisfy.

The passage concerning the generations that have observed the Ushas indicates that the interactions of proto-Venus and the Earth had reoccurred for at least hundreds of years. In the same stanza we see that the present generation 'watched for thee' and gazes 'on thy approach.' Thus it was not dawn, which occurs every day, but a transient phenomenon, the cause of which is approaching the Earth. It also establishes that the time scale of each encounter could be experienced at least once in a human lifetime. In the fifth of the above stanzas, the gods 'start into consciousness' means that the proto-Venus, of which the Ushas were the precursor, was being affected by the gravitational field of the Earth as it approached. The crust was cracking and hot magma and fire were beginning to stream out, increasing its brightness, as described earlier. The hot gases vented at high speeds are the source of the charged particles which cause the Ushas (Auroras). The hymn merely turns the physical situation around, picturing the Ushas as having the power to excite the great god Varuna. The last stanza dispels any doubt as to the subject, as it describes a god as hasting down from the sky, and consuming the sacrificial food. It is easy to recognize our old friend proto-Venus. There is an interesting insight into the attitude of at least one class of the people, in the following quotations.

Arouse, O Ushas, liberal goddess, them that give; the niggards shall sleep unawakened! and even more clearly O shining Dawns, ye liberal goddesses, do ye today suggest to the rich that they shall give bounty!

When the auroras began to increase in intensity, portending a close approach of proto-Venus with its potential for disaster, some of the rich began giving alms to the poor, in hopes of receiving mercy from the deity. At the times of the impending catastrophes, the poor, having nothing to lose, looked forward to the event, because although many people were killed, those that survived took over the cattle, land, and possessions of the deceased.

Baksheesh's roomy chariot hath been harnessed,
And the immortal gods have mounted on it,
The friendly Ushas, wide-spread, from out of darkness
Has risen up to care for the abode of mortals.

The mighty goddess arose before all the creatures,
She wins the booty and always conquers riches;
Ushas look forth, young and reviving ever,

> She came the first hero here to our morning offering.[42]

This provides some insight into the priesthood which composed and nurtured the *Rig Veda*, in that they preserved, in the Vedic hymns, the point of view of the common man, not the rich or privileged class.

The Asvins were twin gods, also called the 'two Horsemen' or 'Sons of Heaven' (diva napata). They are very closely related to the Ushas.

> As harbingers of Ushas, whom they preceded each morning in their golden car drawn by horses or birds, they were known as the twin gods of morning...In general, however, their connection with the early morning sun is quite clear.[43]

Based on this quote it is relatively clear that the Asvins were related to the auroral phenomena of the Ushas. Note that they are called the 'twin horsemen' and that there were two 'sacrificial posts' associated with the description of the Ushas. The deity that they were associated with was not, as interpreted by the above writer, the Sun, but with proto-Venus. Again, because of the two magnetic poles of the Earth, the auroral displays were split into pairs. Therefore, the Asvins, as with the Ushas which followed them, showed similar patterns to the north and south - on opposite sides of the approaching proto-Venus, which was to the east. We believe the following quote refers to the Asvins, explaining that they were similar but not identical.

> Twins ever differ in their strength and vigor: Two, even kinsmen, differ in their bounty.[44]

They took on various forms, that is, birds and horsemen - limited only by the imagination of the people.

> They are represented as young and handsome, agile and brilliant. They received perhaps even more enthusiastic worship than the Ushas, for while like her they were ancient yet ever young, they differed from her in that they were capable of bestowing youth on men.[45]

Strictly speaking the auroras are an atmospheric phenomenon. From the description of the Asvins as harbingers of the Ushas, these were probably the first glowing clouds of plasma that were released from proto-Venus by a significant explosion. The statement that they were agile indicates their shapes changed rapidly as they arced through space between the planets. When they hit the atmosphere, the auroral phenomenon, called the Ushas, was produced. Thus, the Asvins and the Ushas, although propitiated as deities, were not actual planets, but auroral effects visible on the Earth, which were precursors of a close approach of proto-Venus to the Earth.

5
INDRA

Think, in this battered Caravanserai
Whose portals are alternate Night and Day,
How Sultan after Sultan with his Pomp
Abode his destined Hour, and went his way.

The Rubaiyat of Omar Khayyam , Verse 17
Lyrical Translation by Edward Fitzgerald

It was due to the genius of Velikovsky that we have come to understand the cosmic nature of the knowledge stored in ancient myth. But our understanding of the nature of Indra as Mars signals the departure of the V/A theory, from that of Velikovsky. Although he was sure there were multiple interactions between Mars and the Earth, he never really figured out the most amazing aspects of these interactions. It is most likely that no one ever would have gained this understanding if it were not for the *Rig Veda* and Hindu mythology.

Based primarily on these works we claim that the period of planetary chaos began some 6,000 years ago and lasted about 3,000 years. Because he erred by some 2500 years in dating the beginning of what we call the Vedic period, Velikovsky attributed a number of mythical descriptions of Mars, to proto-Venus. For reasons which will become clear in later sections, we will refer to Mars during the Vedic period as *priori-Mars*.

Indra was Mars

The *Rig Veda* corroborates Velikovsky's conclusion that there were two epochs during Vedic times. The first, discussed in the previous chapters, involved the birth of proto-Venus as Aditi, and the subsequent close encounters of proto-Venus with the Earth, originally as Agni (fire), as Mitra-Varuna, and the lesser deities. This was the most devastating period, in which most hominid and animals on the face of the Earth were killed. The *Rig Veda* implies that it lasted less than a hundred years. Although proto-Venus remained in an eccentric orbit for thousands of years, it was no longer the primary threat to the Vedic people, but it still played a pivotal role in the unfolding of cosmic events involving priori-Mars and the Earth.

As amazing as the events involving proto-Venus were, those of the second part of the Vedic period were even more incredible, because they required the synchronization of the orbits of three planets, the Earth, proto-Venus and priori-Mars. This was the period dominated by the worship of Indra, in the *Rig Veda*.

There are more hymns in praise of Indra in the Rig-Veda than any other deity with the exception of Soma. This is primarily due to the large number of close encounters between priori-Mars and the Earth and their long durations. Although Indra was the predominant Vedic deity, the penchant of the rishis to apply different names to the same physical planet still causes great confusion. Other names for priori-Mars in the *Rig Veda* are Hari, Yama, Rudra, and Vishnu.

The Conflict of Agni/Varuna with Indra

The interactions between these deities began very early in the Vedic period and extended into the second period. This is evidenced by many passages in which relationships between Varuna and Indra are described. Some of these are stated directly in the *Rig Veda*, and are recognized by many scholars. Others are identified for the first time herein. The reason why these have not been recognized is the old habit of the rishis of giving different names to one and the same physical body. Assuming our planetary scenario is correct, there could have been one and only one causative agent in the mythical 'birth' of Indra (Mars), and that is proto-Venus. The following quote shows that the rishis understood this:

> Two periods, of different complexions, revolve for their own purposes, and each in succession severally nourishes a son; in one Hari is the receiver of oblations, in the other, the brilliant Agni is beheld. [46]

Here the author explains that there were just two heavenly bodies circulating at the same time, Agni (Venus) and Hari (Mars), and that each had its characteristic period of revolution. The conflict of Varuna and Indra is described in the following passages from the *Rig Veda*:

> Signs of this conflict are already apparent in the Rig-Veda. Between the two great gods, Indra and Varuna, there is cooperation at first, but cooperation soon turns to rivalry, and rivalry to discomfiture of the Asura king. In R.V. 7.82-85 the two gods work together, but their functions are contrasted...In another hymn (R.V.4-42), however, the rivalry between the (two) comes to the surface: Varuna addresses Indra and asserts his absolute sovereignty:
>
> 'I Varuna, am king,' he says, 'To me was the dignity of *asura* first assigned...I Varuna, am Indra [too]. I, knowing the two wide, deep firmly established areas of space in all their grandeur, [knowing] all creatures as their fashioner, I have set in motion both the worlds and maintain them. I made the dripping waters swell forth; in the seat of the law did I establish the heavens. By virtue of the law is the son of Aditi (Varuna) possessed of the law, and threefold has he extended the Earth'.

proto-Venus, had indeed set the worlds in motion, that is, created chaos, when it burst into the inner solar system. It also maintained the pattern of chaotic planetary motion, for some three thousand years, by influencing the interactions between priori-Mars and the Earth.

Older translations use the term 'demons' for *asura*, because they were seen to be in conflict with the *suras*, or gods. This was based on the Sanskrit convention of negation by prefixing an 'a.' In other words *asuras* were 'not gods,' and therefore were demons. This term, which Varuna proudly applies to himself, is currently thought to be derived from the word *asu* which means 'breath' or 'spirit.' We claim that in the *Rig Veda* this word is used to describe bodies with 'atmospheres,' which is closely related to 'breath.' In other words, an *asura* was a body that had visible gaseous envelope or a tail associated with it. But there is another more physical interpretation along the lines of the old translation. The word *suras*, literally means 'glowing' - and because these glowing bodies wreaked havoc, this became synonymous with gods. But the presence of an atmosphere tended to accelerate its forming a crust and thereby suppressed the glow of the body. Thus the interpretation of *asura* as 'not glowing,' fit the observation perfectly. In other words *asura* may have been a cleverly crafted word with *two* meanings, both of which fit the bodies being described.

Varuna staked his claim to sovereignty on law (*rta*, the counterpart of the later *dharma*) whereas Indra appealed only to naked force:

> It is I whom heroes (*narah*), rivaling each other in riding their goodly horses, invoke when they are surrounded in battle. I, Indra, the widely generous, stir up the battle. I, Indra, raise up the dust. I whose might is overwhelming. All this have I done; no power of the gods can restrain me, for I am invincible. Once the Soma and the hymns have made me drunk, then are both immeasurable worlds struck with terror.

We interpret the mysterious *rta* as the gravitational force of a celestial body, which is directly proportional to its mass. Due to its greater mass, proto-Venus impressed the rishis with the intensity of the earthquake activity it generated even when much more distant than priori-Mars. The red planet had to swoop down much closer to the Earth to have the same effect as proto-Venus.

> As we read the hymns of the Rig-Veda Indra's victory is seen to resemble nothing so much as a takeover bid. Attributes that had formerly been proper to Varuna now pass over to Indra...all the gods bestow on him the title of *asura*...As an asura he becomes the creator and sustainer of the universe which he yet transcends, for the two worlds are equal to but half of him;...Creation, however, in India never means creation ex nihilo; it means the ordering of an already existing matter into intelligible form...
> Indra's promotion to the dignity of an asura, however, did not save that class of deity from degradation. Their eclipse in the shape of

> Varuna himself is movingly described in RV. 10.124 where Agni, the divine priest, deserts Varuna for Indra.
>
> Indra speaks: O Agni, come to our sacrifice...too long hast thou dwelt in darkness.
>
> And Agni replies: secretly and hidden do I, a god, depart from him who is no longer a god...Many years did I pass with him. Choosing Indra I leave the father
>
> Indra speaks: 'The asuras have lost their uncanny power (maya) ... ' Varuna set the waters free, for he has no more power over them...These now follow Indra's highest power. [47]

As with the planet proto-Venus, priori-Mars was called by 'manifold' names in the Vedas. The earliest and most popular was Indra. According to the Vedas, 'innumerable' interactions between Indra and Varuna were observed. The atmospheric bridge that formed between the two planets proves that these were true close encounters, not just alignments in which the two planets appeared close. These encounters affected the orbits of both planets. The net effect, as revealed by the hymns, was to transfer orbital energy from proto-Venus to priori-Mars, aiding in the rapid decrease of the eccentricity of proto-Venus' orbit and increasing that of priori-Mars. At one point in this process, a complex interaction occurred between them, as a result of which the final increment of energy was transferred to priori-Mars, causing its orbit to cross that of the Earth. This mythical birth was the result of its changing from a distant point of light, into a 'world' of enormous size. Thus the rishis understood that was not an actual creation 'ex nihilo,' as had just occurred with proto-Venus. Based on the immediacy of its approach to the Earth as stated in the hymns, these events occurred within a short time of one another.

The Birth of Indra

The following quote is a summary of the Vedic descriptions of the birth of Indra and the events which immediately succeeded it:

> In some ways Indra prefigures Krishna, the incarnate god of the Great Epic, for though he was to rise to the rank of king of the gods (in classical Sanskrit, the word Indra means simply 'lord' or 'king') his beginnings are humble...Indra's childhood is beset with difficulties. He is not born in the normal way but emerges from his mother's side... Once born he is wanted by none but feared by all; his mother attempts to hide him...for Indra was a hero of fearful might who threatened to upset the older order of things. At his birth, mountains, heaven, and Earth were seized with trembling...No sooner is he born than he robs his father, Tvashtri, of the Soma and quaffs an enormous draft of it 'worth a hundred cows.' Thus fortified his stature expands to such an extent that it fills the two worlds,

> immediately he slays Vritra, that is, all that obstructs, and gives the waters free course. Thereupon he turns upon his father, seizes him by the foot and smashes him to pieces...Tvashtri, Indra's father, is elsewhere identified with Varuna, the universal king who rules by magic power; between the two throughout the *Rig Veda* there is great tension.[48]

The one difficulty with this passage stems from the implication that there were two agents, a mother and a father, probably introduced to make a better story, when what occurred was really two occultations of priori-Mars by proto-Venus which took place in quick succession, followed by a true close encounter between the two. Although the mother is not named the father is said to be Tvashtri. The fact that Tvashtri is associated with Varuna elsewhere in the Vedas provides the linkage which corroborates the scenario presented below.

Our interpretation of the myth of Indra's birth is that the Vedic people observed the close interaction of proto-Venus and priori-Mars (Indra) which, at the time were relatively close to the Earth, because we are told that the conjunction of the two planets caused the Earth to tremble. Since priori-Mars was much smaller, it could easily have been hidden behind proto-Venus for a period as it approached that planet. When it emerged into view, being smaller, it appeared to come out of the side of proto-Venus, that is, it was 'born.' It probably remained close to proto-Venus as seen from the Earth but still on the far side, and even passed out of view again, thus appearing to be hidden. As it approached even closer to proto-Venus an 'atmospheric bridge' filled the corridor between them. The myth says that the Soma draught was taken from Tvashtri, his 'father,' immediately after he was 'born' from the side of an unnamed mother. We believe this is merely for the story sake, and the fact that the mother's name is not given implies that there was but one other planet involved. It is clear that this process involved multiple apparent encounters between the planets, since after his birth, out of the side of his mother, she tries to hide him. These represent two occludings of priori-Mars by proto-Venus. The 'bridge' implies that the third (apparent) event was the closest approach of the two planets. As the less massive priori-Mars passed by proto-Venus, it experienced an energy boost due to the orbital velocity of proto-Venus, similar to the slingshot maneuver used by NASA to boost the energy of its interplanetary probes. priori-Mars, with its orbit expanded, moved toward the Earth, appearing to gain in stature (size), then passing very close to the Earth, doing battle with Vritra, and 'freeing the waters.' The final step described was a second encounter between priori-Mars and proto-Venus, at which time an electrical discharge took place between the pair and they separated. The term 'smashes him to pieces' is a embellishment in the translation, which should just be 'smashes him.' This could mean that an actual glancing collision took place between priori-Mars and proto-Venus, in which case the more plastic proto-Venus might have had some material splashed into space. The fact that a thunderbolt or electrical discharge passed between the two was corroborated in the myths of other cultures, cited by Velikovsky. This was the reason why he hypothesized that electromagnetic forces cannot be ignored in the close interactions of planetary bodies. It makes physical sense that a very hot body like proto-Venus could accumulate a net electrical charge due to the loss of electrons.

As a result of the same interaction, the orbit of proto-Venus was changed so that it no longer crossed that of the Earth. Because proto-Venus' orbit moved inward, away from the Earth, it was less distorted by the gravitational field of the Earth at each conjunction. When it was closer the distortions had cracked its thin crust, causing great flare-ups which were characterized in the Vedas as Varuna reverting to Agni. Because these eruptions ceased, *Agni* (fire) was imagined to have left Varuna. At the same time Indra began approaching the Earth and experiencing volcanic eruptions due to the tidal force of the Earth disrupting its interior. Thus the notion arose that *Agni* (fire) left Varuna and moved to Indra. (See quote at top of page 58.) This transfer of power from the proto-Venus deity to the Mars deity also occurred in both Egyptian and Greek mythology. The same orbital shift also resulted in Varuna losing his power to 'release the waters' in the Himalayan glaciers, which was an important point to the Vedic people because it was a sign of godly benevolence.

Indra Releasing the Waters

The primary act of 'heroism' for which Indra is famous in the *Rig Veda* is 'releasing the waters.'

> At the time of his birth men were imploring the gods to come to their aid against the demon vritra who had imprisoned the cloud-cattle, thus reducing men to starvation through drought. [49]

Prior to the birth of Indra, Varuna (proto-Venus) had performed this 'miracle.' Most scholars merely repeat the idea that the drought was due to lack of rainfall, and that Indra was a hero because he caused rain to fall. The following metrical translation, based on the *Rig Veda*, is rather cryptic but upon careful consideration, one must have serious doubts that rain storms are being described.

To Indra

I will declare the manly deeds of Indra, the first that he
achieved, the thunder wielder,
He slew the dragon, then disclosed the waters, and cleft the
channels of the mountain torrents.

He slew the dragons lying on the mountain: his heavenly bolt of
thunder Tvashtar fashioned.
Like lowing kine in rapid flow descending, the waters glided
downward to the ocean.

Impetuous as a bull, he chose the Soma, and in three sacred
beakers drank the juices.
Maghaven grasped the thunder of his weapons, and smote to death
the firstborn.

Indra

When, Indra, thou hadst slain the dragon's firstborn and
overcome the charms of the enchanters,
Then, giving life to Sun and Dawn and Heaven, thou foundest
not one foe to stand against thee.

Indra, with his own great and deadly thunder, smote into pieces
Vritra, worst of Vritras.
As trunks of trees, what time the axe hath felled them low on
the Earth, so lies the prostrate dragon.

He, like a mad, weak warrior, challenged Indra, the great,
impetuous, many-slaying hero.
He brooking not the clashing of the weapons, crushed -
Indra's foe- the shattered forts in falling.

Footless and handless still, he challenged Indra, who smote
him with his bolt between the shoulders.
Emasculate yet claiming manly vigor, thus Vritra lay with
scattered limbs dissevered.

There he lies like a bank-bursting river, the waters taking
courage flow above him.
The dragon lies beneath the feet of torrents which Vritra
with his greatness had encompassed.

Then humbled was the strength of Vritra's mother: Indra
hath cast his deadly bolt against her.
The mother was above, the son was under, and like a cow
beside her calf lay Danu.

Rolled in the midst of never-ceasing currents flowing
without a rest for ever onward.
The waters bear off Vritra's nameless body: the foe of Indra
sank to during darkness.

Guarded by Ahi stood the thralls of Dasas, the waters stayed
like kine held by the robber.
But he, when he had smitten Vritra, opened the cave where-
in the floods had been imprisoned.

A horse's tail wast thou when he, O Indra, smote on thy bolt;
thou god without a second,
Thou has won back the kine, hast won the soma; thou hast
let loose to flow the seven rivers.

Nothing availed him lightning, nothing thunder, hailstorm
of mist which he hath spread around him:
When Indra and the dragon strove in battle, Maghaven
gained the victory forever.

Whom sawest thou to avenge the dragon, Indra, that fear
possessed thy heart when thou hadst slain him;
That, like a hawk affrighted through the regions, thou
crossedst nine-and-ninety flowing rivers?

Indra is king of all that moves and moves not, of creatures
tame and horned, the thunder-wielder.
Over all living men he rules as sovran, containing all as
spokes within the felly. [50]

Figure 5.1. A mountain glacier forming a 'horse tail' shape.

Early interactions with proto-Venus had uplifted the Tibetan plateau and the Himalaya mountains, increasing their elevations far above the levels we know today. As a result of their great height the mountains were cold and had trapped all the precipitation in enormous valley glaciers as they do to this day. This caused the land below, where the people lived, to be parched. They knew the water was trapped in the mountains because Varuna (Venus) had released it on previous occasions, and offered sacrifices and praise to the approaching Indra, to induce him to perform this function. The characterization of the evil creatures which withheld the water, as dragons and horse tails is not difficult to understand, in light of the photographs in Figures 5.1, which pictures a horse-tail glacier, and 5.2

which shows that the glaciers look exactly like reptiles laying in the mountain valleys. It is amazing that the Vedic people imagined them in this form even though they did not have the perspective afforded by an Earth orbiting satellite. Glaciers which spread at the end of their course are called horse tails to this day. These glaciers hold an enormous volume of water. When Varuna had 'released the waters' in earlier times, the process took place primarily via earthquakes in the still very young Himalayas. These opened innumerable fissures allowing the upward flow of magma which increased the surface temperatures and causing the glaciers to melt. Geologists today recognize the large number of intrusians in the Himalayan rocks

Figure 5.2 NASA photographs of mountain valley glaciers that resemble serpents.

Priori-Mars did not have nearly as strong a gravitational field as proto-Venus. However, it managed to release the waters of the Himalayas in its own unique way. It came closer to the Earth, and as we will show, remained over the Trans-Himalayas for years at a time, melting subsurface rock and drawing it upward into billions of cracks in the rocks, increasing their temperature, and thus melting the glacier. The Vedic people experienced the continuous earthquakes that were part of this process.

Because the mass of priori-Mars at that time was only about one sixth that of the Earth, the gravitational field of the Earth caused great turmoil within it, and Martian volcanoes expelled many large hot chunks of glowing rock into space. Aided by the gravitational force of the Earth many flew across the gap, striking the Earth, primarily in the Himalaya area. Some of these caused 'thunderbolts,' greatly adding to their destructiveness. These were nothing like ordinary lightning bolts, otherwise they would have had little effect on the massive glaciers. Working in combination with the tidally induced up-welling of magma within the Earth, they broke up the glaciers and began melting the ice relatively quickly, swelling the streams that flowed down from the mountains into mighty rivers. It was no wonder that these people characterized Varuna and Indra as gods - each with different appearances and 'personalities.'

The hymn actually describes the volume of water as flowing over the glacier, and how the 'never ceasing' flow becomes so great that it actually bears off chunks of the 'dragons' body. The long duration of the flow implied by the passage is a clue to the fact that priori-Mars not only melted the glaciers, but it also rained down enormous amounts of water into the area. Not one, but seven rivers eventually resulted from the onslaught of Indra, and the entire climate of the region was changed for years to come.

In the next to last stanza the rishi expresses wonder that Indra, being so successful in the battle, suddenly rushes off as if afraid of something. This lends credence to the idea that during these close encounters, gravitational 'locking up' occurred, during which priori-Mars remained stationary in the sky. Only this makes understandable the surprise at Indra suddenly 'running off,' i.e. when later released from this geosynchronus orbit.

There is considerable evidence in the Vedas and Hindu myths that priori-Mars interacted with the Earth during close encounters in such a way that it remained above the up-lifted Himalayan-Tibetan complex for extended periods of time. This idea is discussed in more detail later. Mount Kailas is declared the heavenly home of Indra in the Vedas and in the Tibetan myths. We infer that this was the point directly above which priori-Mars remained during its encounters with the Earth.

The Many Battles of Indra and Vritra

Just as Indra was observed quaffing drafts of Soma from Tvashtri (Venus) on many occasions, so was this process of releasing the waters repeated on numerous occasions. The great number of planetary interactions between priori-Mars and Earth during Vedic times is one of the most fantastic revelations of mythology. The following metrical translation describes how time after time the glaciers gradually reform only to have Indra come once more and defeat Vritra:

He whose magic powers
From Earth withhold the genial waters;
Of mortal men the foe malign,
And rival of the race divine:
Whose demon hosts from age to age
With Indra war unceasing wage;
Who, times unnumbered crushed and slain,
Is ever newly born again,
And evermore renews the strife
In which again he forfeits life. [51]

The following quote suggests that Indra drew on certain 'weapons' other than thunder and lightning in his battles with Vritra, that is, meteorites of iron:

And soon the knell of Vritra's doom
Was sounded by the clang and boom
Of Indra's iron shower.
Pierced, cloven, crushed, with horrid yell,
The dying demon headlong fell
Down from his cloud-built tower. [52]

Parjanya

Parjanya, Vayu, and the Maruts are the 'storm gods' - lesser deities associated with Indra during his close interactions with the Earth. Parjanya is the god associated with the thunderbolt, although some believe he is just another name for Indra, since his name implies one acting for another, and as illustrated below, has all the same properties ascribed to Indra himself.

> He splits the trees; he destroys the demons that withhold the waters. The whole creation is afraid of his mighty stroke. Like a charioteer urging forward his horses with a whip, the god brings into view his showery scouts. [53]

Thunderbolts were not ordinary lightning bolts, or they would not have been worshipper as a god by these primitive people. They were much more powerful or they could not have broken up the gigantic glaciers. Thunderbolts are also associated with Maruts, as discussed below.

Vayu

Vayu is the god of the winds and, with Indra, he rules over the atmosphere. The following quotes, referring to Vayu, imply that this wind may be more than just an earthly one:

> He won the race for the first draft of the Soma juice, and at Indra's request, allowed him (Indra) to have a quarter of it...
> The two worlds generated him for wealth...
> Vayu is described as being most handsome in form; one who moves noisily in a shining car drawn by a pair of red or purple horses...
> (I celebrate) the glory of Vata's (another name for Vayu) chariot; his noise comes rending and resounding. Touching the sky, he moves onward, making all things ruddy; and he comes propelling the dust of the Earth. The gusts of air rush after him, and congregate upon him as women in an assembly. Sitting along with them on the same car, the god (Indra), who is king of this universe is borne along. Hasting forward...he never rests...His sounds have been heard, but his form is not (seen). [54]

Our interpretation of these lines is that Vayu was the 'wind' which traversed space from priori-Mars (Indra) to the Earth across a temporary 'atmospheric bridge' that formed between the two planets at times when priori-Mars was convulsed by extraordinary tidal forces. We suggest six reasons for this interpretation. First, it is connected to the idea of the passage of Soma between the planets; second, the two worlds generated it; third, it could be seen moving across space from priori-Mars to the Earth, with Indra in the background; fourth, when it reached the top of the Earth's atmosphere it 'touched' the sky; fifth, it carried sounds which could not have propagated across empty space; and sixth, it carried red dust and rain from the Red Planet to the Earth, making everything 'ruddy.'

Additional clues are given by Vayu's epithets:

> Other names by which this deity is known are the following: - Anila, breath; Marut, air that is necessary to life; Sparsana, he who touches; Gandhavaha, he who carries odors. [55]

The Maruts

The Maruts comprise a multitude of gods. In the Vedas they are the constant companions of Indra. They are also said to be the sons of Rudra, a less popular deity associated with priori-Mars, but one which forms an important link to the later Puranic deity Siva (Shiva). This common association of the unique Maruts with Indra and Rudra is one factor linking both these deities to the planet priori-Mars. Other factors are cited in the section on Rudra. The following quote summarizes the Maruts:

> They are youthful warriors, but they are also geniuses of the thunderstorm. For whereas the interpretation of Indra as god of the storm is frequently forced, the close connection of the Maruts with thunder and lightning, wind and rain, cannot be denied...They hold the lightning in their hands and the lightning is their spear. Their golden wheeled chariots gleam in the lightning as they rush forth like boisterous winds...They love to deck themselves with golden ornaments, to scrub each other clean, and to lick each others backs. They are irresponsible and gay, yet wrathful as serpents. They are always present at the slaying of Vritra and they egg great Indra on with their songs. [56]

In this interpretation the author used the terms 'thunderstorm' and 'lightning' because they comprised his experience, but the true meaning was 'thunderbolts.' The following are direct translations from the *Rig Veda*:

> ...that brilliant and mighty band of leading lights...whose arms are decorated with bracelets (quoits) ... given to roaring, potent, dispensing treasures...infinite in greatness.[57]

> This wide-extended Earth is for the Maruts; the spacious heaven is for the spreading winds; the paths of the firmament are provided for their course.[58]

Our interpretation of the Maruts is as follows: Priori-Mars was greatly disturbed by the enormous gravitational field of the Earth when the two planets approached one another, and many enormous volcanic vents were opened on the surface of priori-Mars. At times of alignments with the Moon, Sun, or proto-Venus when near conjunction, the vents erupted with explosive force, expelling great masses of hot rocks at high velocity. Many of these glowing bodies followed a trajectory to the Earth, displaying a certain golden beauty and graceful motion as they arced silently through space. Elsewhere in the vedas, they were described as cinders being expelled from a burning log. When they hit the upper atmosphere, rings of ionized gas formed at the impact points, and they appeared to be donning bracelets. However, as some of them passed through the ionosphere and fell to the Earth they created an ionized path through which the enormous charge of the ionosphere could discharge to the Earth. This caused what the Vedic people called a 'thunderbolt.' It was not mere lightning, with which these primitive people were quite familiar, but an electrical discharge of such enormous power, as has never been experienced in modern times. The 'shorting out' of the ionosphere magnified the energy of the meteorite impact by many orders of magnitude. In *Worlds in Collision* Velikovsky quotes ancient myths in which it is stated that *when some thunderbolts hit the Earth, entire mountain ranges disappeared, and others were thrust upward instantaneously*. The onslaught of these Maruts, concentrated in the Himalayas, broke up the glaciers while at the same time the sustained drawing of magma into fractures in the rocks, by the tidal force of priori-Mars, heated them and melted the ice. The number of these encounters is alluded to in the following excerpt from the *Rig Veda*:

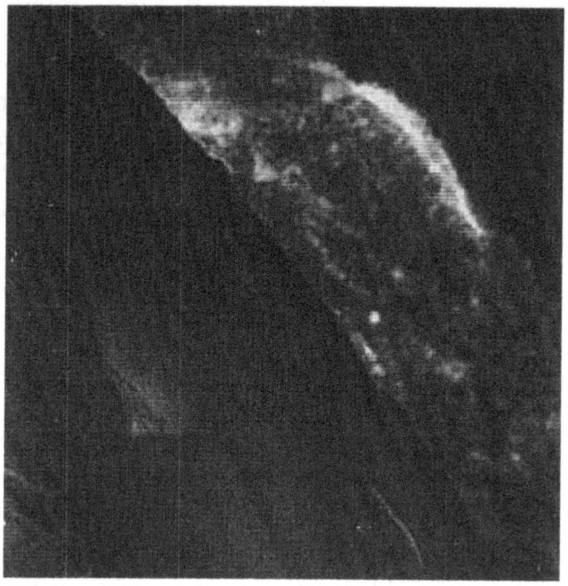

Figure 5.3 The bow-like plume from a volcanic vent on the limb of Io, as photographed by the NASA Voyager probe.

> Indra, unopposable, slew ninety-nine Vrtras with the bones of Dadhyanc. [59]

This implies that there were ninety-nine encounters with priori-Mars, and agrees closely with the number of days in the life of Brahmā, discussed in a subsequent chapter on Hindu mythology. The question of the duration of these encounters, and the profound physical adaptations that were made by both planets in order to accommodate them, will be explained later.

Indra's Bow

The following quote concerns Indra's 'bow,' an attribute which at first reading, seems a bit puzzling. There is no doubt that the analogy is with an archer's bow, since it is mentioned more than once in the *Rig Veda*. Indra is described as holding the bow in one hand. Velikovsky refers to bows in the myths of other cultures, which obviously were describing the same phenomenon.

> With the Bow let us win kine, with Bow the battle,
> With Bow be victorious in our hot encounters.
> The Bow brings grief and sorrow to the foeman:
> Armed with the Bow may we subdue all regions. [60]

The solution to the 'riddle' of the bow of Indra comes from NASA photographs of Io, the closest moon of Jupiter, taken by the Voyager planetary probe. One of these photographs is reproduced in Figure 5.3. These show that Io is currently volcanically active. In fact, the material being spewed from one of the volcanos can be seen at the limb of the moon, where it is visible against the dark background of space. The gaseous material which is flowing out of these volcanos follows a parabolic path and is radially symmetric because there is very little atmosphere on the small moon. When such a volcano happens to be on the limb of a moon (or planet), the shape of the plume viewed in 'profile' looks like a bow. The description of Indra with a bow 'in his hand' merely means that the Vedic people saw the plume from a volcano on the limb of priori-Mars during its encounters with the Earth. Moreover, when priori-Mars suffered acute internal convulsions, this volcano would shoot out a succession of hot glowing rocks at high speed. These were observed coming out of the center of the 'bow,' as would an arrow, thus reinforcing the image.

Deities Synonymous with Indra

Other deity names in the Vedas for Indra are Rudra, Yama, and Vishnu. Although the latter name became the best known deity in the later Puranic period, Vishnu in the Vedas did not have near the stature of the other gods. The concept of Vishnu, the 'sustainer' only evolved after thousands of years, at which time the events reported in the Vedas were seen in a longer term perspective.

Rudra

Rudra is a secondary deity in the Vedas, of which Siva (pronounced Shiva), 'auspicious' was merely an epithet. As we shall see, in the later Puranic period Siva is the third of the well known Hindu Triad, and is sometimes there referred to as Rudra-Siva when he is being particularly nasty. Rudra never attains any real importance in either period.

> Rudra seems ill at ease in the setting of the *Rig Veda*; he is rarely associated with other gods, except the Maruts, also called Rudras, who are his sons. He, the divine archer pursues a solitary course, seeking whom he may devour. He the great Asura of Heaven (RV,2.1.6) and universal lord of heaven, shoots his arrows at whom he will, and his arrows bring death and disease, fever, cough, and poison (AV 11..2.22). So wrathful and unpredictable is he that even the prostrations of his devotees may move him to anger, as may a hymn incorrectly composed...If in the RigVeda, Rudra is terrible, 'spitting like a wild beast (RV, 2.33.11), in the later Vedas and the

> Brahmanas he is far more so. Here he is black, swarthy, murderous, and fearful (AV 11.2.18), a robber, cheat, and deceiver, the lord of thieves and robbers. In the same hymn Rudra is called the father of the Maruts or Stormgods... thou fitly holdest arrows and a bow; fitly thou (wearest) a glorious necklace of every form.
> But there is another side to Rudra: he smites that he might heal: he is wrathful yet he is mild. As a boy bows down to his father as he blesses him, I bowed down to thee, O Rudra, when thou drawest nigh (RV 2.33.12). For he is not only the great destroyer, he is also the divine physician with a thousand remedies at his disposal...In Rudra opposites meet but are not yet reconciled; and this is one of the reasons that this terror inspiring god was able to develop to his full stature in Rudra-Siva, for in Siva all opposites were to meet only to be transcended and reconciled. We invoke with obeisance the ruddy bear of the sky, with spirally braided hair, a brilliant form." "May he who glides away, blue-necked and red-coloured be gracious unto us. " [61]

The most obvious clue to the physical identity of Rudra as an aspect of priori-Mars is through his association with the Maruts, who were also associated with Indra (Mars). Secondly, it is a well known fact, based on interplanetary probes, that about one half of priori-Mars, now known as the northern plains, is full of unusual 'geologic' features such as enormous twisted flood channels, now devoid of water, a great chasm as wide as the Grand Canyon is long, and three thousand miles in length, enormous volcanoes perched atop a continent-sized uplift, and lava flows millions of square miles in size. The other half, now termed the southern highlands, is covered by gently sloping hills with (now dry) river-beds and tributaries that characterized the planet prior to the time when the upheavals that modified the northern plains began. It is the features on the convulsed side of priori-Mars that are referred to in the quotation above, for example, the surface is ruddy, the great chasm is the origin of the blue-neck, and the spirally braided hair are the twisted outflow channels, and the bow is the same as Indra's bow, which we hypothesize is one of many active volcanos. The name Rudra was most likely applied at times when priori-Mars was breaking away from a close encounter with the Earth. It was at these times that the lava 'fountains' settled down, formed a crust, and appeared black and swarthy. As discussed in subsequent chapters, this was a time of great danger on Earth - when earthquakes and floods killed many animals and people. Rudra is the Vedic deity identified with priori-Mars during these dark times.

Yama

The deity Yama is not directly identified with Indra in the *Rig Veda*, but the identification with priori-Mars is easily made, based on our current knowledge of that planet. First, Yama is the twin brother of Yami:

> Tvastri had formed them as man and wife in the womb.

Note Tvashtri is the father in the story of the birth of Indra, from whom Indra stole the draught of Soma. The Yama-Yami duality is consistent with the striking differences between the two morphologic provinces of priori-Mars - one half is Yama, while the other is Yami. Second:

> He is said to have two insatiable dogs, with four eyes and wide nostrils, which guard the road to his abode.

This obviously refers to the two moons of priori-Mars, Phobos and Diemos, pictured in photographs taken by the NASA Viking probe. As with many of the surface features identified in the *Rig Veda*, these moons are far too small to be seen from the Earth without a powerful telescope, in current times, let alone to see craters on them, which were described as 'wide nostrils.' Based on the fact that the larger moon is 40 kilometers in diameter, and it's 'nostril' (crater) is 3.7 kilometers across, the farthest distance at which this could have been observed with the naked eye (resolution 1 minute of arc) is about 12,700 kilometers. Assuming Phobos, the larger moon was in its current orbit around priori-Mars, this could imply that the surface of priori-Mars came within 18,700 kilometers of the surface of the Earth in order that these observations could have been made. However, the orbit of the Martian moons was undoubtedly extended toward the Earth when the two planets were close, so that priori-Mars need not have been quite that close. Based on certain assumptions about the mass of priori-Mars at that time, we estimate that its surface was only 35,000 kilometers above the Himalayas during the 'dance encounters.' Third:

> According to the popular ideas now prevailing, Yama is represented as a green man, clothed in red garments.

This statement not only corroborates the fact that the two halves of priori-Mars have radically different appearances, but introduces the exciting probability that *vegetation still covered the southern half of the red planet within the last six thousand years.* This, combined with the youthful-looking earth-like drainage patterns in the southern hemisphere, suggest that priori-Mars had abundant water and supported life very recently, not billions of years ago, as currently hypothesized by planetary science. This reinforces our hypothesis that priori-Mars was in an interior orbit, similar to that in which proto-Venus is currently located, at the beginning of the Vedic period, some 6,000 years ago. It also provides the probable source of the magic Soma, that is, organic vegetable matter on priori-Mars. The "red garments" refer to the northern plains which were already covered with lava flows.

Fourth, in the Puranas:

> Yama fulfills the office of the judge of dead...all that die appearing before him...The virtuous are then conveyed to Swarga (Indra's heaven) .

This is another characteristic of Yama which links him to the deity Indra. The duty of carrying souls of the deceased 'to heaven' implies that the body alternately came close to the Earth, to pick up the souls, and then moved away until it became a point of light just like the other stars, or deities which resided in heaven.

The Vedic Vishnu

Although today he is the most popular deity in the Hindu pantheon, Vishnu is not regarded as a major god in the more ancient Vedas, based on the small number of hymns devoted to him, as compared to Agni, Varuna, Soma, and Indra. In a few hymns he is afforded what sounds like high praise: Vishnu, in the Rig-Veda, is said

> ...to have established the heavens and the Earth; to have, with Indra, made the atmosphere wide, stretched out the worlds, produced the sun, the Dawn, and Fire; to have received the homage of Varuna, whilst his greatness is described as having no limit. [62]

These attributes are diminished by the fact that Vishnu is usually mentioned in conjunction with the powerful Indra, and does not seem to stand on his own:

> But Indra is associated with Vishnu even in some of those texts in which the latter is most highly magnified: nay in one place, the power by which Vishnu takes his three strides is described as being derived from Indra: in another text, Vishnu is represented as celebrating Indra's praise, whilst in another verse, Vishnu is said to have been generated by Soma.[63]

As mentioned previously, Soma was Indra's favorite drink and imagined to be the source of his strength. The most unique attribute associated with Vishnu in the *Rig Veda* is the 'wide strider:'

> Vishnu strode over this (universe); in three places he planted his step; the world was enveloped in his dust. Vishnu, the unconquerable preserver, strode three steps.[64]

Past interpreters have struggled with the meaning of these three steps:

> One regards him as a god, who, in what are called his three strides, is manifested in a three-fold form as Agni on Earth, as Indra or Vayu in the atmosphere, and as the Sun in heaven; the other interprets Vishnu's three strides as the rising, culmination, and setting of the sun. [65]

The identification of Vishnu with the planet priori-Mars is relatively easy, if one accepts the previous hypothesis that the deity Indra is identical with that planet, because of the common association of Vishnu and Indra in the Vedas. Additional clues arise in the Great Epics and Puranas, discussed in the next chapter, but there is sufficient evidence just in the Vedas.

Assuming that the initial orbit of priori-Mars was inside that of the Earth, the mysterious three strides are relatively easy to discern, and actually aid in confirming the identity of Vishnu as the planet priori-Mars. The first 'stride' was the orbit of priori-Mars at the beginning of the Vedic period, at a distance from the sun approximately where proto-Venus' orbit is today. The second 'stride' was the orbit which brought priori-Mars close to the Earth on numerous occasions. The third 'stride' of Vishnu was the attaining of Mars' current orbit, completely outside that of the Earth's. The deity in the latter orbit is described by the Vedas as being a god of the 'high heaven,' meaning a planet which attains dominance of the mid-night sky. A feat which is not possible for a planet in an orbit inside that of the Earth.

The last stride of Vishnu signaled the end of the Vedic period, the period of chaos in the solar system, however, the end of the Vedic period does not appear to be accentuated by a single event in the Vedas themselves. One oblique reference to the end of this period comes from the teaching that Indra's rule over the gods continued for a hundred divine years; at the expiration of which time he was superseded by another god, or even by a man. Unfortunately universal agreement as to the actual time meant by 'a hundred divine years' has not been reached, but the notion that Indra loses his power, coming from the ages just after the Vedic period, lends credence to the establishment of priori-Mars in its current, 'harmless' orbit at the end of that period

. From a physical point of view, however, one last close encounter with the Earth and nearby proto-Venus must have occurred which supplied the impetus to priori-Mars, resulting in the attainment of its current orbit. The fact that Mar's orbit remains to this day notably more eccentric (0.093) than any of the other planets except Mercury and Pluto, is but one clue to its recent chaotic past. Another lies in the fact that its rotational period is almost the same as that of the Earth. Possible mechanisms by which this final orbit change took place are discussed in subsequent chapters.

The fact that the 'three strides' are the unique attribute of Vishnu in the Vedas, implies that this was a deity conceived at the end of the Vedic period, when the three strides were completed. At that time, the sages obviously consolidated the knowledge of the entire period and combined the various manifestations of priori-Mars, as Indra, Yami, Rudra, etc. into a more omnipresent deity such as Vishnu, who spans the entire period of time, accounting for a number of the earlier deities as 'incarnations.' The identification of these deities with the physical planet priori-Mars seems to have been suppressed by the rishis, but this attitude stems from our modern perspective. We have independently 'discovered' the physical planets due to Brahe, Kepler and Newton in the last four hundred years, and now we are in a position of attempting to marry our world view with that of the Vedic people. We must realize that they had no second world view, that is, the scientific concept of planets did not exist. The names of the deities and the hymns describing them were all they had. Also the major deities of the earliest part of the Vedic period, Agni and Varuna, were rarely mentioned as such in the texts of the later periods, even though many hymns praising them exist to this day in the Vedas. There is evidence that even a few hundred years after the last encounter of priori-Mars with the Earth, it had visible characteristics, such that wise men could still identify it as the once great Indra. At some point in time these ceased and the connection of Indra with the planet was completely lost.

The Vedic deities Rudra, Yama/Yami, and Vishnu are lesser deities associated with the planet priori-Mars. Ample clues to the association of each one with the planet priori-Mars exist in the Vedas, as well as quotes linking them to the primary Vedic deity Indra, which we have already named as priori-Mars. These three deities are transitional, with names appearing in both the Vedic hymns and in the Hindu books which were written later. In this respect they are important because they help make the association between the major deities of the Vedic period and those of the Hindu period. The most important point is that the different names all refer to the one physical body - the planet priori-Mars.

Firmament

6
SOMA

There are more things in heaven and Earth, Horatio, Than are dreamt of in your philosophy.
Shakespeare, Hamlet

'Innumerable' interactions between proto-Venus and priori-Mars were reported in the *Rig Veda*. The primary clue that they were indeed close encounters, as opposed to mere alignments, comes from the description of a white atmospheric bridge between the two planets. This bridge is portrayed as Indra 'quaffing huge draughts of Soma' from Agni, Varuna, or Tvashtri, all names for proto-Venus - the latter is used primarily in the role of the nourisher or Soma-giver. The identification of Soma in this role proves that the Vedic people had had direct experience with it on the Earth. In fact, more hymns in the *Rig Veda* are devoted to this magic, delicious, intoxicating ambrosia than any other deity.

Soma was golden-green in color, sweet to the taste, nourishing, and most probably intoxicating. Its elevation to the position of a 'god' was also due to its delivery from the sky, probably at times of certain celestial alignments. It also implies that this was not a one-time event, but something that occurred on a fairly regular basis over a long period of time. Its pleasant effect on men is extolled the following hymn:

This Soma is a god; he cures
The sharpest ills that man endures.
He heals the sick, and sad he cheers,
He nerves the weak, dispels their fears;
The faint with martial ardour fires,
With lofty thoughts the bard inspires;
The soul from Earth to heaven he lifts;
So great and wondrous are his gifts,
Men feel the god within their veins,
And cry in loud exulting strains:
We've quaffed the Soma bright
And are immortal grown:
We've entered into light,
And all the gods have known. [66]

The identification of the Soma, which fell on the Earth and was consumed by the people, with the proto-Venus deities Agni and Mitra-Varuna is suggested in some passages in the *Rig Veda*. The following quote originated at a time when the eccentricity of the orbit of proto-Venus had already decreased to such a degree that it no longer crossed the orbit of the Earth, and therefore did not threaten the Earth with the 'power' that it once had.

> Away pass Agni, Varuna, and Soma. [67]

As a result it is more difficult for us to believe that proto-Venus was the actual source of the Soma. It is more likely that it fell to Earth during inferior conjunctions of proto-Venus.

> May Mitra-Varuna, sapient pair, bedew our pasturage with oil, with meath the regions of the air. [68]

Soma was declared a god because it had a tangible effect on the people, and rather miraculously was supplied at a time of turmoil, when it was sorely needed. As a result it was highly praised in the myth of the most ancient cultures. As ambrosial rain pours down through the sieve of heaven, the Soma is worshipper as Pavamana (self purifying).

To Soma Pavamana

Through the fine fingers, with the song,
this hero comes with rapid cars,

Going to Indra's special place. (authors note: the Himalayas)

In holy thought he ponders much for the great worship of the gods,
Where the immortals have their seat.

Like a good horse he is led out, when on Earth the path that shines with light.
The mettled steeds exert their strength.

He brandishes his horns on high, and whets them, bull who leads the herd,
Doing with might heroic deeds.

He moves, a vigorous steed, adorned with beauteous rays of shining gold,
Becoming sovran of the streams.

Over the places rough to pass, bringing rich treasures closely packed,
Descends into the reservoirs.

Men beautify him in the vats, him worthy to be
beautified,
Him who brings forth abundant food.

Him, even him, the fingers ten and the seven songs
make beautiful,
Well-weaponed, best of gladdeners.

Although Soma was associated with Agni or Varuna in many hymns, the first stanza only mentions the 'place of Indra' (Mars). The analogy was made of (some) planet as a bull, fertilizing the female Earth with his heavenly semen (Soma), thus making the Earth bring forth abundant plant life and cattle (kine), which were the ultimate measure of wealth in Vedic times. The Soma, 'rich treasures, closely packed' fell in the Himalayas, 'the places rough to pass,' the terminus of the gravitational bridge between priori-Mars and the Earth during the dance encounters. The Soma was then carried by the mountain streams down into the reservoirs.

Soma was important to the people, not just because it produced good health and a pleasant feeling, but it proved to them that this enormous, omnipotent, threatening sky-god had a benevolent nature, to balance its harsh aspect. We are tempted to say that it exhibited exactly those properties that we would expect from a god. But this is putting the cart before the horse. We live in a world that long ago inherited the ideas of gods from ancient peoples, including the Aryans. Whatever forms of worship that existed before them were completely obliterated. These planetary interactions *defined* the nature of gods for all peoples, that is, meting out 'benefits' and 'punishments' with a power far beyond man's ability to influence.

Not only does the whole idea of edible, intoxicating food falling from heaven make a fantastic story, but it provides an important clue for the interpretation of the real meaning of the *Rig Veda*, by which we corroborate the interactions between priori-Mars and proto-Venus, in addition to the interactions of both of those planets with the Earth.

The Reality of Soma

Our hypothesis as to the physical process by which Soma was created and distributed to the Earth is as follows: priori-Mars was covered with live vegetation well into the Vedic period. Many convulsions of its interior occurred, due to alignments with other bodies, such as the Moon. The resulting eruptions launched rock bodies, large and small, and atmospheric gases, including water vapor, across the short distance between priori-Mars and the Earth. Leaves from the vegetation were swept up by the strong winds, heated in the volcanic processes, and carried aloft by a tornado-like vortex that was induced in the atmosphere of priori-Mars at its north pole, the closest point to the Earth. This feature was visible on the Earth and was often referred to as a 'serpent' in the Vedic hymns and myths. This material was transported across the atmospheric bridge to the Earth. Its 'processed' state was probably due to the heat of lava 'fountains' on priori-Mars, its exposure to low pressure during its passage through space, its subsequent heating upon its entry (not *re*entry) into the atmosphere of the Earth, and its condensation on the Earth during the night.

If, as some of the hymns seem to imply, Soma was somehow associated with proto-Venus in the minds of the Vedic people, we interpret this as meaning that the proximity of proto-Venus to the Earth (inferior conjunction) facilitated the release of Soma from priori-Mars. The existence of Soma is one of a number of major revelations that come from the *Rig Veda* but, as pointed out by Velikovsky, it is a consistent entity in the ancient myths of a number of quite distant and distinct cultures. It was called 'manna from heaven' by the Jews, helping to sustain them as they wandered in the desert following their Exodus from Egypt. This is an important data point, since it indicates that this material must have fallen as late as 1400 B.C. In Greek myth the same material was called 'ambrosia.' It fell from the sky at night and was gathered from the ground in the early morning. Such a unique substance could not have been recorded for posterity in such diverse cultures, if it did not actually fall to Earth on a number of occasions and over a wide area.

Regardless of how unbelievable these processes sound, the Soma may have carried the seeds (excuse the pun) of its own proof. The possibility exists that some seeds fell to the Earth with the Soma, and took root here. There is today a fragile plant that can only be found in the high valleys of the Punjab Himalayas, which is supposed to be the source of the deity Soma. In the following quote this is stated with great certainty, because the author believes that the myth only refers to an earthly source for Soma:

> The Soma plant of the Rig-Veda is the Asclepias acida of Roxburgh. It is a creeping plant, almost destitute of leaves. It has small white fragrant flowers collected round the extremities of the branches. Roxburgh says that it yields purer milky juice than any other plant he knows; and that this juice is mild, and of an acid nature. The tender shoots are often plucked by native travelers. It grows on hills of the Punjab, in the Bolan Pass, in the neighborhood of Poona.[69]

This is the plant which the modern scholars say was refined into Soma by the Aryans. If this plant is indeed related to the deity Soma, it is probably the other way around. That is, the Soma that fell to Earth might have carried in it seeds of Asclepias acida of Roxburgh. But where could the original organic material have originated? We know that proto-Venus was too hot then and is still today, for any organic material to grow on its surface. Also, after its initial encounters with the Earth, at the very beginning of the Vedic period, it never again came close enough to transfer any matter to the Earth. Conversely, priori-Mars was in orbit close to the Earth during half of the 3,000 year Vedic period and was covered with vegetation, until so much atmosphere and water was lost that this could no longer be sustained.

It is not unlikely that future missions to the southern highlands of priori-Mars may find freeze-dried samples - fossils of this same plant or actual frozen seeds, buried in the permafrost, which can be proven to be from the same plant by analyzing the genetic material. This is but one of the amazing possibilities suggested in the ancient Vedic hymns.

Certainly, most modern scientists will fail to believe our hypothesis initially, but this attitude will not be new. Disbelief in the Soma myth extends back hundreds of years to past scholars of the

Vedas. One such commentator, who implies his disbelief, in spite of the clarity of the material he is translating, says:

> In the hymns to Soma the pressing of the sacred juice through a woolen filter...is likened to all manner of celestial phenomena with which it would appear to have nothing to do. These comparisons, which sometimes amount to identification, are repeated and intensified in the later literature...the sieve which filters the juice, as we have seen, is likened to the sky; indeed it is the sky. Soma is in the navel of heaven in the woolen filter (R.V. 9.12.4), it 'traverses the lights of heaven, the woolen filter (R.V. 9.37.3), or purifying himself in heaven...he walks with the sun in the filter...(R.V. 9.27.5). In none of this is any incongruity felt, ... And so it is that the golden plant, though born of the mountains on Earth is also the 'child of heaven (R.V. 9.38.5), and though 'running in the vat' it is also 'heaven's Lord who has a hundred rays' (R.V. 9.86.11).[70]

Incredulous of accepting the notion of Soma coming from the sky, he turns the entire meaning around, and suggests that the people got the Soma from an Earth plant, and loved its intoxicating effects so much that they made it a god and then made associations between the processes they used to purify it and heavenly objects such as woolen clouds. This is clearly *not* what is stated in the *Rig Veda*. Nearly all the hymns of its Book 9 are addressed to the deified soma-juice. In this case, as with so many other newly revealed ideas herein, the failure to suspend disbelief will only delay the great advances which they can and will make possible.

The Human Nature of the Vedic People

The next three chapters of this book concern the Hindu Triad, Brahmā, Siva and Vishnu, featured in the Great Epics and the Puranas. Before proceeding to that era, it is of interest to listen to the human side of the Vedic people, revealed in the hymns. The following quotations are translated directly from the *Rig Veda*. The first states the incredulity towards the events that were witnessed, as Indra swooped toward the Earth:

> How is it that, unbound and unsupported, he falleth not although directed downward?
>
> By what self-power moves he? Who hath seen it? He guards the vault of heaven, a close-set pillar.[71]

The following quote makes the analogy between the planets which swoop down towards the Earth, in apparently random fashion, like dice being cast onto a playing board, and resulting in good fortune for some and bad for others.

> Downward they roll on the dice-board, but they bob up on high; handless themselves, they vanquish those that have (powerful) hands. Cast on the board they shine like glowing embers; they are cool no doubt, but how they burn up my heart! [72]

The following quote reinforces the previous thought, and even though it is quoted elsewhere in this book, we repeat here because it is appropriate. It probably refers to the Maruts, because a 'gang' is mentioned:

> The gang of gamesters, pretending to be god-like and true as the sun, ventures the game time and time again; but the dice will not truckle to the rage of a contestant, however truculent; even the king must eat humble pie before them. [73]

In a similar vein, the transient nature of riches is emphasized, since they can quickly be removed by the approaching 'god.'

> The dice their victims hook and tear,
> Disturbing, torturing, false though fair.
> The transient gains they yield today,
> Tomorrow all are swept away. [74]

The poor stood to gain and the rich to lose in the catastrophic encounters between planets. However, some of the rich hoarded food when such an encounter seemed imminent. These people, who would not share their bounty in times of great travail, were greatly despised:

> A man, inhumane and ungenerous, in vain, overstocks himself with provisions. In very truth, this will be the death of him; a man who does not feed his elders and companions, but feasts all alone, is Sin incarnate. [75]

As the ominous Indra approached the Earth, many of the rich began to give alms to the poor in hopes that the threatening god would see this as a benevolent act, and spare the giver his wrath. The poor, on the other hand, knew this to be the case and used the opportunity to extract some 'baksheesh' from the otherwise stingy rich, by recalling for their benefit how destinies vary each time the planet approaches the Earth:

> "Relieve the poor while yet ye may;
> Down future time's long vista look,
> And try to read that darkling book;
> Your riches soon may flit away.

Ye cannot trust their fickle grace.
As chariot wheels in ceaseless round
Now upward turn, now touch the ground,
So riches ever change their place.[76]

These quotes are of great value to our hypothesis, because they emphasize the repetitiveness and *reality* of the catastrophic events that occurred during the Vedic period. No myth, or fairy story, in the sense usually thought of in the modern world, would evoke these gut-level feelings.

Firmament

7
BRAHMA

Koestler explained creativity as the 'bisociation' of two conceptual matrices which are not normally associated, and which may even seem incompatible: 'The basic bisociative pattern of the creative synthesis [is] the sudden interlocking of two previously unrelated skills, or matrices of thought.' The more unusual the bisociation, the more scope there is for truly creative ideas.

Margaret Bodin paraphrasing
Koestler's The Act of Creation, p.210

Of all the revelations of mythical 'gods' in this book the identification of Brahmā is one of the most revolutionary because no Vedic scholar or modern day scientist has even imagined anything close to it. The same explanation also makes possible the identification of the primary deities in the sacred myths of most other ancient cultures, since they all observed the same cosmic phenomenon.

Brahmā was the single most visible, most worshiped feature on the surface of priori-Mars during its encounters with the Earth. Although it took months to rise to its full stature at the beginning of each kalpa (day of Brahmā), it reached staggering proportions. It was an enormous lava fountain that extended upward from the surface of priori-Mars, perhaps a thousand kilometers. It was drawn upward and supported by the tidal force of the Earth, which remained poised directly above it. It was able to rise this high because the outer portions of myriad lava columns hardened into tubes, providing rigidity and allowing liquid lava in their interiors to flow even higher. Brahmā was the central figure on priori-Mars, because this was the location at which the tidal force of the Earth was the strongest. Its great mythical power derived from its enormous size, its complex shapes, colors and textures, and the fiery glow which came from the raw lava at the top of the column. The following quotations contain the clues on which we base our hypothesis.

> There are two states of this Brahmā - one with, and one without shape; one perishable, one imperishable; which are inherent in all beings. The imperishable is the supreme being; the perishable is all the world. The blaze of fire burning in one spot diffuses light and heat around; so the world is nothing more than the manifested energy of the supreme Brahma; and inasmuch as the light and heat are stronger or feebler as we are near to the fire or far off from it, so the energy of the supreme is more or less intense in the beings that are less or more remote from him...Glory to Brahma, who is addressed by that mystic word (Om), associated eternally with the triple universe, and who is one with the four Vedas...He is the invisible, imperishable Brahma; varying in form, invariable in substance; the chief principle, self-engendered; who is said to

> illuminate the caverns of the heart; who is indivisible, radiant, undecaying, multiform... "In the earliest writings Brahma signified a hymn or mantra, whilst Brahmā was the term used to denote a priest or worshipper. It is in the later parts of the Vedas that Brahma is identified with the supreme, and Brahmā becomes his great manifestation. Prajapati, the lord of creatures, was the Creator according to the earlier teaching of the Vedas, and occupied the position in the earlier Pantheon that Brahmā did in the latter. In several texts of the Vedas the two are identified, and thus authority is found for the idea that Brahmā is to be worshiped as the maker of all things. [77]

The invisible, imperishable Brahma (indicated by a short 'a' ending in the nominative case), "without shape," "that illumines the caverns of the heart," comes from the fact that the Vedic people observed the glowing, visible, manifested form of Brahmā (written with a long 'a' ending), gradually rise up out of the interior of priori-Mars at the onset of each encounter. As a result they realized that its source resided permanently hidden within the planet. Brahma was thought to 'arise', i.e. become manifest during the encounters with the Earth in order to create, that is, to shoot thousands of hot bodies into space. The top of this, the largest central fountain of magma, glowed brightly while the smaller surrounding forms, designated by mythical names such as Rama and Krishna were cooler, and therefore did not give off as much light. The glow of Brahmā illuminated them, the ones closer receiving more light than those farther away, i.e. he is said 'to illuminate the caverns of the heart.'

Brahmā, the first of the Hindu gods, was the creator, the father of the gods, the Vedic Prajapati. Later the invisible, formless, sleeping Brahma came to be seen as more important, even to be the invisible force behind all things. *In this concept we see the combining of Brahma and Vishnu, the closest Hindu concept to a supreme being.*

The following passage is attributed to Manu, one of many sages whose names paralleled the names of features observed on the surface of priori-Mars, and who was the hero in the myth of the great flood.

> He who is beyond the cognizance of the senses, subtle, undiscernible, eternal, who is the essence of all things, and inconceivable, himself shown forth.
>
> He desiring, seeking to produce various creatures from his body, first created the waters, and deposited in them a seed.
>
> This (seed) became a golden egg, resplendent as the sun, in which he himself was born as Brahmā, the progenitor of all worlds.
>
> ..That lord, *having continued a year in the egg,* divided it into two parts by his mere thought.

Brahma

In the Maharabata, and some of the Puranas, Brahmā is said to have issued from a lotus that sprung from the navel of Vishnu.

The "seed" was the first manifestation of the subsurface magma, which at first was small, but then grew, just as the fish grew in the waters in the flood myth. It was bright because it was extremely hot. The passage says Brahmā remained in the 'mundane egg' for a year. As it grew it took on various forms. The one described here was one in which it seemed to form two parts, one staying more or less on the surface of priori-Mars and the other extending upward toward the Earth. These shapes resembled the male and female organs, the linga and yoni, during intercourse, and is the reason for the name Purusha (male) used to describe Brahmā in the Vedas. Numerous statuary figures resembling this form can be found in museums and temples of worship in India today. The fact that glowing bodies were 'created' out of this formation reinforced the connection with the sexual organs.

The following quote lends itself to gross exaggerations of both the size and the time scale of the 'egg' feature. Whether this was done to hide the true meaning from the lower castes or because the later writers had lost all cognizance of the physical nature of the events, we cannot be sure.

> The egg referred to above is thus described in the Vishnu Purana - " Its womb, vast as the mountain Meru, was composed of the mountains, and the mighty oceans were the waters which filled its cavity. In that egg were the continents, seas, and mountains; the planets and divisions of the universe; the gods, the demons, and mankind. Brahmā is said to be born; a familiar phrase to signify his manifestation". This wonderful egg, after the creator had inhabited it for a thousand years, burst open, and Brahmā, issuing forth by meditation, commenced the work of creation. Seeing that the Earth was sunk beneath the waters, he assumed the form of a boar, and diving raised it upon his tusks. After this he continued the work of creation. [78]

It is doubtful that the original Sanskrit included specific words meaning 'planets' and 'universe,' but the idea was that it comprised a microcosm of all things in the 'world.' The reference to a thousand years is another example of the embellishment of the lifetimes of the deities by the Hindu commentators who never actually witnessed the events described. It is obvious that this is no chicken egg. Although the features described were physically 'real,' they did not take place on the 'Earth.' The true scale and physical dimensions were impossible to determine because the Vedic people had no means of measuring the distance to priori-Mars.

The notion that Brahmā remained in the 'egg' for a thousand years is typical of the confused time scales associated with a day of Brahma. Although many passages, intentionally or not, tend to obscure these time scales, we believe that a period of a year in the 'egg,' as in the previous quote, is probably accurate. It is consistent with the length of the day of Brahmā which we determine in the subsequent section titled "The Time Scale in the Hindu Myths".

The following is a description of the fully mature form of Brahmā, in which the lava tubes had hardened and his body had become enormously complex:

> In pictures Brahmā is represented as a red man with four heads, though in the Puranas he is said to have had originally five. He is dressed in white raiment, and rides upon a goose. In one hand he carries a staff, in the other a dish for receiving alms. A legend in the Matsya Purana gives the following account of the formation of his numerous heads: - "Brahmā formed from his own immaculate substance a female who is celebrated under the names Satarup, Savitri, Sarasvati, Gayatri, and Bramani. Beholding his daughter, born from his body, Brahmā became wounded with the arrows of love and exclaimed, 'How surpassingly lovely she is!' Satarup turned to the right side from his gaze; but as Brahmā wished to look after her, a second head issued from his body. As she passed to the left, and behind him, to avoid his amorous glances, two other heads successively appeared. At length she sprang into the sky; and as Brahmā was anxious to gaze after her, a fifth head was immediately formed. Brahmā then said to his daughter, 'Let us produce all kinds of animated beings, men, suras (gods), and asuras (demons). Hearing this, she descended, and Brahmā having espoused her, they withdrew to a secluded spot where they dwelt together for a hundred divine years; at the expiration of which time was born Manu, who is also called Svayambhuva and Virj.[79]

We interpret the consorts of Brahmā and Vishnu as bodies of water, or in some cases ('she sprang into the sky'), an enormous cloud of steam resulting from the contact of the tidal ocean, through which the figure of Brahmā arose, and the hot lava out of which it was formed. This explains the observation that Satarupa was born from the body of Brahmā, but at the same time appeared somehow different and more exotic than him. Great flowing rivers on priori-Mars brought water from the heated soil and aquifers which collapsed due to the tidal force as the planet approached the Earth. The water was drawn across the surface of the red planet, not due to the slope of the priori-Mars landscape, but due to the tidal forces from the Earth as the planets converged.

The espousal of Brahmā to his own 'daughter' created a scandal among scholars and interpreters of the Hindu myth - this is discussed in a subsequent section. The statement that they "withdrew to a secluded spot for a hundred years," may mean that the form was obscured by clouds for a short time. Such exaggerations of time periods were common in the Hindu myths, and were probably due to ignorance of the original cosmic encounters, which had ceased by that time. Sarasvati having been produced from Brahmā, was regarded as his daughter; hence her union with him was said to be criminal by the other gods.

> She is celebrated both as a river and a goddess. She was primarily a river deity, as her name 'the watery,' clearly denotes..in two places she is mentioned along with rivers or fertilizing waters: 'Ye opulent waters command riches: ye possess excellent power and immortality...In another place she is said to 'surpass all other rivers, and to flow pure from the mountains to the sea.

These interpretations are consistent with our hypothesis that several of the consorts of the Hindu triad are bodies of water on priori-Mars. In this quote she is described as a river. The giving of the same name to a river on Earth was consistent with the duality of similar features that runs through the Vedas. In addition to references as a river and possibly a lake, Sarasvati is also described as 'mounting upward' in one myth which describes the 'growth' of the four (or five) heads of Brahmā. The implication of that usage is that the water out of which Brahmā was born was exposed to great heat as the lava forms became larger and stronger. As a result, some of this water was converted into clouds of steam which formed beautiful shapes entwining themselves about the form of Brahmā, mythologically causing him to become enamored with her.

The lava which jetted into the air from thousands of cracks in the surface ('Brahmā had a thousand feet') of priori-Mars combined and rose to great heights in the low gravity zone between the two planets. At the top of the lava column the molten material formed a series of stacked horizontal plates or shelves, one above the other with some space in between. These were described in the Hindu myths as the 'heads' of Brahmā. They may also have been displaced horizontally from one another. Their actual shape is belied by the highly anthropomorphised figures of Hindu deities, but the general shape can be gleaned from the symbology of other cultures, such as the Chinese and Egyptians, discussed in subsequent chapters.

The myth of the fifth head of Brahmā is of particular interest, because it appears to be a description of the only sustained, naked fission reaction ever witnessed by mankind:

> His fifth head also having read the Vedas, which the other four heads had delivered, acquired a splendor which neither suras (gods) nor asuras (demons) could endure...Unable to approach or behold it, they determined to apply to Siva for relief...Siva granted their wish and proceeded with them to where Brahmā remained inflated with pride. On seeing Siva, Brahmā did not pay him the customary honours. Siva, seeing Brahmā's fifth head inflicting distress on the universe by its effulgent beams, brighter than a thousand suns, approached him and said, 'Oh! this head shines with too much splendour,' and immediately cut it off with the nail of his left thumb. The Mahabarata says it was because of his seduction his own daughter that Siva decapitated him. This crime was attempted when in a fit of intoxication. [80]

The extreme brightness of the fifth head of Brahmā is a unique observation 'brighter than a thousand suns.' The myth explains whimsically that the brightness of the fifth head of Brahmā as being a result of pride, at having received or learned the complete knowledge of the Vedas. But there is an explanation based on the laws of physics. The lava from many cracks which shot high into the 'air' above priori-Mars remained suspended partly due to the upward tidal force exerted by the more massive Earth. This enormous lava column produced local concentrations of many elements or compounds, just as certain elements are concentrated in highly localized deposits in the crust of the Earth. The unique nearly weightless state in the lava fountain allowed the hottest elements dissolved in the lava to rise to the top of the column, similar to the way volatiles do in a catalytic cracker. In particular, naturally radioactive elements such as Plutonium or Uranium rose to the top of the column because they were hot. We suggest that more than a critical mass of one of these elements accumulated within the top layer or head of the enormous 'lava fountain' and, as a result, a sustained, naked (unshielded) nuclear reaction took place suspended in space far above the surface of priori-Mars. This continued for an extended period of time until the nuclear fuel was expended. This would have emitted copious amounts of radiation, accounting for the fact that the emitted light was so bright that it could not be viewed directly even though the source was some 30,000 kilometers from the Earth. The extremely energetic gamma radiation from this reaction may be manifested in the rocks that were exposed on the surface of priori-Mars at that time.

Associated with this myth is the notion that this reaction cut the circular grooves in the surface of priori-Mars that defined the concentric circular continents and circular seas described above. This seems to be fanciful writing but it cannot be dismissed out of hand.

Brahmā the Creator

Most of the quotations given in this section come from the Original Sanskrit Texts, v.1, and particularly from a section in that reference entitled 'Mythical Accounts of the Creation of Man and the Origin of the Four Castes.' The story of the creation of 'beings' is reiterated in most of the Vedic and Hindu books. Although there are numerous versions of the same story, with minor variations, the basic theme is that the creator, called by the names Purusha or Prajapati in the Vedas, and Brahmā in the Hindu myths, produces different types of 'beings,' apparently from different parts of his body. The author, Dr. John Muir, and many other scholars over the last century have been particularly intrigued by the question of how the four castes of India came about. As a result he devoted an entire section of his Volume 1 to quotes which mention the castes, and their 'creation' by Brahmā.

The great form known as Brahmā is said to have created 'beings' of the four castes of Indian people - Brahmins from his head, Kshatriyas from his chest, Viasas from his thighs and Sudras from his feet. We interpret this to mean that massive hot chunks of rock were periodically blasted out of the Brahmā formation into space. *We claim that the reference to the four castes was a way to describe the colors of Brahmā. It indicates that the highest point of the formationwas raw hot lava, because the Brahmin has the lightest skin; the 'chest' was cooler, but still bright red, the color of the Kshatriya; the 'legs' were dark red, like the skin of the farming caste, the Viasas; and the 'feet' were black, as the skin of the Sudra, or lowest caste.* Therefore this myth indicates that the temperature of the form of Brahmā was highest at the top, the head, and the lower down on the formation the cooler it became.

The darkness of the lower portions of the Brahmā form provide us with yet another clue as to the physical processes which occurred on the surface of priori-Mars during the 'dance encounters' between the red planet and the Earth. Assuming the description is accurate, the entire body of Brahmā cannot have been a 'fountain' of raw exposed lava. *What this myth tells us is that the lava which comprised this and other forms, cooled and hardened on the outside, forming a multitude of vertical 'lava tubes,' through which the liquid lava continued to flow upward.* This process is common on the Earth for certain types of lava as it flows down slopes from an active volcano. The only difference being that the tubes with which we are familiar extend along the surface of the Earth, and not in the vertical direction. The formation of the tubes extended to greater and greater altitudes until a lava tube structure as tall as a thousand kilometers formed. This was a tenuous structure which could not support its own weight, but the tidal force of the Earth aided in its support while the two bodies orbited one another. The colors related in the myth are indicative of the temperature of the shell of this enormous lava tube structure which was observed by millions of people on Earth. Figure 7.1 is an artist's rendition of the column of smoke and fire, otherwise known as Brahmā.

Figure 7.1 An artists rendition of the fully developed form of Brahmā, with five 'heads' and a thousand feet.

In the Hindu myths, Brahmā was associated with the creation of a number of 'creatures,' called gods, fathers, rishis, asuras, Manus, and men, as well as the Vedas themselves. The following quote gives an unembellished idea of the creation which the Vedic people witnessed:

> Now dissolutions and creations are said in the Vedas to take place during (the creator's) sleep and at his waking. "When the sleeper does not see any vision, and when his breath is concentrated in him, then the voice with all names enters into him, the eye with all forms enters into him, the ear with all sounds enters into him, and the mind with all thoughts enters into him. When he wakes, just as sparks shoot out in all directions from blazing fire, so do all breaths according to their several seats issue from this Soul; from the

> breaths spring deities; and from the deities worlds" (Kaushitaki Brahmana, latter part, 3, 3). [81]

The dissolutions referred to were the collapsing of the stationary (lava fountain) forms, at the times when the 'gravitational lock' was broken and a close encounter (a day of Brahmā or a Kalpa) between the Earth and priori-Mars concluded. At this time Brahmā collapsed back down to the surface of the planet, as the internal gravitational field of priori-Mars once again became the dominant force and that of the Earth was removed. These events occurred at the times when Brahmā was said to go to sleep. He was imagined to be in a state of repose, in which all his powers were present, but internalized within the planet. When priori-Mars again approached the Earth, the tidal sea formed and Brahmā rose again through the water. After his gradual 'awakening', all of his powers once again became manifest.

The rock bodies blasted out of the volcanic vents were described as being propelled by the 'breaths' of the god, that is, the expulsion of gases was witnessed by the Vedic people. As the larger vents out-gassed they often disgorged a number of glowing rocks in rapid succession with exactly the same direction and velocity. Many were ejected with sufficient speed with such speed that many escaped the gravitational field of priori-Mars, some falling to Earth and others being injected into orbits around the Earth and the Sun. From the Earth they appeared like sparks shooting out of a fire, but as they approached the Earth their enormous size became evident, that is, they were seen as 'worlds.'

> As these (streams) shine, they are called ambhāmsi. This root *bhā* is used by the intelligent in the senses of pervading and shining, and the male, Prajapati, declares (the fact). [82]

Obviously, the vestiges of the circular continents did not disappear completely in the time between Brahmā's sleeping and subsequent awakening:

> At the close of this day of Brahmā a collapse (pratisancharah) of the universe takes place, which lasts through a night of Brahmā, equal in duration to his day, during which period the three worlds are converted into one great ocean, when the lotus-born god, expanded by his deglutition of the universe, and contemplated by the yogis and gods in Janaloka, sleeps on the serpent Sesha. At the end of that night he awakes and creates anew. [83]

The coils of the serpent Sesha are the vestiges of the ringed continents, which were covered by the shattered debris of the collapsed lava tube forms at the time of the last dissolution and reshaped by the internal gravitational field of priori-Mars during the night of Brahmā, during which time it orbited the Sun independently of the Earth. In spite of these effects, the vestiges of the continents did not disappear completely between encounters. The waking of the creator occurred again each time priori-Mars approached the Earth, Brahmā 'arose' through the tidal sea, and the 'creation' began anew.

> At the beginning of the past (or Pādma) Kalpa, arising from his night slumber, Brahmā, the lord, ... beheld the universe void. He was the supreme lord Nārāyana, who cannot even be conceived by other beings, the deity without beginning, the source of all things, existing in the form of Brahmā...This lord of creatures, discovering by inference, - when the world had become one ocean, - that the Earth lay within the waters, and being desirous to raise it up, assumed another body. As formerly, at the beginning of the Kalpas, he had taken the form of a fish, a tortoise, and so forth, (so now) entering the body of a boar, - a form composed of the vedas and of sacrifice, - the lord of creatures...entered the water...(and) the supreme soul, the upholder of the Earth, lifted her up quickly and placed her upon the great waters. Resting upon this mass of water, like a vast ship, she does not sink, owing to her expansion. Then, having leveled the Earth, ... (and) heaped together mountains according to their divisions. He...created on the surface of the Earth all those mountains which had been burnt up in the former creation. Having then divided the Earth, just as it had been, into seven dvapas, he formed the four worlds Bhrloka and others as before. Becoming next pervaded with the quality of passion..., assuming the form of Brahmā, with four faces, effected the creation. [84]

In the following excerpt, the translator obviously has some difficulty identifying the body on which the events took place. At first he translates it universe, but in the next sentence he places Earth in parentheses, probably because the object was implied in the Sanskrit. But why does he use a different word?

> The universe was formerly waters, fluid. On it Prajapati, becoming wind, moved. He saw this (Earth). Becoming a boar, he took her up. Becoming Visvakarman, he wiped (the moisture from) her. She extended. She became the extended one (prithiv). [85]
> It is possible that the idea assigned to the word Nārāyana, "he whose place of movement is on the waters," may be connected with this passage. See also Genesis 1, 2; "And the Spirit of God moved upon the face of the waters ."[86]

Interpreters have difficulty with terms that seem to imply 'the Earth,' because they did not understand that the events being described actually took place on the surface of priori-Mars which was quite visible during the periods when it remained close to the Earth over the Himalayas. The ancients could see that priori-Mars was a sphere and it never occurred to them that there would be confusion between 'the world' in their myths and the Earth.

Streams of glowing rock bodies were blasted out of volcanic vents into space. These were distinguished both by their appearance and their destinations. The suras (gods, or glowing) were different from the asuras (with breaths, or vapor trails). Thus the same terms that were applied to the planets themselves were also applied to the smaller bodies ejected from priori-Mars. Asuras were cooler bodies ejected from near surface layers that contained water while Suras were glowing hot rocks that came from deeper layers. The latter are now referred to by astronomers as asteroids while those which still retain some of their water are called short period comets.

The 'embodied spirits' (spirits given bodies by Brahmā) were also categorized according to their destinations. Those bodies which traversed the short distance and fell on the Earth were termed 'men,' those that fell on the Moon were called the 'fathers,' and those that stayed in orbit about the Earth or the Sun were called 'gods,' because they remained in the heavens.

The lava fountain features which rose up on the surface of priori-Mars gradually changed as new eruptions emerged and old ones became occluded. It was the shapes and relative positions of these bodies to one another which prompted many of the myths and Avataras. Other 'actors' on this celestial stage were bodies of water, tornado-like 'serpents,' and beautiful 'female' vapor forms which entwined themselves about the lava-fountain 'heroes.'

Perhaps the hymn which best expresses the sheer size and complexity of the fully developed Brahmā is the 90th hymn of the 10th book of the *Rig Veda* Sanhita - called the Purusha Sukta, or the hymn to Purusha, an earlier Vedic name for Brahmā. It has always been of particular interest to Indian scholars because it contains the earliest mention of the four castes of the Indian people, for which they have sought the origin:

> Purusha has a thousand heads, a thousand eyes, a thousand feet. On every side enveloping the Earth he occupies (it) by a space of ten fingers. Purusha himself is this whole universe, whatever has been and whatever shall be. He is also the lord of immortality, since with another he expands. Such is his greatness, and Purusha's is superior to this. All existences are a quarter of him; and three-fourths of him are that which is immortal in the sky. With three quarters Purusha mounted upwards. A quarter of him was again produced here. He was then diffused everywhere over things which eat and things which do not eat. From him was born Viraj, and from Viraj, Purusha. When born, he extended beyond the Earth, both behind and before. When the gods performed a sacrifice with Purusha as the oblation, the spring was its butter, the summer its fuel, and the autumn its (accompanying) offering. This victim, Purusha, born in the beginning, they immolated on the sacrificial grass. With him the gods, the Sādhyas, and the rishis sacrificed. From that universal sacrifice were provided curds and butter. It formed those aerial (creatures) and animals both wild and tame. From that universal sacrifice sprang the rich and sāman verses, the metres, and the yajush. From it sprang horses, and all animals with

> two rows of teeth; kine sprang from it; from it goats and sheep. When the gods divided Purusha, into how many parts did they cut him up? what was his mouth; what arms had he? what two objects are said (to have been) his thighs and feet? The Brahmin was his mouth; the Rajanya was made his arms; the being called the Vaisya, he was his thighs; the Sudra sprang from his feet. The moon sprang from his soul (manas), the sun from his eye, Indra and Agni from his mouth, and Vayu from his breath. From his navel arose the air, from his head the sky, from his feet the Earth, from his ear the four quarters: in this manner formed the worlds. When the gods, performing sacrifice, bound Purusha as a victim, there were seven sticks (stuck up) for it (around the fire), and thrice seven pieces of fuel were made. With sacrifice the gods performed the sacrifice. These were the earliest rites. These great powers have sought the sky, where are the former Sādhyas, gods. [87]

The mythical idea behind this passage is slightly different from that of the previous quotes but the same physical occurrences are being described. In it Purusha was imagined to be the victim of a great sacrifice performed by the 'gods' and the seven rishis. As a result of this sacrifice, all kinds of 'aerial' creatures were formed. The descriptions of the various things produced from his sacrificed body, although not as detailed as in the later Hindu books, generally corroborate that the head was the brightest, because the things associated with it are light (Brahmins) or fire (Indra and Agni). These created entities are all allegorical. Even the *Vedas* are said to have issued from Purusha.

The seven sticks, posts to which the sacrificial animal was bound, were columns of magma and volcanic ash streaming upward from the seven volcanos frequently referred to as the seven rishis. Normally, as photographed on Io by the Voyager satellite, the emissions from a volcano spread out in all directions and fall back to the surface in a shape resembling an umbrella, but in the bizarre gravitational field between the two planets, the tendency was for the effluents to stream straight up into the area of null gravity, just as did the form of Purusha.

In the earlier portion of the passage were attempts at describing Purusha's size and shape. First, the thousand heads, eyes and feet were an attempt to show that this form was not a single, smooth form, but contained enormous detail, which could be seen from the Earth. In this particular description the 'four heads' were not mentioned. Next 'he occupies (the Earth) by a space of ten fingers,' may mean the glowing form was as wide to the eye of the observer as are ten fingers (two hands) held side by side at arms length. The statement about expanding because he was with another, corresponds to other mythical descriptions of Purusha, in which he took on the imagined form of the male sex organ, prithiv, the lingus, and the yoni. The yoni was the breach in the surface of priori-Mars and the lingus was the lava protrusion 'mounting upwards,' which got larger and larger. Many statuettes of Brahmā or Purusha in this simple form still exist to this day, but they look more like eggs than like the male member. The formation became more and more complex as the lava tubes hardened and developed into the full form of Prajapati (Brahmā). In addition there were a multitude of smaller lava forms surrounding the central figure. These smaller flows were pictured as 'the sacrificial grass' in this passage.

Internal convulsions, triggered by sudden alignments with other heavenly bodies, shot streams of large bodies from volcanic vents and calderas shattering the hardened part of the lava forms. Most of them were launched into orbit forming 'aerial bodies' and did not return to priori-Mars. The author attempts to describe the vast amount of material seen being ejected into space

> ...all existences are a quarter of him and three quarters remained [on priori-Mars], diffused everywhere...and is immortal in the sky...
>
> with the half he produced the whole world; what became of the other half of him? [88]

These are attempts of various observers to describe the enormous mass of material that was blasted into space during their lifetimes. Considering that the Vedic period was almost three thousand years in duration, the total amount of material lost from priori-Mars as a result of these convulsions is almost beyond our ability to comprehend.

The aerial bodies were also described as various creatures depending on their appearances -brilliance, shape, and velocity. The form of Viraj was a surface form, similar to but somewhat smaller than Purusha which stood next to him, and was sometimes called his 'son.'

The following version is of interest because it gives a much more physical description of the sequence of events observed by the Vedic people:

> At first this (universe) was not anything. There was neither sky, nor Earth nor air. Being non-existent, it resolved 'let me be.' It became heated. From that heat smoke was produced. It again became heated. From that heat fire was produced. It again became heated. From that heat light was produced. It again became heated. From that heat rays were produced. It again became heated. From that heat blazes (or fogs) were produced. It again became heated. It became condensed like a cloud. It clove its bladder. That became the sea. Hence men do not drink of the sea. For they regard it as like the place of generation. Hence water issues forth before an animal when it is being born...This was then water, fluid. Prajapati wept, (exclaiming), 'For what purpose have I been born, if (I have been born) from this which forms no support?.' That which fell into the waters became the Earth. That which he wiped away, became the air. That which he wiped away upwards became the sky. From the circumstance that he wept (arodit), these two regions have the name of rodasi, (worlds). They do not weep in the house of the man who knows this. This was the birth of these worlds. He who thus knows the birth of these worlds, incurs no suffering in these worlds.

> He obtained this Earth as a basis. Having obtained (this Earth as a) basis, he desired, 'May I be propagated.' He practiced austere fervour. He became pregnant. He created Asuras from his abdomen. To them he milked out food in an earthen dish. He cast off that body of his. It became darkness. He desired, 'May I be propagated.' He practiced austere fervor. He became pregnant. He created living beings (prajh) from his organ of generation. Hence they are the most numerous because he created them from his generative organ. To them he milked out milk in a wooden dish. He cast off that body of his. It became moon-light. He desired 'May I propagate.' He practiced austere fervor. He became pregnant. He created the seasons from his armpits. To them he milked out butter in a silver dish. He cast off that body of his. It became the period which connects day and night. He desired 'May I be propagated.' He practiced austere fervor. He became pregnant. He created the gods from his mouth. To them he milked out Soma in a golden dish. He cast off that body of his. It became day. These are Prajapati's milkings. [89]

This was a description of the growth of Prajapati (Brahmā) at the beginning of one Kalpa (day of Brahmā). First the heating up of the planet, pictured by the Vedic people as being due to the will of the (as yet) invisible Brahma. This occurred in graphically described stages, producing smoke, fire, light, heat rays, and blazes. As a result of this heating up, ground water was released and rushed toward the point nearest the Earth due to the enormous tidal force. Prajapati was then born; that is, he first appeared in the midst of the water, and commiserates because he has no 'support,' as on dry land. The creation of the worlds, in the myths, applies to the 'Earth,' the air, and the sky. This implies that as a result of observing priori-Mars up close, they understood that it was a sphere, that the atmosphere around it was a spherical shell, and that the sky above it was an infinite sphere. (Given this viewpoint sages of that time undoubtedly figured out by inference that the Earth on which they lived was also a sphere.) Prajapati was pictured as reaching down and finding the surface of priori-Mars beneath the waters as a basis (a support). He then willed that he be propagated, that is, have offspring. The term "He practiced austere fervour" can be translated 'He heated up.' With each state of growth he became larger and hotter. Initially, he is the shape of the lingus, or organ of generation - a relatively simple shape. The lava fountain continued to harden and extend upward. Finally it reached the maximum size and complexity that could be maintained, but this form was not permanent. When alignments of priori-Mars and the Earth with other bodies such as the Moon, Sun, and proto-Venus occurred, explosions in the interior blasted out particularly large bodies into space from some of the same vents that fed the great lava-fountain Brahmā. They shattered the hardened lava tubes, causing the entire structure to collapse on the surface of priori-Mars. This was described as the 'casting off' of that body. This also led to the notion that the 'creator' was sacrificed in order to create the expelled bodies. But the lava fountains continued to flow upward, so a new form started to build up and the process repeated. A particularly meticulous translator says that the sense of the original Sanskrit is that of 'a serpent shedding his skin.' The material which fell to the surface also got hotter at each stage of growth as indicated by the colors of the 'dish' and 'bowl' which went from darkness, milk in a wooden dish, moonlight, butter in a silver dish, and finally Soma in a golden dish.

In a particularly allegorical passage, the Vedas themselves are said to have been created by Brahmā. This is another facet of the duality between features or phenomena observed on priori-Mars and things which existed on the Earth. The most detailed description of the 'heads' of Brahmā is given in terms of the four books of the Veda (Rich, Yajush, Saman, and Atharvana):

> From the eastern mouth of Brahmā, who sprang by an imperceptible birth from that divided egg (Manu, i. 9, 12), there suddenly issued first of all the rich verses, resembling China roses, brilliant in appearance, internally united, though separated from each other, and characterized by the quality of passion (rajas). From his southern mouth came unrestrained, the yajush verses of the colour of gold, and disunited. From the western mouth of the supreme Brahmā appeared the sāman verses and the metres. From the northern mouth of Vedhas (Brahmā) was manifested the entire Atharvana of the colour of black bees and collyrium, having a character at once terrible and not terrible, capable of neutralizing the arts of enchanters, pleasant, characterized by the qualities both of purity and darkness, and both beautiful and the contrary. The verses of the Rich are distinguished by the quality of passion (rajas), those of the yajush by purity (sattva), those of the sāman by darkness (tamas), and those of the atharvana by both darkness and purity. [90]

The visual properties of the four heads of Brahmā observed by the Vedic people varied in texture, color, brightness (sattva) and motion (rajas). *We suggest that these different properties came from the differences in the elements concentrated at different altitudes by fractionation in the different lava columns, combined with the differences in the local micro gravity.*

This gives an additional idea of the complexity of the phenomena observed on priori-Mars during the encounters. The emissions of each 'head' of Brahmā were characterized by different degrees of form, motion, and brilliancy. Although it is impossible to make any sensible comparison between the books of the Veda, and the physical elements which comprised these celestial forms, it does help us understand the large scale of these features and the enormous amount of detail that was visible to the Vedic people at that time, because the planet priori-Mars was so close to the Earth.

The Time Scale in the Hindu Myths

This section is particularly important because we wish to determine when these planetary encounters occurred and how long they lasted. Finding the answer is not easy, because of the apparent inflation of the true time intervals by the Hindu scholars. The search is tedious, but the result we obtain is clearly supported by Greek mythology - an encouraging fact.

Each day of Brahmā was divided into four Yugas, whose durations became progressively shorter. These same Yugas were repeated during each day of Brahmā. Each Yuga was characterized by the natures and trajectories of the bodies that were ejected during the period. This division was unique to the Hindu myths, which were recorded much later and are not present in the *Rig Veda*. Although the additional information is welcome, it carries with it a 'down side.' The Hindus, either due to misunderstanding of the ancient Sanskrit, or to deliberately obscure the ancient events to the masses, or to attribute what they considered lifetime more worthy of a god, greatly inflated the length of the day of Brahmā and the Yugas within it, to meaninglessly great durations. The currently accepted scheme of time is as follows:

> 12,000 divine years are equal to a period of four Yugas, which is thus made up, viz: [91]

Krita Yugas with its mornings and evenings	4800 divine years
Treta Yugas with its mornings and evenings	3600 divine years
Dvapara Yugas with its mornings and evenings	2400 divine years
Kali Yugas with its mornings and evenings	1200 divine years
making	12,000 divine years.

The basis for the definition of divine years is not known. However, the situation becomes worse, when the following statement is invoked:

> One year of mortals is equal to one day of the gods. [92]

The application of this notion results in the following astronomical numbers.

> The Vishnu Purana supplies further explanation of the great mundane periods, in the third chapter of the first book, and in the first chapter of the sixth book, as translated by Professor Wilson in his notes to page 50:
> ...As a day of the gods is = to one year of mortals, the 12,000 divine years must be multiplied by 360, the assumed number of days in a year...i.e. 4,320,000 human years are = 1 day of Brahmā, and his night is of the same duration. Within that period of a day of Brahmā, 14 Manus reign. [93]

We claim that the cycles of creation and dissolution described in the Hindu myths could not have been experienced if they were hundreds or thousands of times the life of a single person. People must have witnessed an entire day and night of Brahmā during their lives. In order that these periods be observable, it would make more sense if the inverse of the above statement were true, that is, if *a day of men were equal to a year of the gods.* The justification for this interpretation comes from our claim that priori-Mars remained stationary over the Himalayas, i.e. in a geosynchronous orbit. Thus in one day it would go through all four 'seasons' much in the way the Earth does in a period of a year. If we *divide* the periods mentioned above by 360 instead of multiplying by this amount the 12,000 divine years would be only 33 of our years, giving approximately 13.2, 9.9, 6.6, and 3.3 years for the duration of the four Yugas mentioned above - a much more reasonable period. But this still does not explain the reign of fourteen Manus.

There is another possible source of ambiguity in the translations, one which also exists in modern english. The myths state that a *Kalpa* or 'day of Brahmā' is equal in duration to a 'night of Brahmā,' which seems fairly clear. But in statements translated 'One year of mortals is equal to one day of the gods,' it is not clear whether just the daylight part of the day is meant or a full day and a full night. This leads to the possibility that the Yugas had durations closer to 6.6, 4.95, 3.3, and 1.65 years; that is, a day of Brahmā was about fifteen years and a night of Brahmā was of the same length. This makes more sense because it pretty much agrees with the statement that 14 Manus reigned, because Manus refer to 'men.'

We believe that the last statement, seemingly just thrown in as an after-thought, is in actuality a clue to the true length of a day of Brahmā, that is, fourteen years. Since the night of Brahmā was equal to a day of Brahmā, and his life was 100 such day-night periods, this gives a total length of about 2,800 years for the duration of the Vedic period. As will be seen in Chapter 11, this agrees quite closely with the length of time that the Olympian Gods are said to have reigned in Greek mythology (3,000 years.)

To bolster our claim of much shorter actual times, we offer the following quote, which at least indicates that the matter is open to a wide range of interpretation:

> Of this elaborate systems of Yugas, Manvantaras, and Kalpas, of enormous duration, no traces are found in the hymns of the [more ancient] *Rig Veda*. Their authors were, indeed, familiar with the word Yugas, which frequently occurs in the sense of age, generation, or tribe. ...the word is combined with manushy, mnush, manushah, or jannm, Yugas seems to denote "generations", of men...In i. 158, 6, (Rig-veda), it is said that the rishi Dirghatamas became worn out in the tenth Yugas; on which Professor Williams remarks (R.V. vol. ii, 104, note):
> The scholiast understands Yugas in its ordinary interpretation; but the Yugas of five years is perhaps intended, a lustrum, which would be nothing marvellous.

The following quote gives a sense that these periods were known from human experience, which could not have been the case even with the periods in the quote on page 98:

> The first book of the six, into which the work is divided, is occupied chiefly with the details of creation, primary (sarga), and secondary (pratisarga); the first explains how the universe proceeds from

> Prakriti, or eternal crude matter; the second, in what manner the forms of things are developed from the elementary substances previously evolved, or how they re-appear after their temporary destruction. Both of these creations are periodical; but the termination of the first occurs only at the end of the life of Brahmā, when not only all the gods and all other forms are annihilated, but the elements are again merged into primary substance, besides which only one spiritual being exists. The latter takes place at the end of every Kalpa, or day of Brahmā, and affects only the forms of inferior creatures and lower worlds, leaving the substances of the universe entire, and the sages and the gods unharmed. [94]

We claim that the 'secondary creation' describes the process which occurred each time priori-Mars and Earth entered into a new 'dance encounter.' The 'elementary substances' were the collapsed stuff on the (roughly spherical) priori-Mars from the previous encounter, as it approached the Earth, without much structure. Then when priori-Mars was captured, a tidal sea formed on the side facing the Earth, Brahmā gradually rose up through it, 'created' both flying objects (glowing meteors), and innumerable fountains of magma on the surface of that planet, as it hovered above the Himalayas. At the close of the encounter (the day of Brahmā), the 'inferior creatures and lower worlds,' meaning the multitude of smaller hardened lava features, or stationary deities, and the circular continents formed on the surface of priori-Mars facing the Earth, were obliterated, but the meteoroids and comets, called suras and asuras, which were launched into orbits around the sun or the Earth, remained.

The primary creation refers to the entire life of Brahmā, that is, the entire period of interaction of the planets described in the Vedas. That period concluded when priori-Mars ascended to its current orbit. Once this happened, the bodies placed in orbit around the Earth during the Vedic period eventually burnt up in our atmosphere or crashed to Earth. This was when 'all the gods and other forms' were annihilated. Of course we now recognize that there are still a considerable number of the large bodies currently orbiting the Sun. These circle invisibly in orbits which come extremely close to the orbit of the Earth. The existence of these bodies has recently been 'rediscovered' by modern day astronomers, who call them 'near Earth asteroids,' but have no idea as to their true origin and how recently they were injected into these orbits.

The following quote from the Rig-Veda, referring to the Vedic name for priori-Mars, Indra, is of interest here, not only because it indicates that a unique time scale exists for the gods, but also because it describes in very understandable, physical terms the coupling of the Earth and priori-Mars.

> He made the heaven... and the earth as a likeness of heaven; he placed bounds to the air and pillars to the heavens; like two wheels upon one axle he set heaven and earth apart, and fixed them both. He placed the moon in the sky...Indra rejoices far out beyond stream and land. Both the worlds (earth and heaven) form ideas of his sublimity, but they cannot comprehend it; his half equals both of

them; when he grasps both these unbounded worlds together, they are but a handful; as a skin his power rolls heaven and earth together; they both roll after him (by the necessity of nature) as the wheel after the horse. His days do not pass in human fashion;* neither years nor moons make him old; the course of days do not make him to fade...

[footnote - Job 10.5. Are thy days as the days of man? Are thy years as mans days?]. [95]

The two planets orbiting one another were the Earth and priori-Mars. Hovering above the Himalayas, some twenty times the diameter of the full Moon, priori-Mars rotated on its axis once a day. Wise men realized that the Earth was a 'world' just like priori-Mars and imagined the pair as two great wheels upon one axle. The axle was implied by the form of Brahmā which extended downward toward the Earth. This was called the 'axis mundi' or axis of the worlds in other cultures.

Note that the author includes a seemingly related passage from the Bible in a footnote. This statement implies the relationship that we favor - that *one day of men equals one year of the gods.* This is the opposite of the interpretation in the Hindu myths. We believe that the translation of the ancient form of Sanskrit might have been in error, and that error was propagated due to a lack of understanding of the driving events.

8
SIVA

But helpless Pieces of the Game He plays
Upon this Chequer-board of nights and days;
Hither and thither moves, and checks, and slays,
And one by one back in the Closet lays.

The Rubiat of Omar Khayyam
Translation by Edward FitzGerald

Siva (pronounced 'Shiva') is the third deity of the Hindu triad - Vishnu, the sustainer; Brahmā the creator; and Siva, the destroyer.

> According to the teaching of Hinduism, death is not death in the sense of passing into non-existence, but simply a change into a new form of life. He who destroys, therefore causes beings to assume new phases of existence - the Destroyer is really a re-Creator. [96]

In the planetary scenario which we have interpreted from the Hindu myths, a cyclic pattern of encounters between priori-Mars and the Earth is evident. The sustainer, Vishnu, was the deity embodying its recurring nature. Brahmā was called the Creator because out of this enormous 'column of smoke and fire' were 'created' all sorts of embodied spirits, that is, spirits to which he had given bodies. At the close of a day of Brahmā, a collapse (pratisancharab) took place, which lasted through a night of Brahmā, equal in duration to his day, during which the lotus-born god slept on the serpent Sesha. The stationary 'beings' on the surface of priori-Mars, including Brahmā, collapsed in a pile of blackened rubble. This dissolution of such heroic gods was attributed to another powerful deity, Siva. These major collapses occurred when a 'dance encounter' was terminated and priori-Mars was drifting away from the Earth. These events produced great stresses on the Earth and priori-Mars, causing continuous quaking and ejection of large rocks the volcanos on priori-Mars.

But it was not only at the end of a day of Brahmā when Siva came into play. In addition to these great destructions at the end of each encounter, lesser ones occurred much more frequently, due to alignments of the Earth-Mars pair with other heavenly bodies. Great rock bodies were ejected from numerous volcanos on priori-Mars, not just from the form of Brahmā itself. In fact an important clue which points to the alignments of the priori-Mars-Earth pair with other heavenly bodies as the driving source of these eruptions involves the explosion of Olympus Mons, the largest volcano in the entire solar system. This is clearly described in the following myth from the Hindu Mahabharata:

> As Siva was seated on the Himalayas, where he had been engaged in austerities, Uma, attended by her companions and dressed as an

> aesthetic, came behind him and playfully put her hands over his eyes. The effect was tremendous. Suddenly the world became dark, lifeless, and destitute of oblations. The gloom however is as suddenly dispelled. A great flame burst from Mahadeva's [Siva] forehead, in which a third eye, luminous as the sun, was formed. By fire from this eye the mountain was scorched, and everything on it was consumed. Uma hereupon stands in a submissive attitude before her husband.

Although we have identified several of the consorts of the Hindu triad as bodies of water on priori-Mars, this is not true of Uma, said to be the female power behind Siva. We have good reason to believe that Uma was the Moon. Her aesthetic dress implied the ashen gray color of the Moon. Uma is also 'the personification of divine knowledge, distinguished by severe austerity, adored by the worlds - the most excellent of goddesses.'

The statement that Siva was 'seated' in the Himalayas is one more confirmation of our scenario of 'dance encounters,' in which priori-Mars remained stationary over the Transhimalayas for extended periods of time. Since the Vedic people were almost directly below priori-Mars, in the foothills of the Himalayas, the passing of the Moon behind priori-Mars meant that there was an alignment of Earth, priori-Mars, and the Moon at this time. Thus the eruption described in the myth occurred right at the time of an alignment of the Earth-Mars pair with the Moon. The rapidity of the change in the tidal force increased the severity of the reaction. Since the two planets revolved about one another with a period of one day, the times that they were exactly aligned with other bodies came and went in less than an hour. This rapid change caused violent trauma within priori-Mars.

This myth makes quite clear the fact that the Vedic people saw the face of Siva on priori-Mars. The face is still discernable to this day, although not nearly as impressive as it was during the Vedic period. It is shown in Figure 8.1. The 'eyes' were the erupting volcanos Arsia and Ascraeus Mons. The 'magnificent nose' (pictured as the snout of a lion in Egyptian myth) was the Tharsis Bulge, an almost circular feature some 1500 kilometers in diameter now ringed by several faults, which, although now only some seven kilometers above the surface, was greatly distended forming 'a magnificent nose' during the Vedic period. The 'mouth' of this face was formed by the tip of the Valles Marineris which cuts into the Tharsis Bulge.

The notion of Siva's eyes being covered was due to dust which was kicked up from the surface of priori-Mars as a result of powerful quakes, obscuring all detail and creating a featureless appearance. This was the initial result of the Moon approaching alignment with priori-Mars as seen from the Earth. The phrase "destitute of oblations" implies that the many glowing volcanic vents, usually visible on the surface of priori-Mars during these encounters, were hidden from view by the dust clouds. The idea that these fires represented 'oblations' comes from the custom of the Vedic people to burn wood to propitiate their deities.

The myth goes on to tell that a third 'eye' then erupted in Siva's forehead, belching flames. This was the eruption of Olympus Mons, the largest volcano in the solar system, which, as can be seen in Figure 8.1, is in the position of the 'forehead' of the large raised face-like formation spanning several thousand kilometers. The presence of the largest volcano at this location was the original impetus of the custom still practiced in India, and probably carried over into the Roman and Byzantine Catholic ceremony of Ash Wednesday, in which a mark is placed on the forehead of the faithful.

This important myth explains clearly the mechanism of the episodic eruptions of priori-Mars. The tidal forces on priori-Mars, which were already enormous, suddenly spiked when the priori-Mars-Earth pair passed rapidly through alignments with the Moon, causing great stresses in the interior of priori-Mars. During such alignments all the volcano on priori-Mars erupted violently, belching large hot rocks from the interior, many of which fell to Earth causing death and destruction. When the planet was convulsed the volcano 'eyes' shown brightly, the Tharsis Bulge 'nose' became more distended, and on some occasions, liquid iron from the outer core of the planet was drawn out of the 'mouth' into space. By virtue of the alignment described in the quote, Uma is described as representing the 'energy' or 'active power' of Siva.

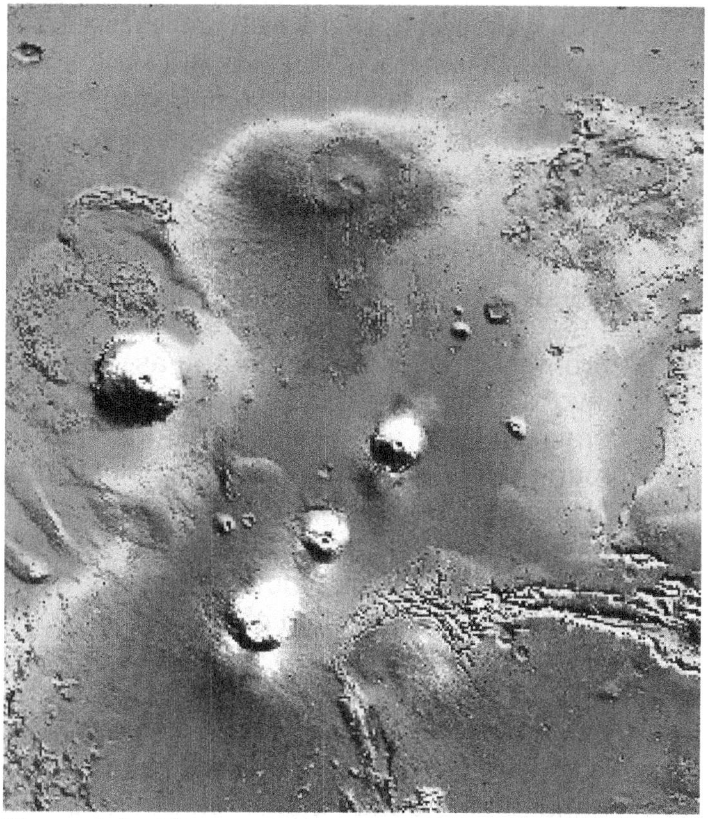

Figure 8.1 Olympus Mons appears as a third eye in the forehead of a face-like figure called Siva. The outer two volcanoes in the row of three form the eyes and the Tharsis Bulge forms the off-center nose or beak, which is cut by the extreme western end of Valles Marineris..

Although priori-Mars may have been continuously losing some atmosphere and water vapor to the Earth throughout the 'dance encounters,' these losses were greatly enhanced at the times of these alignments. The convulsions expelled great masses of rockof all sizes, and along with them choking red dust, atmospheric gases and water, much of which bridged the distance to the Earth and fell over a wide area. The potential size of the largest bodies that were launched is indicated by the overall size of the caldera of Olympus Mons which is some eighty kilometers in diameter, containing a half dozen smaller discernable calderas, indicating the numerous bodies that were ejected at different times.

Space probe measurements of the Tharsis Bulge show a very broad gravity anomaly to this day - representing a mass concentration so great that the moments of inertia of priori-Mars are the most asymmetrical of any planet in the solar system. It is not a large volcano, but an enormous uplifted region, much larger in proportion to priori-Mars than the Tibetan-Himalayan complex is to the Earth. In the figure it is marked by a series of concentric circles, resulting from its collapse.

The Vedic people, who actually experienced and recorded these monumental events, were too close to the subject; that is, they were just reacting to what happened and did not have the luxury that the later Hindus had of seeing the entire period in retrospect. The Hindus developed the idea of cyclic creation and destruction. Thus, although the deity Rudra of the Vedic period is obviously the precursor of Siva, based on his swarthy, unkempt appearance and insane 'actions,' his 'theological' function, as the necessary destroyer, was not so clearly defined. The Vedic people just prayed that this terrible deity would spare their lives. As examples of the visible characteristics of Rudra-Siva from the Vedas:

> From him who is as brilliant as the sun, who shines like gold...We invoke with obeisance the ruddy boar of the sky, with spirally braided hair, a brilliant form...Far be from us thy cow-slaying and man-slaying weapon...Thou fitly holdest arrows and a bow; fitly thou wearest a glorious necklace of every form...May he who glides away blue-necked and red colored be gracious unto us...Reverence to the blue-necked, to the thousand-eyed, to the bountiful, and to the lord of spirits, and to the lord of thieves. [97]

In the later Ramayana, the following description of the less-likeable side of Siva is given:

> He roams about in dreadful cemeteries, attended by hosts of ghosts and sprites, like a madman, naked, with dishevelled hair, wearing a garland of dead mens skulls and ornaments of human bones, ...insane, beloved by the insane, the lord of Bhutas (spirits), beings whose nature is essentially darkness.

The description of Rudra-Siva as 'wandering' is particularly enlightening. Obviously this was considered unusual, even insane behavior. This supports our hypothesis that the *usual* behavior of the planet priori-Mars was to remain fixed in the sky over the Himalayas, while the two planets remained locked together in their 'cosmic dance.' When priori-Mars was released from its orbit around the Earth, it was seen to drift in longitude by the Vedic people - thus the notion that he 'wandered.' Chances are also good that priori-Mars was misshapen, wobbled erratically, and Siva appeared to have spittle running from his mouth, as it moved away - contributing to the impression of its being 'insane.'

Siva adopted the garb and lived the life of an ascetic:

> He is represented in human form, living in the Himalayas along with Parvati...wearing around his black neck a serpent, and a necklace of skulls, and furnished with a whole apparatus of external emblems, such as a white bull on which he rides, elephant's skin, rattle, noose, etc. He has three eyes, one being in his forehead...he has a crescent on his forehead. [98]

These descriptions reinforce the idea that Siva was the name given to priori-Mars at the times of its convulsions. priori-Mars was very hot and glowing a deep red when seen at night. There were still many fissures through which raw lava and flames could be seen. By day it was grey or even black due to the ashes and shattered lava-tube formations that were strewn over much of the northern plains. This appearance was the origin of the notion of spiritual cleansing by covering ones self with ashes. The mention of the third eye and crescent in the middle of Siva's forehead leaves little doubt that Siva was this large face-like combination of features. The crescent was the shadow of Olympus Mons which appeared at lower sun angles.

9

VISHNU

The most beautiful thing we can experience is the mysterious. It is the source of all true art and all science. He to whom this emotion is a stranger, who can no longer pause to wonder and stand rapt in awe, is as good as dead: his eyes are closed.

Albert Einstein

The major books, on which modern Hindu mythology are based, were produced in the thousand years following the Vedic period. They are, in chronological order: the two Great Epics - the Ramayana and the Mahabharata; the Puranas, or 'old traditional stories,' eighteen in number; and the five principal Tantras.

> These books differ from the Vedas in this respect, that whilst the older scriptures (the great epics) treat of the religion common to the Hindus of that period, each Purana is chiefly concerned with some one god whose excellences are extolled whilst others are spoken of in a depreciatory manner. [99]

This implies that during the period of the great epics, up to 0 B. C., as in the earlier Vedic period, the religion was more or less focused so that everyone worshiped the same deities. By the time the *Puranas* were composed, perhaps 500 A.D., the culture probably expanded and became dispersed over a much larger area, so that different regions drifted into an emphasis on one particular deity of the pantheon in the *Puranas*. The gods established in the great epics are the subjects of the Puranas, and are propitiated to this day. *The three deities of primary importance in the Hindu pantheon are Brahmā, Vishnu, and Siva, known as the Hindu Triad.* We have discussed Brahmā and Siva first because they are more easily identifiable, whereas Vishnu is a more abstract deity. In the Puranas there was a noticeable transfer of actions originally ascribed to Brahmā, over to Vishnu. This further clouds the issue of their identification. In spite of these complications, the myths involving the Hindu deities are the source of much information on the encounters between priori-Mars and the Earth.

The reason for our interest in the books of this period, is that, although they contain some distortions of what was actually observed in Vedic times, they are very rich in detail, some of which is either missing or very obscure in the *Vedas* themselves. As we shall see, these details contain valuable information about the planetary interactions which are the subject of the V/A theory - indeed, some of this detail is more fascinating than anyone, even Velikovsky, has previously imagined.

Just as Indra, the preeminent deity in the Vedic period, was associated with priori-Mars, the preoccupation with Martian deities continued as the central theme of the *Epics* and *Puranas*.

Although the body we now call Mars had already ascended to its current, harmless orbit, it was nevertheless the last of the gods which threatened the Earth, and the events were still fresh in the 'folk' memories of the older generations who had experienced the encounters. These memories were passed down in each family, and in each village, and constitute, we believe, the source of the additional detail which was incorporated into the Hindu myths.

The Association of the Hindu Triad with priori-Mars

Brahmā, Siva, and Vishnu were all associated with the planet priori-Mars.

> Brahma is regarded as the supreme Being, the God of gods; of whom Brahmā, Vishnu, and Siva are manifestations. It is true that, in some verses of the Vedas, attributes ascribed to him are also ascribed to other deities, and in some of the Puranas various gods are said to be identical with the supreme Brahma. [100]

As we will see there are many quotes which interrelate these deities, although the argument as to which is superior depends on the source. In the *Padma Purana* we find the following quote stating that Vishnu is supreme but at the same time implying the common identity of the three:

> In the beginning of creation, the great Vishnu, desirous of creating the whole world, became threefold; Creator, Preserver, Destroyer. In order to create this world, the Supreme Spirit produced from the right side of his body himself as Brahmā; then in order to preserve the world, he produced from his left side Vishnu, and in order to destroy the world, he produced from the middle of his body the eternal Siva. Some worship Brahmā, others Vishnu, others Siva; but Vishnu, one yet threefold, creates, preserves, and destroys: therefore let the pious make no difference between the three.

Each of the *Puranas* champions a particular one of the deities and the claim is made that their god is the one who performed a unique act, while exactly the same claim is made in another book for another of the triad. They are often described as interacting, that is, having discussions, worshiping one another, being in conflict with one another, and even being absent while another is present. This has probably caused scholars of these books to hesitate identifying them as being associated with the same physical entity. In the minds of the worshiper, they are separate, even though we maintain that they were all aspects of the planet we call priori-Mars today.

Initially, the many details concerning these gods, their associated animals (vahans), consorts, adversaries, conflicts, and manifestations, seem overwhelming. But we have gleaned much of the knowledge that these stories were meant to convey. The richness of this detail makes it clear that: (1) priori-Mars was extremely close to the Earth on numerous occasions; (2) these encounters each lasted years; (3) amazing distortions and upheavals took place on priori-Mars during these encounters; and (4) all the people that lived in the Eastern hemisphere observed some of these convulsions during their lifetimes because priori-Mars remained fixed in the sky above the Himalayas for extended periods of time. As unbelievable as this hypothesis first sounds to those with some knowledge of planetary science, the many corroborating details, both mythological and physical, constitute what to us is an incontrovertible case.

One challenge here is to understand the rationale by which this one body was seen as manifesting the three deities Brahmā, Vishnu, and Siva. This is similar to the problem of understanding the different deity names associated with proto-Venus in the Vedic period, where it was clearly Agni when 'on fire,' and later became a very different deity, 'Mitra-Varuna,' as a result of cooling off and acquiring an atmosphere. In contrast, the names of the Hindu triad are invoked one, two or all three at a time in some myths. This implies that perhaps surface features associated with all three may have been simultaneously visible. This is consistent with our identification of Siva, which was identified with an enormous 'face' formed by several volcanos and the Tharsis bulge, and Brahmā, formed by a massive column of partially frozen lava, but there is no indication that Vishnu was so identified. Obviously the latter deity was more subtle in nature.

A second means of differentiating these deities is by the animals, or animal features, with which each of these gods was associated:

> Each god is represented as having special fondness for some bird or animal on which he is supposed to travel, and which therefore is called his Vahan or vehicle. The bull is Siva's; and the image of his favorite bull, Nandi, is seen in front of many of the shrines sacred to Mahadeva. [101]

The vahan of Brahmā was a goose, and the vahan of Vishnu was half man and half eagle. These are quite likely references to the same feature on the surface of priori-Mars, the Valles Marineris. Of what possible use could this be to our study? Although it does not permit the identification of each deity, it provides additional support for the priori-Mars hypothesis via photographs of priori-Mars returned by space probes. As shown in the photo mosaic in Figure 12.3, there are large terrain features on the surface of priori-Mars to this day which have a strong resemblance to the animals mentioned in the Hindu myths. It only requires a little non-scientific imagination to pick them out, because they stand out prominently against the background terrain under the right lighting conditions. The fact that these images correspond to those animals described in the Indian texts and those of other cultures, supports the notion that priori-Mars was the source of these myths and that it was very close to the Earth in Vedic times.

A third strategy of identification of the deities of the triad is by color, as suggested in the following quote:

> Brahmā then looked intently at Siva, who by thought summoned Vishnu into their presence. As the three deities looked at each other, from their three refulgent glances sprung into being a virgin of celestial loveliness, of hue cerulean, like the petals of a blue lotus,

> adorned with gems, who bashfully bowed before Brahmā, Vishnu, and Siva. On asking her who she was, and why she was distinguished by the three colors, black, white and red, she said: 'From your glances was I produced; do you not know your own omnipotent energies?' Brahmā then...(said) ... O goddess, divide thyself into three forms, according to the colors by which thou art distinguished.' She then, as Brahmā had requested, divided herself into three parts; one white, one red, and one black. The white one was Sarasvati of a lovely, felicitous form, and the cooperator with Brahmā in creation; the red was Lakshmi the beloved of Vishnu, who with him preserves the universe; the black was Parvati, endowed with many qualities and the energy of Siva.

The colors of regions on the surface of priori-Mars that are observed today agree with those in the passage, but beyond that, they do not correlate with the animal images. This may be because dust storms have rearranged the coloring of the terrain over the last few millennia.

Although all the above statements from the myths generally correlate with what is observed on the surface of priori-Mars today, they do not allow us to uniquely identify Vishnu. The more important associations were based on their actions, because the Vedic people were not merely observers of remote surface feature. Their lives were directly threatened by the actions of these deities. Our approach in the following sections will be to pick out the myths which contribute the most information vis-a-vis the physical scenario of such close encounters between priori-Mars and the Earth. It is the physical events in which we are really interested, and the descriptions of the gods and their exploits are only a means to determine them. It turns out that the true 'identity' of the Hindu triad naturally falls out in the process.

The Avataras of Vishnu

The *Avataras* are stories which describe the deity as taking on different 'forms' or 'appearances.' The term Avataras is applied to Vishnu, while the forms of Brahmā are called *incarnations*. Some scholars have interpreted these as cosmic events, but there are probably very few Hindus in the world who think of their deities as planets or their adjuncts. We claim that none of these sacred myths deal with actual men or gods which stepped foot on the Earth. They do describe real physical events, but ones that took place in the heavens, on the 'astral plane,' the surface of priori-Mars, in clear view of millions of Vedic people.

> This deity is worshiped not only under the name and in the form of Vishnu, but also in one of his many incarnations. Whenever any great calamity occurred in the world, or the wickedness of any of its inhabitants proved an unbearable nuisance to the gods, Vishnu, as Preserver, had to lay aside his invisibility, come to earth in some form, generally human, and, when his work was done, he returned

> again to the skies. There is no certainty as to the number of times he has become incarnate. Some Puranas describe ten Avataras, as they are called; some mention twenty-four; one, the Kalki is still future."Some of these Avataras are of an entirely cosmical character; others, however, are probably based on historical events.[102]

This passage describes the periodic return of priori-Mars to the vicinity of the Earth. Mythologically, the god Vishnu is seen being summoned due to some wicked behavior of mankind. This is a hint of the action attributed to Vishnu. As the sustainer, he ensured that priori-Mars would again return to the vicinity of the Earth, even though it moved far away for years at a time.

The Vamana or Dwarf Avatara - A non-Avatara?

Each Avatara is a myth explaining the circumstances under which the deity took on a particular form. We suggest that the forms described in the true Avatara were the shapes of the lava which rose through the tidal sea at the beginning of each 'dance encounter.' The forms are sometimes the shape of animals, such as a fish or a tortoise, but others were interpreted as 'heroes' such as Rama, or Rama-with-Axe. The Dwarf Avatara is considered the tenth Avatara in spite of the fact that there has been some controversy among scholars as to whether it should actually be categorized as one. Our interpretation of the physical events being described explains why we believe it is not a 'true' Avataras in the sense stated above

Nonetheless, it is of great value because it describes the crucial interaction between proto-Venus and priori-Mars, discussed in a previous chapter as the birth of Indra. Moreover, it illustrates the way the same event, described in a simple 'reporting style' in the *Rig Veda*, becomes embellished in the Puranic stories, resulting in more detail of scientific interest:

After Bali (an *asura*, actually proto-Venus) had been killed by Indra, he was restored to great strength by the Brahmins (by sacrifice). As Bali approached, the gods (*suras*) forsook their capital, which was taken over by Bali. The gods called on Vishnu to overcome Bali and regain their power over the three worlds. Vishnu appears as a dwarf and visits Bali at Indra's heaven.

> Acquainted with his duty, Bali placed upon his head the auspicious water with which the Brahmin's (Vishnu) feet had been washed, and said, 'Welcome to thee O Brahmin! What can I do for thee? Ask of me, student, whatever thou desirest. Son of a Brahmin, I conclude thou art a supplicant; ask a cow, pure gold, an embellished house, food and drink, etc. The dwarf concludes a speech with the semblance of moderation as follows: "I ask from thee a small portion of ground, three paces measured step by step. I desire no more from thee. A wise man incurs no sin, when he asks [only] as much as he needs." The king, though astonished at the smallness of the request, takes a vessel of water in his hand, and ...fulfills his

> promise. With two steps Vishnu strode over the universe; there was nowhere for him to take the third. " Asura, three paces of ground were given to me by thee; with two paces the entire world has been traversed; find a place for the third. As thou hast not given what was promised, it is my pleasure that thou should dwell in the infernal regions" ... Bali goes to Patala. First Vishnu's wife and then Brahmā intercede on the demons [asuras] behalf, who in reply, promises that Bali shall again become powerful; but that in the mean time he must dwell in Sutala, where "by my will, neither mental nor bodily pains, etc. afflict the inhabitants."
>
> Another version teaches that Vishnu gave Bali the choice of going to heaven, taking with him five ignorant people, or of going to hell with five wise. He chose the latter, for there is no pleasure anywhere in the company of the ignorant; whilst a bad place with good company is enjoyable. [103]

We believe that this is the Hindu version of the Vedic myth about the birth of Indra, which describes the critical interaction in which sufficient orbital energy was imparted to priori-Mars to bring it out to the orbit of the Earth. This placed Vishnu in the position of most powerful deity, from the point of view of the Hindus. Vishnu was pictured as cleverly outsmarting the asura - by taking such a giant stride. In fact, the most famous characteristic of Vishnu in Hindu myth were his 'three giant strides.' We interpret these strides as (1) priori-Mars initial orbit, which was inside that of the Earth, similar to the current orbit of proto-Venus, (2) the orbit into which it was injected by proto-Venus, that intersected that of the Earth throughout the Vedic period, the subject of this myth, and (3) the current orbit of priori-Mars, in the far cold reaches of the solar system. Vishnu (Mars) was the sura, because of the volcanic fire, and Bali (Venus) was the asura. Vishnu appeared as a dwarf because priori-Mars was behind and considerably smaller than proto-Venus. The Vedic notion of the transfer of Soma from Mitra-Varuna to Indra was changed to a more physically correct picture here, in which water from Vishnu falls on the head of Bali. The term 'infernal regions' implied either that proto-Venus' orbit after the encounter carried it closer to the Sun or below the horizon.

The True Avataras

The true Avataras described transient volcanic forms that arose on the surface of priori-Mars as lava was drawn up through the tidal sea. The tortoise and boar Avataras were three dimensional shapes which formed on priori-Mars early in their respective encounters. As each encounter progressed, these were obliterated as the lava fountains became larger and more energetic.

Even though referred to as the Avataras of Vishnu, they seem to more closely describe the forms taken on by Prajapati or Brahmā as he arose at the beginning of each Kalpa (day of Brahmā). These include the fish, tortoise, and boar Avataras and other more mysterious ones such as Rama and Rama-with-axe. These forms all 'came up through the great waters.' Indeed, the deity's origin in the water is one of the most pervasive themes in the Avataras.

> Vishnu is commonly called Narayana by those who make him the supreme object of worship, though this was originally a name for Brahmā. [104]

The meaning of this epithet is explained as follows:

> He, (Brahmā), desiring, seeking to produce various creatures from his body, first created the waters and deposited in them a seed...This seed became a golden egg, resplendent as the sun in which he himself was born as Brahmā, the procreator of worlds. The waters are called nra because they are the offspring of Nara; and since they were formerly the place of his movement (ayana), he is therefore called Narayana. [105]

This implies that the 'waters' were thought to have been created by the invisible Brahma and subsequently became the place of his creative actions. Physically, the waters in the northern hemisphere of priori-Mars were drawn into a sea at its north pole by the tidal force of the earth at the beginning of each 'dance encounter.' Another description of this same type of event clearly indicates that Brahmā proceeds out of the body of Vishnu:

> Vishnu, as he meditated ... sank into a mysterious slumber; and as in his sleep he imagined the production of various things, a lotus sprung from his navel. In the center of this lotus Brahmā appeared; and Vishnu, beholding the production of his body, was delighted.[106]

The association with water is reinforced by the presence of a lotus, but the shape of the lotus was obviously chosen because it best described the form of the feature that arose on the surface of priori-Mars:

> According to the Brahmā Purana, another name, Apava (who sports on the waters), is given to Brahmā, which has a similar, though not identical, application with Narayana. Apava, according to that Purana, divided himself into two parts, male and female, from whom proceeded Vishnu. [107]

The obsolescent term 'sporting' has a sexual connotation of a male erection. The description of the lava fountain feature on priori-Mars in terms of the male (lingus) and female (yoni) is a common theme, facilitating our understanding of their appearance.

> The following description of Vaikuntha, the heaven of Vishnu, is from the Mahabharata. It is made entirely of gold, and is 80,000 miles in circumference. All its buildings are made of jewels. The pillars and ornaments of the buildings are of precious stones. The crystal waters of the Ganges fall from the higher heavens on the head of Druva; from thence they fall into the hair of the seven rishis; and from thence they fall and form a river. There are also five pools containing blue, red, and white lotuses. On a seat glorious as the meridian sun, sitting on white lotuses, is Vishnu; and on his right hand Lakshmi, who shines like a continued blaze of lightning, and from whose body the fragrance of the lotus extends 800 miles. [108]

This is a description of priori-Mars as it appeared during the 'dance encounters.' The vast transient floods that occurred in the initial phases as the tidal force of the Earth rapidly drew most of the water in the northern hemisphere toward the north pole of priori-Mars, were seen as rivers and were given names that paralleled the names of rivers on Earth. The massive floods left the 'outflow channels' which are still visible on the surface of priori-Mars.

The Matsya or Fish Avataras

> The earliest account of what was afterwards regarded as an Avataras of Vishnu is found in the "Satapatha Brahmāna." It will be noticed that though in this passage a wonderful fish is described, it is not said to have been an incarnation of any of the gods. The Mahabharata says that Brahmā assumed this form; whilst the Puranas teach that the fish here spoken of is Vishnu. This transfer of work from one deity to another is not a matter of much surprise, when we remember how frequently it is declared that all the various gods are but forms of one supreme being. [109]

One of the best known myths concerning Vishnu (or Brahmā?) is the story of Manu and the flood. Vishnu is pictured as coming back to the Earth, because when Brahmā fell asleep, a demon Hayagriva "came near him and stole the Vedas, which had flowed from his lips." Vishnu, although always the same, takes on the form of a tiny fish. A pious man, called Manu, found the small fish in the river and it implored him, to put it in a safer place. But the little fish outgrew the bowl in just one night. Patiently responding to repeated imploring of the rapidly growing fish, he moved it to a cistern, a pool, a lake and finally to the sea, which then was the only body of water that could contain it. Realizing that the fish could be no other than Hari (Vishnu), Manu propitiated him. The myth, as written in the Bhagavata, continues as follows:

> The lord of the universe, loving the pious man who thus implored him from the sea of destruction caused by the depravity of the age,

thus told him how he was to act: 'In seven days from the present time, O thou tamer of enemies! the three worlds will be plunged in an ocean of death; but, in the midst of the destroying waves, a large vessel sent by me for thy use shall stand before thee. Then thou shalt take all medicinal herbs, all the varieties of seeds, and accompanied by the seven saints, encircled by pairs of brute animals, thou shalt enter the spacious ark, and continue in it secure from the flood, on one immense ocean without light, except the radiance of thy holy companions. When the ship shall be agitated with an impetuous wind, thou shalt fasten it with a large sea serpent on my horn; for I will be near thee, drawing the vessel, with thee and thy attendants. I will remain on the ocean, O chief of men, until a night of Brahmā shall be completely ended. Thou shalt then know my true greatness, rightly named the Supreme Godhead; by my favour all thy questions shall be answered, and thy mind abundantly instructed.'

Hari, having thus directed the monarch disappeared; and Satyavrata humbly waited for the time which the ruler of our senses had appointed. The pious king having scattered toward the east the pointed blades of the grass darbha, and turning his face toward the north, sat meditating on the feet of the god who had borne the form of a fish. The sea overwhelming its shores, deluged the whole Earth, and it was soon perceived to be augmented by showers from immense clouds. He, still meditating on the command of Bhagavat, saw the vessel advancing, and entered it with the chiefs of Brahmins, having carried into it the medicinal plants, and conformed to the directions of Hari. The saints thus addressed him: 'O king, meditate on Kesava; who will surely deliver us from this danger, and grant us prosperity.' The god being invoked by the monarch, appeared again distinctly on the vast ocean in the form of a fish, blazing like gold, extending a million leagues, with one stupendous horn; on which the king, as he had been before commanded by Hari, tied the ship with a cable made of a vast serpent, and, happy in his preservation, stood praising the destroyer of Madhu. When the monarch had finished his hymn, the primeval male, Bhagavat, who watched for his safety on the greater expanse of water, spake aloud to his own divine essence, pronouncing a sacred Purana [the 'Matsya Purana'], which contained the rules of the Sankhya philosophy; but it was an infinite mystery to be concealed within the breast of Satyavrata; who, sitting in the vessel with the saints, heard the principle of soul, the Eternal Being, proclaimed by the preserving power. Then Hari, rising together with Brahmā from the destructive

> deluge, which was now abated, slew the demon Hayagriva, and recovered the sacred books. [110]

Other versions of this myth mention Manu being instructed to take animals with him on board the vessel. Aside from the name Manu this story is practically identical to the flood of Noah in the Bible. This lends credence to the argument that the first part of Genesis was derived from more ancient myths, but nevertheless contain proto-historically accurate information.

This myth explains that at the onset of each new dance encounter, a 'seed' or 'egg' - in this case a fish - began to grow in the tidal sea that gathered at the north pole of priori-Mars. The various shapes which it was imagined to take on, were the 'Incarnations of Brahmā' which later came to be called the 'Avataras of Vishnu.' The hardened column of lava known as Brahmā had fallen asleep (collapsed) at the end of the previous encounter, and the Hindus imagined that the return of priori-Mars was precipitated by Vishnu, the sustainer, in response to the theft of the Vedas.

The arc was a large land feature that extended above the tidal sea. The seven rishis or saints were seven volcanoes, which rose above the waters. They were not *on* the arc, but "accompanied the ark." This is supported by the statement that the ark would float "on one immense ocean without light, except the radiance of thy holy *companions*." These seven permanent volcanos on priori-Mars became activated during every 'dance encounter,' and therefore were the origin of the special emphasis given the number seven in the myths of many ancient cultures.

The serpent, which connects the fish's horn and the ark was a continuous giant waterspout formed in the incredibly bizarre environment of high surface heat and low pressure at the point on priori-Mars closest to the Earth. The serpent is another ubiquitous element in ancient mythology because of its continuous presence on priori-Mars during the encounters.

The myth apparently describes two appearances of the fish, the first in which he grew from a very small size to an enormous size and warned Manu about the flood, and the second in which he returned to save Manu when the flood got really bad. Its temporary disappearance may have been due to clouds, which are mentioned at this juncture, obscuring the scene for a period of time. When Manu realizes the fish is Vishnu, he says, "when thou hast shown us to our amazement the appearance of other bodies, not in reality existing, but successively exhibited," informing the reader that the forms taken on by the lava flowing from the interior varied as the encounter progressed.

When priori-Mars became captured in orbit, major adjustments took place in response to the tidal force of the Earth. The near surface waters were the quickest to react, and the myth points out how they rapidly accumulated to form a body of water that covered practically everything on the side facing the Earth. But the tidal force also acted on the interior of priori-Mars, causing melting of rock in the crust and/or mantle and, once melted, started drawing it inexorably through the crust which had been cracked and weakened by continuous marsquakes. Eventually the lava reached the surface and became visible in the midst of the water.

The tiny fish, which appeared in the waters and which grew as the waters increased until 'blazing like gold and extending a million of leagues,' was molten rock that was drawn up through the water toward the Earth. This was anthropomorphized as a fish in this story because its shape resembled one, but this general process occurred each time priori-Mars approached the Earth. Different shapes arose each time, giving rise to other Avataras. For instance, in the myths quoted previously, the rising ball of lava was seen as a golden egg in the water, out of which Brahmā was born. This is revealed here for the first time in the modern era.

Lest those devoted to the dogmas of planetary science as currently taught split their sides in laughter at this point, let us point out that the breakup of a secondary body in orbit around a larger primary body is commonly discussed in planetary astronomy books under the subject of the Tidal Stability, or Roche Limit. Simplistically, this is the distance, usually stated as 2.4 times the radius of the primary body, inside of which a secondary body with no internal cohesiveness would be pulled apart by the tidal force of the primary. For example, the ring systems that circle the giant ice planets are thought to be the remains of a satellite which had either broken up or which never formed, because it was within this limit. Of course, a solid rock body could still remain intact if within this distance because of its internal strength. We are not aware of a single case in which consideration is given to the potential ramifications of a *differentiated* planet orbiting a larger one. The understanding of such a configuration is greatly complicated by the internal makeup of the secondary planet, which incorporates mobile liquids, both on its surface and in the interior. What was witnessed by the Vedic people, was the slow, step by step, dissolution of priori-Mars by the Earth - even though it was outside the Tidal Stability Limit. In the Fish Avataras, and others of a similar nature, we have a situation in which the structural strength of priori-Mars was initially sufficient to hold it together, except that the liquid materials - first surface water and subsequently higher density magma - flowed due to tidal stress, forming the features described in the myth. In descriptions of later encounters we find clear observations that priori-Mars gradually suffered more and more profound damage as a result of tidally induced internal trauma.

The Kurma or Tortoise Avataras

This excerpt illustrates how the deities of the Hindu triad were said to communicate, defer to, and propitiate one another. Even Indra, the primary deity of the Vedic period is involved.

> The account of the churning of the ocean, to which frequent reference is made in the Hindu scriptures, is found in the Ramayana, and several of the Puranas; though there are some discrepancies, they agree in the main. The reason for this great act is as follows:- A saint named Durvaras, a portion of Siva, was travelling, when he met a celestial nymph with a sweet-smelling garland, which at his request she gave him. Excited with the scent, he was dancing, when he met Indra, seated on his elephant. To please the mighty god, the saint presented him with the garland, who placed it upon his elephant's head. The elephant in his turn, becoming excited, seized the garland with his trunk and threw it upon the ground. Durvaras, seeing his gift slighted, cursed the god in his anger, and told him that his kingdom should be overwhelmed with ruin. From that time Indra's power began to wane; for though he sought forgiveness, the Brahmin's anger was not to be

> appeased. As the effects of the curse were experienced by the gods, they, fearing they should be overcome by the asuras, fled to Brahmā for help. He told them he could not assist them; that Vishnu alone could do this, whom he advised them to seek. Brahmā conducted them to Vishnu; and, having sufficiently lauded him, caused him to lend a willing ear to their request. [111]

As priori-Mars approached, its rotation became unstable, that is, it wobbled around until its north pole became aligned toward the Earth. Indra's elephant was the large permanent ice cap at its south pole, which can easily be imagined to be an elephant with its trunk raised. Its 'becoming excited' was a mythical way of describing the unstable motion. In response to the entreaties of the gods (suras) to help them gain control over their adversaries, the asuras, Brahmā deferred to Vishnu. Vishnu's answer, as put into verse, in the Vishnu Purana is:

> Your strength shall be restored, ye gods--
> Only accomplish what I now command:
> Unite yourselves in peaceful combination
> With these your foes; collect all plants and herbs
> Of diverse kinds from every quarter; cast them
> Into the sea of milk: take Mandara
> The mountain for a churning stick, and Vasuki,
> The serpent, for a rope; together churn
> The ocean to produce the beverage--
> Source of all strength and immortality---
> Then reckon on my aid: I will take care
> Your foes shall share your toil, but not partake
> In its reward, or drink the immortal draught...
>
> Then by the god of gods advised, the host
> United in alliance with the demons.
> Straightway they gathered various herbs, and cast them
> Into the waters; then they took the mountain
> To serve as a churning-staff, and next the snake
> To serve as cord, and in the oceans midst
> Hari himself, present in tortoise form,
> Became a pivot for the churning staff.
> Then did they churn the sea of milk...

The poem describes a number of forms including 'the sea of milk,' a 'serpent,' and a mountain used as a 'stirrer.' Also present was Dhanvantari, the gods' physician with the cup of nectar that both the suras and asuras sought. The Vishnu Purana continues, explaining that Vishnu then cast himself into the form of a beautiful woman and so distracted the asuras (demons), at which time the gods quaffed the magic potion, maintaining their power over the asuras as a result of the strength it gave them. The tortoise was Vishnu and this is the Tortoise Avatara.

> In the [more ancient] Satapatha Brahmāna are these words: "Having assumed the form of the tortoise, Prajapati (Brahmā) created offspring. As the worship of Brahmā became less popular, whilst that of Vishnu increased...the names, attributed, and works of the one deity appear to have been transferred to the other. [112]

The older Brahmana version gives a different account of how the suras took advantage of the asuras. This reveals some additional physical clues:

> The assembled gods were stationed by Krishna at the tail of the serpent Vasuke (the churning rope) and the daityas and danavas at its head and neck. Scorched by the flames emitted from his inflated hood, the demons were shorn of their glory; whilst the clouds, driven by his breath towards his tail, refreshed the gods by vivifying showers. In the midst of the milky sea, Hari [Vishnu] himself in the form of a tortoise, served as a pivot for the mountain as it was whirled around. [113]

This version affords clues to the nature of the serpent. Its tail remains fixed to the tortoise, which was at the north pole, about which the entire planet was spinning. The tortoise was the lava rising through the tidal water at the north pole facing the Earth. The churning took place about the tortoise shell and an off-axis mountain was the stirring stick. The heat from the rising lava, located on the spin axis, generated a vortex or tornado in the rising column of hot air. The clouds near the surface were drawn into the vortex along with the hot gases from the lava. It carried the fiery products of the eruptions on the surface high into the sky where they were sprayed around, causing it's 'inflated hood.' The mention of a serpent, or a rope, is common to a number of Hindu myths because this tornado was almost always present due to the conditions at the north pole. The Vedic people had a 'birds eye view' of the entire process, because it occurred at the nearest point to the Earth.

In this Avatara we again find water at the north pole. This arrangement makes sense from a physical point of view because *it allowed all the water in the northern hemisphere of priori-Mars to be drawn into a single tidal sea on the side facing the Earth. The stirring analogy described in the three above quotes implies that the spin axis of priori-Mars remained oriented toward the Earth.* This configuration is consistent with the smoothness and depressed level of the northern plains of priori-Mars, the direction and magnitude of the outflow channels, and the multiple shorelines, the Oceanus Borealis, detected from Viking photographs of the northern hemisphere of priori-Mars. [114] The physical principle by which this configuration was maintained throughout each 'dance encounter' is explained in the sister volume of theis book entitled *Chaos*, by Angiras.

Circular Symmetry of Martian Continents

The various layers in the geologic section of priori-Mars were 'peeled' open by the continual tidal force exerted by the Earth while the planet was spinning with its axis pointed toward the Earth. It is quite natural that the resulting distortions of the planet would be symmetrical with respect to the approximate line of centers between priori-Mars and the Earth, just as the tortoise shell in the above myth was aligned with the spin axis of priori-Mars. Exactly such a description exists, although it has not been correctly interpreted to date. The circular symmetry of the 'continents' on priori-Mars is clearly described in the Bhagavata. It is also clearly stated that these same features were recreated at the beginning of each encounter between priori-Mars and Earth:

> Priyavrata, being dissatisfied that only half the earth was illuminated at one time by the solar rays, "followed the sun seven times round the earth in his own flaming car of equal velocity, like another celestial orb, resolved to turn night into day. Brahmā, however, came and stopped him, saying this was not his province. The ruts which were formed by the motion of his chariot wheels were the seven oceans. In this way the seven continents of the earth were made.
>
> These seven continents are called "Jambu dvipa, Plaksha dvipa, Salmali dvipa, Kusa dvipa, Krauncha dvipa, Sake dvipa, and Pushkara dvipa. They are surrounded severally by seven great seas, of salt water, sugar cane juice, wine, clarified butter, curds, milk, and fresh water" (V.P. ii.2,4)...Jambu dvipa is in the centre of all these continents...In the centre of Jambu dvipa is the golden mountain Meru, 84,000 yolangas high, and crowned by the great city of Brahmā. There are in this continent six cross-ranges of boundary mountains, those of Himavat (Himalaya), Hemakuta, and Nishadha lying south of Meru; and those of Nila, Sveta, and Sringin, situated to the northward. Of these, Nishadha and Nila are nearest to Meru
>
> Jambu dvipa is surrounded by a sea of salt water; and that sea again is bounded on its outer side by the dvipa or continent of Plasksha running all round it. According to this scheme the several continents and seas form concentric circles, Jambu dvipa being a circular island occupying the centre of this system. Plaksha dvipa is of twice the extent of Jambu dvipa...Thus each continent (ring) is of twice the extent of the one immediately inside it...Pushkara is surrounded by a sea of fresh water equal to itself in compass...On the other side of the sea is beheld a golden land of twice its extent, but without inhabitants. Beyond that is the Lokaloka mountain,

which is ten thousand yolangas in breadth, and as many thousands in height. On all sides it is invested with darkness. This darkness is encompassed by the shell of the mundane egg". In the following chapter, however, (seventh) of this same book, the shell of the mundane egg is said to be outside of the seven spheres of which the system is compassed.

These seven spheres have been described by me; and there are also seven Patalas: this is the extent of Brahmā's egg. The whole is surrounded by the shell of the egg at the sides, above, and below, just as the seed of the wood-apple (is covered by the rind). [115]

The amount of detail given about this 'landscape' implies that it was real, that the writer had a birds-eye view of the scene, and that what he was describing was relatively stable. Note that mount Meru is at the center, with the city of Brahmā on top. However, because the commentator had no idea that the texts were referring to a 'world' other than the Earth, he generally dismissed these passages as indicating the ignorance of the Vedic people concerning the Earth around them. His comments are as follows:

It will clearly appear from the contents of the present chapter that the authors or compilers of the Puranas in reality knew nothing of any part of the world except that immediately around them. Whenever they wander away beyond their own neighborhood, they at once lose themselves in a misty region of fiction, and give the most unbridled scope to their fantastic imaginations. [116]

The description goes into considerable detail (not shown), giving divisions, or Varshas, within the continents, and describing the appearance and even the lifestyles of the different beings who supposedly inhabit each of these regions. The ancient Hindus imagined different 'heavens' existing on the various continents and divisions, and, based on color, imagined that different tribes or castes of people living on earth, were destined to pass over to specific places on priori-Mars after their death. Some excerpts give the idea of their musings:

I have attained, even in my embodied condition, to the heavenly state; or to the holy Northern Kurus, or to the Amaravati (the city of Indra). [117]

The country to the north of the ocean, and to the south of the Himadri (or snowy range), is Bharata Varsha, where the descendants of Bharata dwell". "In Bharata Varsha, and nowhere else, do the four Yugas, Krita, Treta, Dvapara, and Kali exist. Here devotees perform austerities, and priests sacrifice...In this respect

> Bharata is the most excellent division of the Jambu dvipa; for this is the land of works, while the others are places of enjoyment...Heaven is of three kinds, in the sky, earth, and in the abyss. Here the Varshas are called terrestrial heavens.

We interpret the reference to the 'land of works,' as stemming from the observation of fires, from volcanic vents, apparently only in the Bharata dvipa, which were interpreted by the Vedic people as sacrificial fires built by the inhabitants who were thus thought to be deeply religious.

The Identification of Vishnu

The myths involving Vishnu supply important physical insight into the processes at work on priori-Mars during the 'dance encounters.' However, Vishnu remains the most obscure of the Hindu deities. In contrast, Brahmā and Siva are relatively easy to pin down. Most of the stories now attributed to Vishnu, such as the Avataras, were originally the incarnations of Brahmā, and as such make more sense to us, because they describe the unique shapes of the molten rock that rose through the tidal sea, each of which subsequently grew into the thousand kilometer high form of Brahmā.

Even the notion of Vishnu as the 'sustainer,' the deity which ensured the repeated return of priori-Mars to the Earth, seems to have been borrowed from that of the supreme, eternal, invisible Brahma, from which the manifest Brahmā was generated during each encounter.

We have explained the 'three great strides' of Vishnu - his most famous feat. Perhaps the function of the sustainer can be considered as a subset of these strides. Perhaps the best match for Vishnu is with Indra, the Vedic name for priori-Mars. A prime example being the similarity of the Vedic myth called 'the birth of Indra' and the Hindu myth called 'the dwarf Avataras.'

A Summary of Vedic Hymns and the Hindu Myths

Details of many planetary interactions were recorded in the Vedas and in the Hindu classics and Puranas. Although they introduce some distortions, the Hindu myths supply important details not in the Vedas. Our interpretation of these myths, along with findings of archaeology and anthropology of the East leads us to the conclusion that the Vedic period lasted some 3,000 years - longer than the time from Christ to the present, from the beginning of the Bronze Age, usually given as 3300 B.C. (radiometric date) or 3,900 B.C. (corrected date, see Chapter 17), down to 687 B.C.

With the perspective of time, the Hindus of the classic period recast the Vedic 'deities' and associated with each one more 'comprehensive' functions along with new names. They recognized the continuity of priori-Mars throughout the entire Vedic period as Vishnu; the periodic 'wakings,' 'creations,' and 'sleepings' of the primary volcanic form on priori-Mars, as Brahmā; and the 'destructions' of this world of 'creatures' on priori-Mars by Siva, as a necessary part of a cycle that repeated itself over and over. The older, Vedic counterparts of the Hindu triad were Indra, Purusha, and Rudra.

These Hindu myths focus primarily on the events during which priori-Mars interacted closely with the Earth. proto-Venus is identified as a protagonist of priori-Mars, for example, as Bali in the Dwarf Avatara, but the actual devastating interactions of proto-Venus with the Earth, which occurred early in the Vedic period, are given much less attention than in the *Rig Veda*.

We have divided the Vedic period into two major sub-periods, a short but most destructive one dominated by proto-Venus, followed by one lasting some three millennia, dominated by priori-Mars. Domination in the minds of the Vedic people meant only one thing - the direct uncompromising threat to their lives. Real awe, and worship were 'inspired' by a god which, unlike the more subtle gods of our day, actually bore down on the Earth, appearing larger and larger, causing violent and continuous quaking and 'moaning' of the Earth, great floods, thunderbolts, storms, auroras, and consequently greater destruction than anything imaginable in modern times.

In the Hindu versions of these events, the sustaining deity, Vishnu, was priori-Mars, whether near or far - from the beginning of the Vedic period to the end. In Vishnu was recognized the fact that in spite of its orbit changes; in spite of its changes in appearance; in spite of other names it was given due to the events which occurred; there was one physical body that gave continuity, indeed reality, to events so diverse that otherwise might seem irrational. The so-called Avataras (manifestations) of Vishnu graphically describe the sequence of events observed on priori-Mars as the magma from the interior was gradually drawn up through the tidal sea, and therefore should be called the Avataras of Brahmā. The ultimate form which repeatedly arose was so enormous, so impressive, and so full of fire and action, that it was given the name Brahmā, and worshiped as a separate deity, although 'born out of the body of Vishnu.' This deity was evoked when priori-Mars was captured and dissolved when it was released. Thus the day of Brahmā, and the night of Brahmā. 'Credit' for many of the feats originally attributed to Brahmā is transferred to Vishnu in later books devoted to him, adding confusion on the one hand, but at the same time corroborating the fact that these were merely aspects of the same physical entity, priori-Mars. We estimate that Brahmā must have been a thousand kilometers high in order to command the rapt attention and provide such a richness of detail described in the Hindu myths. The same impressive column of smoke and fire was the principle deity of all of the ancient cultures that existed during the Vedic period.

The third deity of the 'Hindu triad' was Siva, who was associated with the periodic collapse of the forms on the surface of priori-Mars, at the end of the kalpa, or 'day of Brahmā.' Rudra-Siva was described as an insane, wandering, ascetic with a blue neck and twisted locks of hair. His appearance was that of the planet priori-Mars after the gravitational lock was broken, the form of Brahmā that had 'laid down and gone to sleep.' Siva's destructive power was not limited to the forms on priori-Mars, but was feared by the Vedic people also, since his activation resulted in great trauma for the Earth itself, involving floods, continuous earthquakes, myriads of falling meteorites (Maruts), thunderbolts and, at the onset and termination of each encounter, spin axis changes of the lithosphere.

Because of the many encounters and their duration, many secondary lava fountain shapes gave rise to many stories. It is not possible to imagine the transient forms that inspired them. We have discussed the ones that are the simplest and easiest to interpret in light of the V/A scenario.

Each one discussed herein contributes clues, which help to understand the nature of the physical events which took place. The many similarities between different myths also indicates a set of events which reoccurred each time the two planets came together:

(1) priori-Mars wobbled as its spin axis was forced to become reoriented toward the Earth;

(2) The waters in the northern hemisphere were drawn into a tidal sea at the north pole of priori-Mars, which remained facing the Earth during the 'dance encounters,'

(3) The subsurface rock gradually melted and was inexorably drawn from the interior of the planet up through the gathered sea;

(4) The magma shot high into the air, solidified into tubes and continued to grow into enormous and grotesque shapes, extending toward the Earth. The greatest of these arose during every encounter, and was called Brahmā by the Hindus;

(5) The continuously changing shapes of these forms which gave rise to the many myths, the most famous of which is the four-headed deity Brahmā himself;

(6) Glowing streams of hot rocks, ranging in size as large as tens of kilometers, accompanied with 'storms' of smaller rocks, red dust, water, and atmospheric gases were ejected from priori-Mars at times of sudden alignments with the Moon and other heavenly bodies. Their velocities, combined with a slingshot effect from the nearby earth sent them into orbits around the Earth and even the Sun - myriads of which eventually impacted the Moon, and the Sun;

(7) The release of priori-Mars from its orbit around the Earth resulted in the collapse of the lava-fountain forms, creating great stress on both planets.

All of these events were seen by millions of people and are described in the ancient myths of many cultures. These myths describing them resonate with our collective subconscious. It now remains for the scientists, who believe they know all about the history of the Earth and the planets, to acknowledge this great store of true knowledge and to revisit the currently accepted uniformitarian scenarios which dominate the earth and planetary sciences.

10
EGYPTIAN MYTHOLOGY

Do you guess that I have some intricate purpose?
Well I have, for the Forth-month showers have, and the mica
on the side of a rock has.
Do you take I would astonish?
Does the daylight astonish?
Does the early redstart twittering through the woods?
Do I astonish more than they?
This hour I tell things in confidence,
I might not tell everybody, but I will tell you.
Walt Whitman
Song of Myself

Egyptian mythology is the most mysterious ancient source of clues to the cosmic chaos suggested by the V/A scenario. Its information comprises the most diverse forms of any culture. The spoken language is essentially lost, and difficult to reconstruct, because of the purely symbolic nature of its 'writing.' A symbol may be worth a thousand words, but only to those who know what the symbol 'stands for.' We are able here to explain the meaning of some hieroglyphs that have never been correctly identified because Egyptologists had no idea of what the ancients actually saw and worshiped. The hieroglyphics and art corroborate our scenario. The paintings of the Egyptian deities are much more informative than, say, the Greek gods, which all took strictly human forms. Not only do the bizarre shapes of the heads of the Egyptian deities convey information; even when normal, they are often supplemented with symbology representing the physical appearance of the heavenly body, on top of their heads, as in Figure 10.1.

The hieroglyph for Osiris (in Egyptian, *'Asar' or 'un-nefer'*) has several eyes, indeed, its

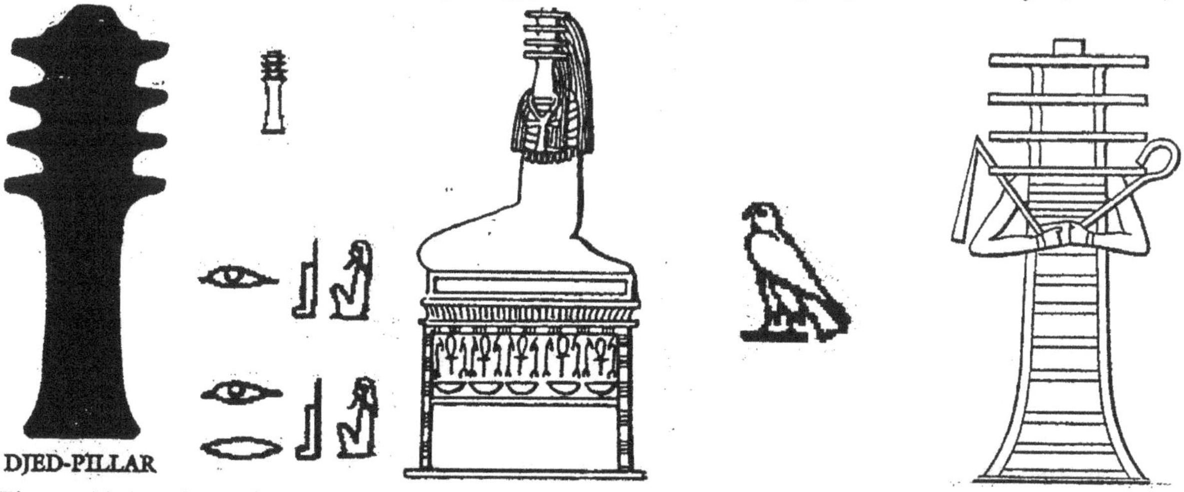

Figure 10.1 A few of the many symbols for Osiris from left to right. A carved djed-pillar, three hieroglyths for Osiris (the lone Tet is from earliest times), Osiris Un-nefer with djed-pillar head, Horus (Osiris' son), and the djed-pillar with arms holding the scepter and whip of Osiris .

Figure 10.2 Thoth and Horus praising Osiris. From the papyrus of Nesi-ta-neb-ashru. From: *Osiris & The Egyptian Ressurection* V1, E.A.W. Budge, Dover, NY, 1973, p. 42-43.)

literal translation is *many-eyed*, but no one currently understands its origin. The V/A scenario explains that this stemmed from the time when proto-Venus had a thin crust. Each time it approached the Earth, the tidal distortion of the still-liquid planet caused its crust to crack in many places, allowing the bright radiation from the interior to come through. Each of these bright spots in the surface was seen by the Egyptians as an eye - thus the name *many-eyed*. Additional corroboration of our interpretation comes from the fact that the *Rig Veda* used exactly the same epithet for Varuna. Thus the earliest manifestation of Osiris was proto-Venus.

As the crust formed on proto-Venus, quenching the bright glow of fire, the power seat of Osiris was transferred to priori-Mars, where he was imagined to be embodied in the column of smoke and fire. The hieroglyph representing this impressive feature is the *djed pillar* or *tet*. The association of Osiris with this hieroglyph is unmistakable. Many paintings associate Osiris with this common hieroglyph, which Egyptologists have had great difficulty interpreting, terming it 'life.' We claim that the Tet symbol is a quasi-graphical representation of the great lava fountain that rose up on the surface of priori-Mars during the dance encounters. Some highly stylized paintings of the Tet have intermixed attributes which are those of Osiris. Some of these are shown in Figure 10.2. Also shown is a small carved figure of the pillar. The multiple horizontal 'layers' at the top of this figure corresponded to the four 'heads' of Brahmā in the Hindu myths.

Horus was a later deity associated with priori-Mars. The Tharsis Bulge was quite distended during the Vedic period, and was 'sliced open' at the bottom by the end of the Valles Marinaris. As a result, it resembled a bird beak. The eyes associated with the beak were the volcanos Arsia and Ascreaus Mons. This same combination of features was called the face of Siva in Hindu mythology. Further corroboration of the beak interpretation comes from the description of Horus as a 'child,' because he was always sucking on his finger:'[118]

> Horus, the son of Osiris and Isis appears in Egyptian texts usually as Heru-p-khart, or "Horus the child," who afterwards became the

> "avenger of his father Osiris", and occupied his throne, as we are told in many places in the Book of the Dead. In the pyramid texts the deceased is identified with Heru-p-khart, and a reference is made to the fact that the god is always represented with a finger in his mouth.

This is obviously a reference to the linear feature Vallis Marineris which originates in the lower side of the 'beak' at the same angle that a child's finger would be placed in its mouth. (See Figure 8.1 on page 103.) The hieroglyph for Horus is a standing bird because priori-Mars remained stationary in the sky throughout each dance encounter.

Other hieroglyphs not understood are those currently translated 'the world of the horizon,' and 'the world of the two lands,' that is, the 'two lands' which emerged from the primal waters. We interpret both of these symbols as referring to priori-Mars; which remained stationary on the eastern horizon during the dance encounters and was characterized by a surface dichotomy between the northern plains, covered with lava flows, and the southern highlands, which were still covered with green vegetation.

Figure 10.3 The Egyptian goddess Hathor (the abode of Horus) with characteristic 'horned-sphere' head-piece.

As each dance encounter progressed, priori-Mars spherical shape became more and more distorted due to the tidal tug of the Earth and periodic convulsions in its interior. As a result, there were a number of Egyptian deities, the shape of whose heads mimic the distorted shapes of priori-Mars. These include a frog, a jackal with a long snout and an ibis, with the most elongation of all. Two of these, Thoth and Horus, are shown in Figure 10.2. Hathor was probably the planet priori-Mars itself, as indicated by its Egyptian name which means 'the abode of Horus.' The symbol above her head in Figure 10.3 indicates that priori-Mars had a wake. This may have been the case, but this is perhaps a clue that there is another possibility, one which is discussed below.

Ptah was 'the lord of the two lands,' one dominated by Osiris and the other by Set. The land in the northern plains of priori-Mars was prone to enormous volcanic activity and so displayed many visible points of light that looked like stars at night. This was where the column of smoke and fire, signified by the djed pillar or tet, embodying Osiris in the latter part of the Vedic period, was located. Because of these lights, this hemisphere was associated with life (Shu). When in the shadow of the Earth, the many volcanos on priori-Mars appeared as stars. This was the Egyptian *Duat*, 'the Circle of the Gods.' Just as the Hindus, the Egyptians divided this circle into (twelve) zones, and probably named the 'stars' in each.

The Egyptian deity Anubis has the head of a jackal. An image of this deity is shown in Figure 10.4. Aside from the extreme distortion of the planet implied by this form, what is particularly interesting is the fact that the 'back' of this head is colored green, while the most distorted

Figure 10.4. The Egyptian deity Anubis, whose face is black and dog-shaped and whose 'hood' is green. (From *The Egyptian Book of the Dead*, E.A.W. Budge, Dover, 1967.)

part, (deviating the most from a spherical shape) is black. *This is a very graphic clue indicating that at the time of the earlier dance encounters the southern highlands of priori-Mars were still covered with green vegetation, while the side facing the earth, which was traumatized by continuous volcanic eruptions, was black.*

It was in homage to Horus that the great sphinx (Greek name) at Giza was built. It faced the east where Horus, the son of Osiris, remained stationary in the sky for fourteen years at a time. *The fact that the Egyptian name for the sphinx is Horus-am-Akhet, meaning 'Horus on the Horizon,' is a direct corroboration of our thesis.* It was not built to honor the rising Sun as is currently supposed. All the heavenly bodies including the Sun move across the sky, appearing first in the east and disappearing in the west due to the rotation of the Earth. *Priori-Mars, when locked in its dances with the Earth is the one body which did not. It remained stationary in the eastern sky, day and night for fourteen years at a time, and the sphinx was placed facing it.* Although the head of the sphinx is now an image of a pharaoh, many scholars believe that it was originally that of a lion, matching its body - a shape which closely matched the body of priori-Mars. The lions face was probably the same as the face of Siva, while the mane was the southern highlands. The importance of the east is also evidenced by the belief that the buried pharaohs exited their tombs through a faux door, provided for this purpose, which faced the east.

Osiris was the most prominent Egyptian deity in the early portion of the Vedic period. He encompassed two stages - the hot and glowing proto-Venus in the early part of the Vedic period, and in the later portion, the column of smoke and fire on priori-Mars. He was credited with bringing the knowledge on which the great Pharonic dynasties were built. In spite of the fact that he represented proto-Venus early-on, we know of no overt association between him and the latter two overturnings of the Earth, of which the Egyptians had full knowledge. This was related to Solon, the Greek, by the Egyptian priests, according to Plato's *Timaeus and Critius*. Perhaps the Egyptians believed that these were a necessary aspect of the historical sequence which led to the birth of their great civilization and did not wish to offend Osiris. Quoting from this work, the Egyptian priest says:

> Oh Solon, Solon, you Greeks are all children, and there is no such thing as an old Greek...
> You are all young in mind...you have no belief rooted in old tradition and no knowledge hoary with age. And the reason is this. There have been and will be many different calamities to destroy mankind, the greatest of them by fire and water, lesser ones by countless other means. Your own story of how Phaethon...burnt up things on the earth...is a mythical version of the truth that there is at long intervals a variation in the course of the heavenly bodies and a consequent widespread destruction by fire of things on earth. On such occasions those who live in the mountains or in high places suffer more than those living by rivers or by the sea; as for us, the Nile, our own regular savior, is freed to preserve us in this emergency. When on the other hand the gods purge the earth with a deluge, the herdsmen and shepherds in the mountains escape, but those living in the cities in your part of the world are swept into the sea by the rivers; here water never falls on the land from above either then or at any other time, but rises up naturally from below.

In the V/A scenario, two close passages of the flaming proto-Venus near the beginning of the Vedic period caused both types of damage. Its heat burned the surface of the earth and its tidal force, acting on the lithosphere, actually caused it to overturn, with the position of the north and south poles being reversed each time. These 'overturnings of the Earth' caused enormous tidal waves which washed completely over some of the continents.

Going back even further in time, Egyptian priests told Herodotus around 475 B.C. that their kingdom was 341 generations old. Based on Herodotus' estimate of thirty years per generation, this means it began some 12,770 years before the present. The priests also told him that in the entire memory of their people, *the Earth had overturned four times.* Mexican hieroglyphics also describe 'four motions' and 'four prehistoric suns' or 'world ages.' Based on this, the Egyptians had the most ancient 'memory' of any people.

The Legend of Osiris - Part One (Venus)

Osiris was the pre-eminent Egyptian deity. When he was born of the goddess Nut, a voice was heard to proclaim that the lord of creation was born. In the course of time he became the king of Egypt and devoted himself to civilizing his subjects and teaching them crafts. Once Egypt was peaceful and flourishing he set out to instruct other nations. During his absence, Isis ruled his kingdom well. When he returned, Typhon plotted with seventy-two comrades to slay him. At a banquet they got him to lie down in a beautiful chest, which was immediately closed by Typhon who drowned Osiris by placing it in the Nile. This happened when he was in the twenty-eighth year of his reign or of his age. (The legend of Osiris is continued below.)

Although seemingly an impossible task, we believe it is possible to interpret this story in terms of the immediate heavenly bodies in the V/A scenario. The voice heard at the time of the birth of Osiris was the disturbance of the Earth at the time the traveler impacted Jupiter. "In the course of time" implies that it took several years for proto-Venus to approach Earth after its 'birth' out of Jupiter. To become lord of the Earth meant to become the primary deity that controlled the destiny of the Earth.

Osiris was identified with fire or 'glowing' - this was the origin of the word we now interpret as 'god' in almost every ancient culture, and the word from which we derive the term 'glory' in the Judeo-Christian religion. Early in the Vedic period his power was associated with the newly born proto-Venus. The sealing up of Osiris in the fine chest when in the twenty-eighth year of his reign implies that a tenuous crust covered proto-Venus in only a few decades after reaching the vicinity of Earth, obscuring its glow. To the ancient Egyptians, this was seen as a loss of power and the 'first' death of the deity Osiris.

The following attributes of Osiris are consistent with the V/A interpretation of the Vedic deity Varuna which we associate with proto-Venus in Chapter 2.

(1) Osiris commanded great respect because he governed by subtle persuasion. This parallels the Vedic respect for Varuna and disdain for the noisy war-like blustering associated with Indra (Mars). This was due to the greater mass of proto-Venus which caused much stronger earthquake activity than priori-Mars, even though it was farther away and because of its beauty after it had acquired its 'shining robe' (atmosphere).

(2) Egyptian myth conceived Osiris as being the husband of Isis, whom we interpret as the Moon, whose name is literally translated 'ancient.' In Egyptian myth, when Osiris 'traveled' he left Isis to administer his kingdom. This implies the comings and goings of proto-Venus at inferior conjunction, while the Moon remained in orbit around the Earth.

(3) He lived only a short time - 28 years. The period during which proto-Venus came close enough that the tidal force of the Earth was sufficient to crack its newly formed crust and elicit fire from its interior was amazingly short. Once the raw aspect of fire no longer occurred on proto-Venus, Osiris (fire) was seen mythologically to pass to priori-Mars, where fire regularly appeared in the form of enormous lava fountains.

(4) The hieroglyphic for Osiris is literally translated 'many eyed.' This is exactly the same as one of the epithets for Varuna in the Vedas.

(5) The earliest figure thought to represent Osiris dates from the time of the Egyptian King Narmer - who is dated at about 3300 (uncorrected radiometric date) or 3900 B.C., *the very date we propose as the beginning of the Vedic period.*

The most beautiful images associated with the Egyptian deities began to be created early in the third millennia B.C. a few centuries after the two destructive encounters between proto-Venus and Earth. These adorn the walls of their temples and burial places. The Egyptians concentrated on the glorious and positive nature of their deities, because they felt that their great culture had been given to them by the gods.

Osiris' protagonist, Set, the counterpart of Typhon in Greek mythology, was the serpentine smoke column which emanated from proto-Venus when it first approached the Earth. As in many cultures, it was seen, not as emanating from proto-Venus, but as a serpent-like creature attacking

Osiris. Although in modern times, we would describe the smoke cloud as the tail out gassing from the super-hot proto-Venus, the notion of it as an adversary is ubiquitous in ancient mythology. This must have been a striking image.

Figure 10.5 At left, a portion of the Papyrus of Nekht, showing Osiris Khenti Amenti wrapped in white and Queen Maat holding the sign translated 'life'. At right, the surface feature on proto-Venus which may be the source of the ankh symbol. (From *Osiris, & The Egyptian Ressurection* V1, E.A.W. Budge, Dover,1973)

A number of paintings of Osiris provide clues to his early association with proto-Venus. His earliest known image is called Osiris Khenti Amenti. It is depicted in the Papyrus of Nekht, a royal scribe, shown in Figure 10.5. The figures on the pedestal at the left, are the goddess Maat, 'queen of the two lands,' and Osiris Khenti Amenti. We interpret this pair of deities in the same way as we did Mitra and Varuna in the Vedic hymns. Queen Maat represents the solid planet as does Mitra. Her identification as the planet proto-Venus is consistent with the fact that she is considered one of the most ancient of the Egyptian deities. Osiris, always shown wrapped in white, is analogous to Varuna who 'wore a golden robe.' The white wrapping is commonly interpreted as a mummy shroud, but we interpret it as the bright atmosphere which the ancient people saw surrounding proto-Venus at night in the

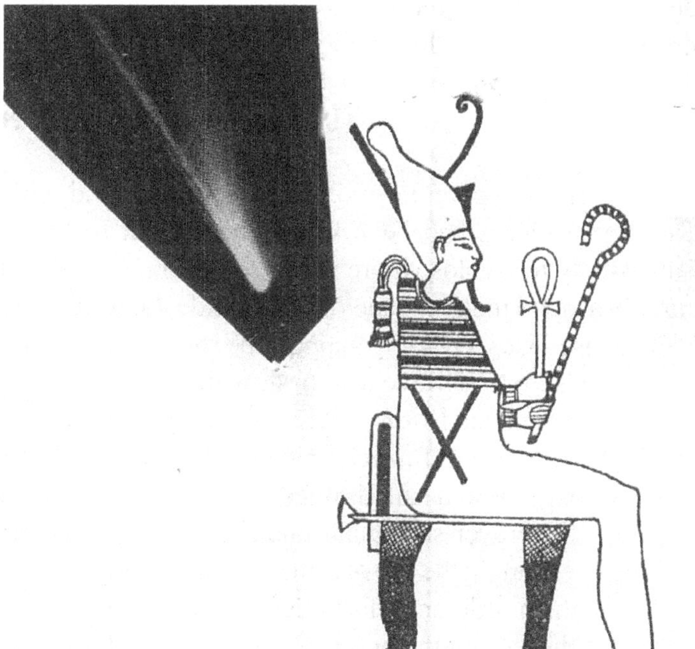

Figure 10.6 Osiris wearing 'the crown of the two kingdoms', and (insert) the comet Mrkos, which resembles the crown.

years when it was near favorable conjunctions. Note that Queen Maat holds in her hand a strangely shaped symbol, which has its own hieroglyph 'ankh' commonly translated 'life.' This may be an abstraction of a feature on the surface of proto-Venus which was seen by the ancient Egyptians before the planet became enshrouded with sulfuric acid clouds - one which is still present to this day, and has been captured in radar images. It is shown at the right in Figure 10.5. The notion that it is a surface feature is consistent with its being carried by Queen Maat who represents the solid planet. If the ankh hieroglyph was derived from a surface feature, translations of a number of passages will need changing. We suggest that the same feature is referred to in the Rig-Veda, where an epithet for Varuna is 'Pasi, the noose-carrier,' since the ankh when turned upside down resembles a noose.

We suggest that the strange shapes pictured on the heads of the Egyptian gods were not 'crowns,' but were stylized depictions of the appearance of the heavenly bodies as they actually appeared. This notion is supported by the fact that no crowns resembling any of those shown in ancient Egyptian art have ever been found among the artifacts in excavated tombs. Thus the shape above Osiris Khenti Amenti represented the smoke being given off by the 'comet' proto-Venus. The thick portion depicting the primary smoke tail and the others representing flaming promenances are similar to those pictured as feather by the Mexicans in the god Quetzlcouhatl - the feathered serpent. Similarly, the 'crown of the two kingdoms' worn by Osiris in Figure 10.6 is probably a representation of proto-Venus at a later stage with a broad dust tail and a thinner ionized tail, characteristic of 'normal' comets, as illustrated in the insert by a photograph of the comet Mrkos. The currently accepted interpretation of this double crown is that the two promenances represent the combined crowns of upper and lower Egypt, showing his sovereignty over both.

The Legend of Osiris - Part Two (Mars)

The Egyptian legend of Osiris did not end with his death at the hands of Set. When Isis found that Osiris had been killed she set out to find his body which had been carried by sea, supposedly to Byblos where it had come to rest in the branches of a tamarisk tree. The tree grew rapidly around the chest enclosing it inside the trunk. The king of the country had the tree cut down and the used it as a pillar to support the roof of his house. Isis found it and prevailed upon his queen for permission to retrieve the chest, which she brought back to Egypt. She hid the chest while she went to find her son Horus, but Set, one night hunting by moonlight found the chest and recognizing the body, tore it into fourteen pieces which he scattered. When Isis heard of this she took a boat made of papyrus and sailed round the world gathering all the fragments of Osiris' body except the phallus. Wherever she found one she placed a monument. But now Horus had grown up, and having been taught the use of arms by Osiris, who returned from the other world, he went to do battle with Set, the murderer of his father. The fight lasted many days and Set was made captive. But Isis, to whom the care of the prisoner was given, so far from aiding her son Horus, released Set. Horus in his rage tore from her head the royal diadem; but Thoth gave her a helmet in the shape of a cow's head.

Our interpretation of this myth is as follows: The transportation of the chest containing Osiris on the water reflects presence of a glowing mass in the tidal sea that formed on priori-Mars at the beginning of each 'dance encounter.' The tamarisk tree was the lava-tube figure which grew from that mass. This was called Brahmā in the Hindu myths. The top of that form was very hot, thus it embodied the primary characteristic of Osiris - *fire*. The enclosure of the chest in this tree means that the deity associated with fire, Osiris, was imagined to have moved from proto-Venus to the planet priori-Mars.

That is, Osiris was initially associated with proto-Venus but was subsequently transferred to priori-Mars, paralleling the transfer of Agni (fire) from Varuna to Indra in the Vedas, also of Zeus, originally associated with Jupiter, to proto-Venus and then to priori-Mars in Greek myth (next chapter). Thus Osiris became the column of smoke and fire on the surface of priori-Mars.

As discussed in Chapter 8, streams of large glowing rocks were expelled from the vents which fed the lava fountain. When this occurred the hardened lava-tube formation shattered and collapsed. The largest events probably occurred at the times of alignments with proto-Venus, which was Set because it still had a tail. The exceptionally large bodies comprising one such stream were observed arcing out toward the Moon. They were interpreted as pieces of the body of Osiris, scattered by Set. Because the bodies were ejected one after another in a 'stream' through a common vent they followed the same trajectory.

The idea that Isis, the Moon, tirelessly searched for and retrieved the pieces of Osiris' body implies that their orbit carried them outward from the Earth-Mars pair to the Moon, where they impacted. The Egyptians watched these impacts on the Moon and recorded them in their mythology.

Isis (the Moon) was pictured going round the Earth in a boat of papyrus collecting the scattered parts of Osiris' body. The 'boat' in which Isis is said to have traveled was the illuminated crescent of the Moon as it circled Earth. The association of this unusual shape to that of a boat is easily seen in paintings of Egyptian ceremonial ships, an example of which is shown in Figure 10.7.

Believe it or not, the monuments that Isis placed are still visible from the Earth to this day, although some overlap others and obscure the actual number. They are the lunar maria, the dark areas which are only found on the side of the Moon facing the Earth.

To this day, the Egyptians celebrate a festival in early spring which is called expressly 'The Entrance of Osiris into the Moon.'[119] The specific mention of the number fourteen in the Egyptian myth may have been an actual estimate of the number of bodies that were seen to impact the Moon or it may have been an abstract clue as to the length of the 'dance encounters,' since we have independently concluded, based on the Hindu myths, that their duration was fourteen years and the time between them was also fourteen years. (It would be interesting to examine the lunar gravity data to see if fourteen distinct 'mascons' can be discriminated.)

Figure 10.7. The boat on which the Egyptian deities travel has the shape of a crescent.

Figure 10.8 Osiris on a 'raft' floating on a lake with lotus plants. (From Osiris, & The Egyptian Ressurection V1, E.A.W. Budge, Dover,1973)

This beautiful myth is consistent with the Hindu myth, in which Brahmā created four classes of embodied spirits, gods (suras), demons (asuras), fathers, and men, categorized by their appearances and destinations. *The class of interest in connection with the Osiris myth is the 'fathers,' because their home was said to be on the Moon.* The Egyptians imagined that the original Osiris, who was associated with proto-Venus, died after 28 years and was resurrected under the auspices of Horus (Mars). This reinforces the brevity of the period in which proto-Venus wreaked its havoc on the Earth.

Osiris, when embodied in the lava fountain formation on priori-Mars, was obviously 'sacrificed' (collapsed) many more times. Apparently the story of Isis collecting his parts stemmed from a unique event involving a stream of particularly large bodies, the trajectories of which could be followed all the way to the Moon. There is indirect evidence that great storms of rocks and dust, repeatedly blasted from priori-Mars, accumulated on the Moon throughout the Vedic period, both before and after this event. This comes from the fact that *the entire near side of the Moon is covered with as much as two kilometers of 'regolith' and because it is measurably lopsided, with unaccounted for mass on the side toward the Earth.*

The Egyptians did not believe that, as a result of the scattering of his parts, Osiris perished permanently. Instead he repeatedly returned from the other world, as the lava fountain was quickly rebuilt after each convulsion of the interior . Figure 10.8 and 10.9 show anthropomorphized images of Osiris after his 'transfer' from proto-Venus to priori-Mars. In the first he is pictured in his 'raft' on a lake with lotus plants in the water, and in the other, as Osiris-Res (Osiris the Riser) lying on the ground with a snake below or beside him. These images parallel the Hindu myths which describe Brahmā as rising out of the water, being

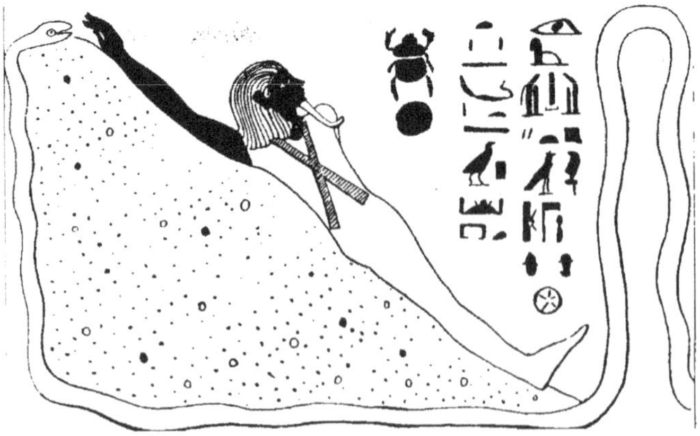

Figure 10.9. Osiris the Riser, reclining on a snake. (From Osiris, & The Egyptian Ressurection V1, E.A.W. Budge, Dover,1973)

born in a lotus growing from the navel of Vishnu, and reposing on the coils of a great snake during the 'nights of Brahmā.'

A primitive stone stela from the Early Bronze II or III Age (3050 to 2300 B.C. radiocarbon dates), was unearthed at Arad in the northern Negev desert. It is shown in Figure 10.10. It depicts two anthropomorphized forms - one standing and one laying down. This connotes the erect lava-tube formation on priori-Mars during the dance encounters and the collapsed form, as it was between encounters. What is of interest in connection with the previous explanation of the Egyptian djed-pillar is the 'head' of these figures. They are again depicted as a series of parallel lines, further reinforcing the notion that this was the shape of the formation observed on priori-Mars.

Figure 10.10. A stela from Arad depicting a deity in two positions, standing and laying down. (From Archaeology of the Land of the Bible, A. Mazar, Doubleday, NY, 1992, Fig. 4.24, p.137)

When a pharaoh died, the Egyptians believed he was taken up into the sky to be with the gods. Referring to this process, the Ancient Egyptian Pyramid Texts from the pyramid of Unas:

> The gods who are in the sky are brought to you, the gods who are on the earth assemble for you, they place their hands under you, they make a ladder for you that you may ascend on it into the sky, the doors of the sky are thrown open to you, the doors of the starry firmament are thrown open for you.

We suggest that the 'ladder' referred to in this passage is describing the same formation as the Tet. In this instance, the series of parallel lines pictured at the 'top' of the Tet are the rungs of this celestial ladder. A startling corroboration of this interpretation comes from Graham Hancock in *Fingerprints of the Gods*, who after quoting the above passage, paraphrases some of the other 'utterances' in the texts as follows:

> The ascending pharaoh was identified with, and frequently referred to as 'an Osiris.' Osiris himself, as we have seen, was frequently linked to and associated with the constellation Orion. Osiris-Orion was said to have been the first to have climbed the great ladder the gods had made. And several utterances left no doubt that this ladder had not extended upwards from earth to heaven, but

downwards from heaven to earth. It was described as a *rope-ladder* and the belief was that it had hung from an 'iron plate' suspended in the sky.

Lacking any knowledge of a body in the sky, from which the ladder could extend, Mr. Hancock imagines that the constellation of Orion was the home of the skygods - a very unimpressive deity to be the focus of such a detailed description. We claim that during the dance encounters priori-Mars remained fixed in the heavens above the eastern hemisphere, regardless of the rotation of the Earth and appeared over ten times the diameter and over a hundred times the visible area of the Moon. The ladder was the thousand kilometer high column of smoke and fire on the planet priori-Mars. The most telling aspect of this interpretation of the Pyramid Texts is the description of *the ladder hanging down from heaven toward the earth.* This is exactly the way that the lava fountain formation represented by the Tet appeared to the people on earth. *It pointed from priori-Mars down toward the Earth and the parallel horizontal lines at the bottom could easily be pictured as the rungs of the ladder.*

Also of interest is the reference to the 'heaven' as an 'iron plate.' There is considerable evidence that the Egyptians realized that iron meteorites came from priori-Mars. They saw the massive eruptions that occurred periodically on priori-Mars and connected these logically to the meteorites that subsequently fell to earth. As discussed previously these eruptions were described in the myths of many cultures as the process by which the supreme deity created 'embodied entities.' Therefore the iron meteorites were directly related to the gods and iron (*bja*, 'celestial metal') was considered sacred because its only source was from the heavens.

Ra

Some 'utterances' from the Pyramid Texts referring to Ra are of interest:

Utterance 261 ... The King is a flame, moving before the wind to the end of the sky and to the end of the earth ... the King travels the air and traverses the earth
Utterance 332 ... the king says, 'I am this one who has escaped from the coiled serpent. I have ascended in a blast of fire having turned myself about. The two skies go to me.
The door of the sky at the horizon opens to you, the gods are glad at meeting you ... May you sit on this iron throne of yours, as the Great One who is in Heliopolis.
The earth speaks, the gate of the earth god is open, the doors of Geb are opened for you ... May you remove yourself to the sky upon your iron throne.

These ancient Egyptian texts contain references to many of the same features on priori-Mars that we have found in the Vedas and Hindu myths. Priori-Mars 'travels the sky and traverses the earth' as it established its position over the Himalayas and moved across the land again at the time each 'dance encounter' was broken off. The description of a deity that 'escaped from the coiled serpent' is almost identical to the that in the Hindu myth in which Brahmā laid down to sleep at the end of each day of Brahmā, on the coils of the serpent *Sesha*. At the beginning of each dance encounter, the same deity was seen to rise up from these coils, and thus 'escape' them. For the 'door of the sky' to 'open up' means priori-Mars had returned to its fixed position over the Himalayas, which placed it permanently on the eastern 'horizon.' The 'iron throne' was the place of Osiris on the central mountain on priori-Mars. The 'earth speaks' refers to the continuous groaning of the Earth in response to the changing tidal stress induced by the approach of priori-Mars.

The mythical association of the transformed Osiris with Horus (Mars) is given more credence by the Papyrus of Anhai in which the Mert goddesses of the north and south praise Osiris:

> Thou risest [in] beauty in the Horizon of Eternity.

The presence of these goddesses of vegetation connects Osiris with the crops of the country. The tidal effects of priori-Mars and proto-Venus when close to the Earth during the dance encounters had a profound effect on the distribution of water around the world and consequently on agriculture.

One of the *Pyramid Texts* which has proved particularly intractable is translated as follows by R. O. Faulkner:

> 502 ... The phallus of Babi is drawn back, the doors of the sky are opened, the king (Ra) has opened the doors because of the furnace heat which is beneath what the gods pour out.

Although Egyptologists have great difficulty making sense of this passage, the V/A theory explains it handily. The 'phallus of Babi' was the erect, rigid lava fountain on priori-Mars. When alignments of the Earth-Mars pair with other heavenly bodies occurred, the entire structure, symbolized here as an erect phallus, collapsed, that is, was 'drawn back.' The Egyptians saw that it was the heat below the surface of priori-Mars which launched the bodies that 'the gods poured out' and the doors being opened meant that those bodies were ejected out into space.

Isis and Nephthys

The two female deities, Isis and Nephthys (Nept-het), are often mentioned together in Egyptian myth. We claim that Isis ('ancient') is the Moon but we have not found such a sure identification for Nephthys. The fact that the sky goddess Nut gave birth to only four deities; Osiris, Isis, Set, and Nephthys, seems to place her in important company, but does not guarantee that each is a separate and distinct heavenly body - Osiris being a good example. On the one hand her feline nature leads us to believe she might correspond to one of the consorts of the Hindu triad, like Lakshmi, which we have identified with bodies of water or vapor clouds which arose on priori-Mars during the dance encounters, their sinuous movement giving the notion that they were dancing around the more rigid male deities such as Brahmā. On the other hand her 'dark side,' alluded to by her marriage to Set, the primary protagonist of Osiris, suggests she might just be the dark or earth-lit portion of the Moon leaving Isis to be the illuminated portion.

Regardless, the study of Isis and Nephthys provides an important clue. Representatives of these two goddesses were always the chief mourners at ceremonial Egyptian funerals, where they performed the mysterious symbolic ritual called 'the opening of the mouth.' The body in its humanoid coffin was placed upright at the entrance to the tomb by a priest. *A hook, made of iron, was then used to touch the corner of the mouth.*

The specification of an iron hook for this ceremony is particularly meaningful. We claim that toward the end of each dance encounter, liquid iron from the outer core of priori-Mars was drawn out into space, forming bizarre shapes, among them the 'headdress' and 'beard' worn by the pharaohs and the sphinx. Some of it flowed out through the western-most end of the feature we call today the Vallis Marineris. This was just at the opening of the beak of Horus, where the Valles Marineris cut into the distended Tharsis bulge. Thus the placing of the iron implement at the mouth on the coffin, was a direct mimicking of the events that were observed on priori-Mars.

The Eye of Ra

Additional details of the physical processes observed on priori-Mars come from the legend of Ra and Isis. As the god Ra (an ancient Egyptian name for priori-Mars) aged and became infirm, Isis fashioned a serpent from the dust which, having been placed in his path, bit Ra and poisoned him. When he was close to death, Isis, in the form of an enchantress, heeded his call for help and administered to him. In order to treat him, she asks only that he state his name. Then Ra hid himself from the gods and the throne in the Boat of Ra was empty. When it was the time for the heart to come forth the great god was made to yield up his name, and Isis, the great lady of enchantments, said:

> Flow on, poison, and come forth from Ra; let the Eye of Horus come forth and shine outside his mouth...Let Ra live, and let the poison die; and if the poison live then Ra shall die.

We have previously established the idea that the tidal force of the Earth, combined with episodic convulsions due to alignments with other heavenly bodies, gradually drew magma from the interior of priori-Mars and exposed deeper and deeper layers of the planet as each dance encounter progressed. We have also established that alignments of the Moon with priori-Mars were the primary impetus for the convulsions within priori-Mars as it circled the Earth once per day.

As a result of this hypothesis, our interpretation of this quote is that priori-Mars became weakened (infirm) and distorted as each dance encounter drew to its end, due to many such internal convulsions. Eventually the liquid iron of the outer core was drawn out through the Valles Marineris, taking on bizarre forms, such as the 'hood' worn by the pharaohs. This was the impetus for the mythical idea that Horus dribbled from his mouth. When the Moon, and particularly the Sun and Moon combined, became aligned with priori-Mars, more severe convulsions took place. But every fourteen or fifteen years proto-Venus reached its closest point to the orbit of the Earth and its tidal force combined with the others literally pulled priori-Mars apart.

In this myth the alignment of the Moon is implied by the involvement of Isis. The marsquakes raised dust over the entire planet, obscuring all the natural features ('Ra hid himself'). It was then that the 'heart of Ra came forth.' In the quote, the 'heart' is also described as 'the eye of Horus' or 'the eye of Ra,' which is represented by a unique and mysterious hieroglyph that has never been correctly interpreted to date. It is shown at the top of the composite Figure 10.11.

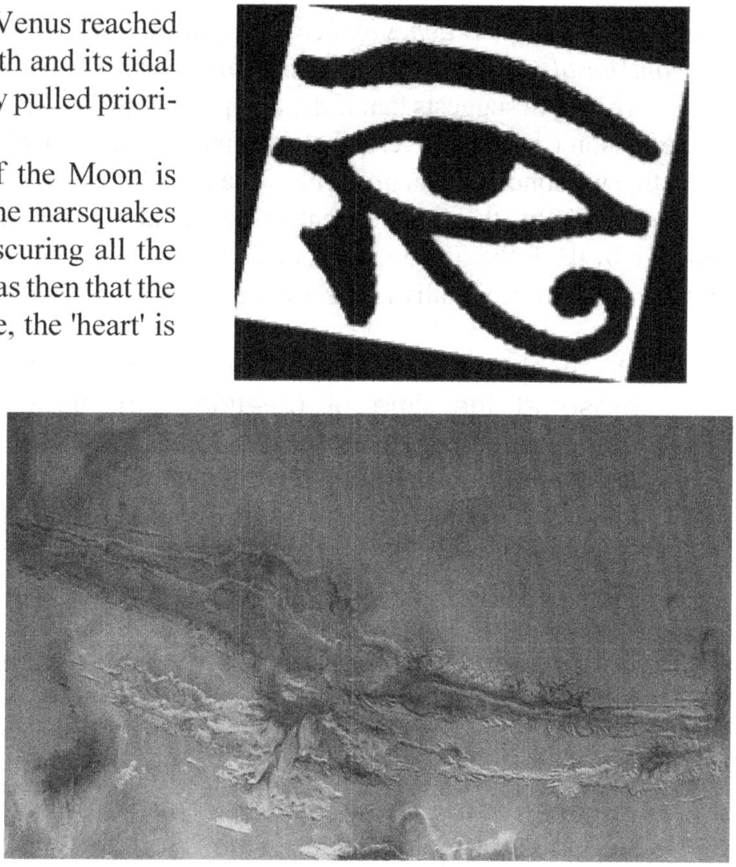

Figure 10.11 Below: A photo of Valles Marineris as it appears today. Top: The hieroglyphic for the 'eye of Ra.'

We claim that the 'eye' stemmed from the lithosphere being peeled back at the Valles Marineris, like two eyelids, exposing the glowing solid iron core of priori-Mars still within the planet.

What was seen as an eye when still within the opening became a mouth when the glowing solid core actually exited the planet into space, i..e. to 'come forth and shine.' We suggest that what was being described was *the solid iron core of the planet priori-Mars actually being drawn out of the semi-rigid shell or lithosphere of the planet to the point at which it became a separate glowing body in space. Figure 10.11 shows the remarkable resemblance of the Valles Marineris with the hieroglyph for the 'Eye of Ra.' Note that the lines extending below the canyon are exactly in the same configuration as those below the eye in the hieroglyph. It is easy to picture the solid core protruding through the (now closed) opening. This view is upside down compared to normal pictures of this canyon. The reason is that during the dance encounters the north pole of priori-Mars pointed toward the Earth. Therefore, people saw everything upside down. This orientation is a further corroboration of the V/A scenario.*

The liquid iron was drawn out from the interior first, forming the serpent in this myth. The fact that this comes from the funeral ceremony implies this event was linked to the 'death' of a deity, which we claim was priori-Mars. The exiting of the core signaled the end of each dance encounter, but the Egyptians knew from experience that priori-Mars would return fourteen years later, which it did a total of ninety-nine times. *When the 'poison' (the core) failed to return, the true 'death' of Ra took place, that is, its disembodied lithosphere, which we now call priori-Mars, was permanently relegated to the cold reaches of the solar system. This was the terminal event that sent priori-Mars, once and for all, from the vicinity of the Earth, ending some 2800 years of chaos in the solar system.*

The myth suggests that if the two parts of priori-Mars come back together 'Ra shall live' but if 'the poison (the core) lives,' that is, remains outside its lithosphere, then priori-Mars 'will die.' From this we conclude that on earlier occasions the eye of Ra came out of the planet and went back. This was probably the crucial event in the breaking off of the dance encounters. Further support is supplied by the following quote. As with many books on ancient myth, the translator has no inkling that his exposition actually is describing something that happened to priori-Mars:

> Re [Ra] first addresses Nu, the primeval material out of which he arose at the time of creation. In his statement he mentions how mankind emerged from the tears of his eyes ... and now they are conspiring against him. He wants to know Nu's opinion before he kills the entire human race. Nu's reply is that the Eye of Re, the solar eye, will be the instrument to terrorize and slay mankind. Re now becomes aware that men know he is angry over their plot and discovers that they have fled into the deserts of Egypt. The gods in unison urge Re to take vengeance on the conspirators.
>
> The symbol of the Eye of Re is complex but an underlying feature of it is that it can form an entity independent of the sun god himself - even to the extent of going off on journeys to remote regions and having to be enticed back. Here the Eye of Re becomes his daughter, the goddess Hathor ... In the myth of cataclysm ... Hathor becomes a deity of invincibly destructive powers, pursuing men in the desert and slaughtering them ...
>
> The rest of this compilation, following the punishment and near extermination of the human race, is concerned primarily with the rebirth and ascension of the sun god, and therefore of the monarch, into heaven on the back of the 'Divine Cow.' Both the shrine of Tutankhamen and the tomb of Sety I have depictions of the Cow 'Mehet wer' or 'Great Flood' who forms the celestial firmament, identifying the sky goddess Nut. [120]

We believe the second paragraph describes the separation and return of the solid core of priori-Mars from its lithosphere. The third associates this process with the departure of priori-Mars from a dance encounter. The translator of this myth probably confuses the destruction of North Africa by proto-Venus with later events involving the core of priori-Mars. But there is little doubt that the core was a marauder which was greatly feared for the relatively short time that it was separate from the shell of priori-Mars. It was very hot and its orbit may have brought it closer to the Earth than priori-Mars. It may have only made one orbit of the Earth before rejoining its 'parent body.' The passage implies that the iron core departed from priori-Mars on more than one occasion but soon returned ("rebirth"). Their recombining is translated as "the ascension of the 'sun god' into heaven on the back of the divine cow". The core is the sun god because it was glowing, and it has joined the lithosphere but has not yet reentered the interior. The 'celestial firmament' was the planet priori-Mars, hovering in the sky above the Earth, not the sky as we understand it today. The process of reforming into a normal planet took place once it was far from the Earth, so that its own gravity became the dominant factor. How long this process took is unknown. It may have taken months or years.

Figure 10.12. Protome showing a lion holding the head of a Nubian in its jaws.

Another clue to these bizarre events comes in the form of a protome from ancient Egypt shown in Figure 10.12. This depicts a lion holding the head of a Nubian in its jaws. Archaeologists suppose that the lion is a symbol of the power of the pharaoh subjugating a Nubian, a people who were one of the traditional adversaries of Egypt. Our interpretation is that this is a stylized representation of the 'eye of Ra,' the solid core, coming forth from the interior of priori-Mars at the end of one of the dance encounters.[121]

Heaven and Reincarnation

The ancient eastern religions are all based on the repetitive 'dance encounters' between priori-Mars and Earth. Because the people of the Vedic period observed similar forms rising up, taking on life, during each encounter and 'lying down' at each termination, appearing to die, they came to believe that the same thing happens to human beings when they pass away. This was the source of the notion of reincarnation characteristic of the eastern religions.

Another religious belief from this period was that of 'going to heaven' or 'going to the gods,' when one dies. This was reinforced by the proximity of priori-Mars to the Earth during these encounters. The people believed that when they passed away, their spirits would go to nearby priori-Mars, which many ancient cultures referred to as 'heaven' or 'the heavens.' The notion of priori-Mars as 'heaven' also explains the mysterious word *firmament*, which implies something solid or firm up in the air. For example, the Germanic root **hibin-* meant "the stony vault of heaven." This no longer made sense to the generations born after Vedic period ended and is now interpreted as the 'vault or arch of the sky' - a poor substitute for its original meaning. Many cultures believed that priori-Mars was only a 'way station,' and that their ascension only became complete when the dance encounter was broken off and priori-Mars drifted away to join the 'stars,' thus carrying their spirits 'to the gods.' This was the particular function of Yama, a priori-Mars deity, in the Vedic hymns.

An interpretation of the recently assembled funerary writings from the tomb of Pepi I at Saqqara, by Catherine Berger of the French Archaeological Mission indicates that the deceased king has a choice of being reborn each day, crossing the sky in a celestial boat, or turning each night with the circumpolar stars. All three of these were attributes of the planet priori-Mars as seen from the Earth during the dance encounters. The daily rebirth referred to the passage of the planet through all the phases, similar to the Moon's, in a period of one day. The celestial boat is the crescent shape of the illuminated portion of priori-Mars. The 'circumpolar stars' were not stars at all. They were active volcanos on the surface of priori-Mars, which appeared like stars when priori-Mars was in the shadow of the Earth. Because of priori-Mars' rotation these points of light were seen circling its polar axis, thus the name 'circumpolar' was quite appropriate.

11
GREEK MYTHOLOGY

Men stumble over the truth from time to time, but most pick themselves up and hurry off as if nothing happened.

Winston Churchill

Greek myths lack the time ordering of the *Rig Veda*, and like the later Indian works, some have been embellished to the extent that the driving events are difficult to discern. Nonetheless we have uncovered some new clues and a corroborated a number of elements of our scenario. The most important revelation is the straightforward confirmation that the Olympian gods reigned for 3000 years. This is directly stated in almost every work on Greek mythology. Their reign corresponded to what we call the Vedic period, which we previously estimated at about the same length based on the very cryptic description of the life of Brahmā. Other important confirmations in Greek myth include the burning of the Earth by proto-Venus, the Mitra-Varuna duality, the transfer of the primary deity to the temporal seat of power based on fire, the influence of proto-Venus on the capture and release of priori-Mars, and the process by which the dance encounters were terminated. We also explain a number of the Greek myths associated with priori-Mars, such as Zeus, Prometheus and Atlas. The latter leads to an explanation of Atlantis and, finally, we find corroboration of the sequence of cosmic events, derived in the previous chapter from the Egyptian myths, which led to the break off of the dance encounters.

The Birth of proto-Venus in Greek Myth

In Greek myth the Titans, led by King Cronos, had ruled the Earth until led a rebellion that put the Olympian gods in power. The rebellion of Zeus was initiated by the birth of the goddess Pallas-Athene out of the head of Jupiter. In one version of her birth, it is said that no one knew the names of her parents, but Zeus gave her legitimacy by claiming that one day he swallowed his wife, Metis, when pregnant. In another form, Zeus was warned by Gaea and Uranus that if he allowed the child in the womb of Metis to be born it would turn out to be more powerful than he, so he swallowed Metis to prevent this from happening. This caused him such a headache that his howling ('like a thousand wolves') prompted another god, Hephaestus, to help him by splitting open his head with an axe. Out of the gaping wound sprang Athene fully armed and brandishing a sharp javelin.

> Great Olympus was profoundly shaken by the dash and impetuosity of the bright-eyed goddess. The Earth echoed with a terrible sound, the sea trembled and its dark waves rose.

This parallels the disturbance of the Earth at the birth of Aditya in the *Rig Veda*. The rebellion of the Olympian gods was triggered by the impact of the traveler which produced proto-Venus, described in Greek myth as the birth of Pallas (young)-Athene. She was the embodyment of proto-Venus. This signaled the beginning of the Olympian period, which was the Greek name of what we refer to as the Vedic period. *An important confirmation of the V/A scenario comes from the fact that Greek myth claims the Olympians reigned over the Earth for 3,000 years, since we have deduced independently from the Aryan/Indian myths that this was the duration of the Vedic period.*

In another version of her birth from Crete it is said that the goddess Pallas-Athene had been hidden in a cloud and it was by striking this cloud with his head that Zeus had caused her to emerge. Our interpretation of this myth is that Metis, if she was indeed seen, was the 'traveler' which hit Jupiter, the original power seat of Zeus. An enormous glowing cloud of plasma many times the size of Jupiter 'rebounded' from the site of the impact. This was her 'armor' and also the 'cloud' against which Zeus struck his head, out of which Pallas-Athene contracted.

The mention of her being born with a sharp javelin is particularly important. The javelin was a super high velocity jet of material which shot hundreds of thousands of kilometers into space from the impact site on Jupiter. Evidence cited in subsequent chapters from the middle ages and from Galileo data, indicate that *the jet continued to shoot out of the impact crater into space for millennia.* This suggests a phenomenon of which modern science has absolutely no knowledge and one which has a bearing on many current mysteries about the Jupiter system and the main asteroid belt.

Greek myth also reinforces the notion that the Earth was shaken by these events, in spite of its great distance from Jupiter. The sound came from the interior of the Earth which was suddenly put under great tidal stress due either to the rapid passing of the massive traveler or its impact on Jupiter. (Physicists searching for the elusive 'gravity waves' would have had an exciting day.)

The fact that King Cronos was associated with the planet Saturn implies that before the impact that created proto-Venus, Saturn was the brightest planet in the heavens. We infer from this that Saturn had absorbed the impact of a high energy body at an even earlier time, which most likely produced its ring system. As a result it was still glowing brightly while Jupiter was in a more benign state, similar to that of Uranus and Neptune today.

The parallel between the births of Pallas-Athene and Aditi are striking. Both struggle with the identity of the ‘second’ parent. This stemmed from the fact that the traveler was not visible. Both emerged from a glowing cloud as a bright white-hot, star-like, body, which shone with such a bright light that it was millions of times brighter than Jupiter itself. The extremely high temperature of proto-Venus at that time is implied by the term 'bright-eyed.’

Greek Gods and Goddesses Associated with proto-Venus

As Pallas-Athene approached the Earth, a few years later, King Cronus is said to have released his monstrous half brothers, normally sequestered in the mountains, to battle her. These were described as '*huge twisted dark creatures taller than trees*.’ They were the dark comet tails of proto-Venus. Just as in the *Rig Veda*, they were pictured, not as smoke emanating from proto-Venus, but as denizens of darkness attacking the glowing goddess. At that time there was a great rushing and tumult of the skies. The people on Earth heard mighty thunder and saw mountains shatter. The

Earth quaked and tidal waves rolled as these gods fought. By the classical era, this myth was one 'recollection' the Greeks had of the proto-Venus encounters, but they no longer understood its true meaning.

The myth which best describes the devastation caused by close passes of proto-Venus to the Earth is the story of Phaëthon, the son of Helios. He attempted to drive the chariot of the Sun but could not control the horses which broke loose from their normal course. They raced aimlessly and dipped too close to Earth. The heat was so great that northern Africa, which was a savanna at the time, became a desert and remains so to this day. Entire rivers like the Nile were dried up in an instant and clouds of steam rose everywhere. All volcanos erupted. Although the popular interpretation names the Sun, the chariot was actually the blazing proto-Venus which passed close to the Earth. In Greek, Phaëthon means 'the blazing one,' a good parallel with the Vedic god Agni (literally *fire*). Although this is one of the most widely known myths in the world, no one really understands its profound implications.

Velikovsky[122] quotes dozens of ancient documents from Egyptian, Jewish, Greek, Chinese, and Mexican sources, describing two 'overturnings' of the Earth and the attendant destruction and chaos. The following is Ovid's version of the legend of Phaëthon:

> The Greek legend says that the young Phaëthon, who claimed parentage of the Sun, on that fatal day tried to drive the chariot of the sun. Phaëthon was unable to make his way "against the whirling poles" and "their swift axis" swept him away. Phaëthon in Greek means "the blazing one".
>
> The chariot of the sun, driven by Phaëthon, moved "no longer in the same course ... rush aimlessly, knocking against the stars ... borne along just as a ship driven before the headlong blast, whose pilot has let the useless rudder go and abandoned the ship to the gods and prayers.
>
> The earth bursts into flame, the highest parts first, and splits into deep cracks, and its moisture is all dried up. The meadows are burned to white ashes; the trees are consumed, green leaves and all, and the ripe grain furnishes fuel for its own destruction ... Great cities perish with their walls, and the vast conflagration reduces whole nations to ashes.
>
> The woods are ablaze with the mountains ... Aetna is blazing boundlessly ... and twin peaked Parnassus ... Nor does its chilling clime save Scythia; Caucasus burns ... and the heaven-piercing Alps and cloud-capped Apennines.
>
> Then also Libya became a desert, for the heat dried up her moisture ... The Don's waters steam: Babylonian Euphrates burns; the Ganges, Phasis, Danube, Alpheus boil; Spercheos' banks are aflame. The golden sands of Tagus melt in the intense heat ... The

> Nile fled in terror to the ends of the earth ... the seven mouths lie empty, filled with dust; seven broad channels, all without a stream ... Great cracks yawn everywhere ... Even the sea shrinks up, and what was but now a great watery expanse is a dry plain of sand. The mountains, which the deep sea had covered before, spring forth ... If we are to believe report, one whole day went without the sun. But the burning world gave light.

This is a graphic description of the devastation by fire that was caused by the close approach of the young proto-Venus to the Earth. *We take this description literally.* The myth does not state that Phaëthon was the Sun, only that the Sun is his parent.

As proto-Venus swept past the Earth and their relative positions changed, the resulting torques changed, producing 'all manner of twistings.' The destruction caused by proto-Venus took two forms. The most immediate was a scorching of a wide band of latitudes that extended across Africa and the entire Asian continent. We claim that the effect of this scorching was to kill most of the organic material along its path to such an extent that its path is still obvious to this day. See Figure 14.4 on page 190.

Typhon was the serpentine tail of proto-Venus when it approached the Earth early in the Vedic period and encountered the geomagnetic field. It was imagined to be a serpent-like monster with wings and multiple tails attacking Pallas-Athene, since the blazing proto-Venus appeared to be in its mouth. When this tail dissipated it was thought to have fallen to Earth and become imprisoned underground.

Later Athene was characterized as a battle goddess that exercised great restraint - never going to war unless forced. Her silver throne had golden basket work and a crown of violets made from blue lapis lazuli set above it. However Ares, her primary protagonist at this stage complained to Zeus that she was:

> ... a rash and foolish goddess who delights in disobeying the will of Zeus.

This more benevolent form of proto-Venus, resulted from the altering of proto-Venus' eccentricity, bringing its orbit completely inside that of the Earth, along with the formation of a thin crust and the acquisition of an atmosphere. This description parallels that of Mitra-Varuna in the Vedas. The golden basket-work was the body of proto-Venus with light showing through the many cracks in the tenuous crust, while the crown of violets was its newly acquired atmosphere. The notion of intractability may have resulted from the fact that the orbit of proto-Venus was still highly eccentric, making it impossible for the astronomers of the time to predict its motion.

In Greek myth, Athene had a sister which expressed a more peaceful, beautiful aspect of proto-Venus. This was Aphrodite, the equivalent of Varuna in the Vedic hymns. The dual aspect of proto-Venus is alluded to in the myths as follows:

> Next to Athene sat Aphrodite, Goddess of Love and Beauty. Nobody knew who her parents were either ... Aphrodite had a magic girdle, which she would wear whenever she wanted to make anyone love her madly.

Aphrodite is commonly associated with the Moon in mythology because it is thought to be the most beautiful and romantic body in the heavens, but we disagree with this interpretation. We suggest that Aphrodite was the beautiful aspect of proto-Venus, after the flames had been suppressed by the formation of a thin crust, and the magic girdle was the beautiful atmospheric robe, which was visible from the Earth at night. However, the two aspects of proto-Venus were intermixed, because although proto-Venus as Aphrodite had a peaceful countenance, Ares (priori-Mars) often 'rushed off' to do battle with her, (or to make love to her because of that irresistible girdle) at which time the warrior-goddess Athene emerged as the tenuous crust cracked and flaming magma was exposed. The conflict of Athene/Aphrodite with Ares is discussed below.

The Conflict of the Gods

The Greek account of the conflict between priori-Mars and proto-Venus, given by Homer in the Iliad, is paraphrased by Velikovsky.[123] He states that a battle between the Trojans and the Achaeans took place on Earth while the gods Ares (Mars) and Pallas-Athene (Venus) fought in the cosmos above. Each army chose one of the celestial combatants as their champion. The details of this battle give a number of clues to the identity of the Greek gods, but not without considerable ambiguity. Pallas-Athene (Maiden Athene) and Ares were obviously the primary combatants. Aphrodite attempted to enter the battle but is warned by Zeus not to become involved. Zeus 'presides' over the entire conflict.

Ares is said to "darken the battlefield." Pallas-Athene retreats then turns and hurls a spear against his "nether-belly." Both gods are said to produce loud bellows or roars, then "over against her spouted Ares" and "together they clashed with a mighty din." At this point Hera "spreads a thick mist about herself." Ares strikes Pallas-Athene with his long "spear" and, although giving ground, she then heaves a gigantic "stone" at him "loosening his limbs."

At one point, priori-Mars probably eclipsed the Sun, darkening the earthly battlefield. The 'bellows' were due to the rapidly changing tidal force exerted by the two nearby planets on the interior of the Earth. Apparently enormous electrical discharges occurred between the priori-Mars and proto-Venus when they approached one another. This is understandable, since the V/A theory proposes that proto-Venus had been literally incandescent for years before these encounters and as a result was likely to have accumulated a net electrical charge.

We propose that Hera was another goddess associated with the body of proto-Venus. Thus the common interpretation that she was a goddess of the 'world' is really that she was goddess of another 'world,' which was clearly seen as such by the ancient people. The fact that some scholars claim the name is derived from the Sanskrit word for the sky may also be true, since the myth describes her capturing a 'sky' or atmospheric robe. Some say that she was the wife of Zeus, with whom she continually squabbled. But she sometimes borrowed the girdle of Aphrodite in order to arouse his passion and thereby get her way. This relationship is consistent with the transference of Zeus to priori-Mars, since priori-Mars and proto-Venus were the primary combatants in our scenario.

Ares, (Mars), was pictured as either rushing off to engage in combat with Athene or to make love to Aphrodite, both of which were aspects of proto-Venus. The physical meaning is clear in either case, that is, when priori-Mars was engaged in a 'dance encounter' with the Earth and proto-Venus reached its closest point to the Earth it caused priori-Mars to break off the encounter and 'go after' her. The important implication of this 'rushing off to do battle' theme is that the gravitational force of proto-Venus at 'favorable' inferior conjunctions was a significant factor, at least in the release and probably also in the capture of priori-Mars. It is likely that the effect of proto-Venus in facilitating the capture and release of priori-Mars from the 'dance encounters' continued throughout the Vedic period. Indeed, it was probably a crucial factor in this process.

The myth in which the 'local hero' rushes off occasionally to do battle with an 'intruder' is found in a number of cultures. The notion of a deity 'rushing off' is consistent with our claim that (a) priori-Mars orbited the Earth for extended periods of time, (b) broke off the encounters when proto-Venus, the ‘intruder’, approached the Earth and (c) returned numerous times.

Greek Deities Associated with priori-Mars

Although it is obvious that Zeus was embodied in Jupiter at the time of the birth of Athene, it is difficult to understand the stature of distant Jupiter still being considered significant during the period when priori-Mars and proto-Venus were close to (and were having devastating effects on) Earth. Since Zeus is described as the deity that led the battle against the Titans, we suggest that the primary power of Zeus was transferred from Jupiter to proto-Venus at the beginning of the Vedic period. Later he is described as living on Mount Olympus, implying that his mythical power base was transferred again - this time to the column of smoke and fire on priori-Mars. Reinforcing this is the notion that Zeus rained down thunderbolts and storms on the Earth and was without contest the most powerful of all the Olympians. The transference of his power to priori-Mars parallels the transference of Agni (Vedic) and Osiris (Egyptian) from proto-Venus to priori-Mars early in the Vedic period. In all three cases this was a transfer of the deity to the ‘seat of power’ on priori-Mars, which was the great flaming column of smoke and fire, identified as Brahmā in Hindu mythology.

Just as the triad of Hindu deities conversed and even disagreed, so does Zeus curse Ares (Mars) as the "Most hateful of all gods that hold Olympus." It is necessary to see through these ambiguities, and not get 'hung up' on them.

Atlas

Greek mythology contains another deity that is associated with the column of smoke and fire on priori-Mars. In fact, no deity description fits our scenario more perfectly than Atlas. He was supposedly the commander-in-chief of the armies of the Titans, who lost to the Olympian gods led by Zeus. When the Olympian Gods took over the Earth, Atlas was condemned by the council to carry the 'heavens' on his shoulders until the end of the world.

Greek Mythology

When priori-Mars approached the Earth it grew larger and larger. Upon attaining its geosynchronous orbit above the Himalayas it appeared to 'stop.' It was enormous - over twenty times the apparent diameter, and four hundred times the area of the Moon. The people did not understand what kept it from continuing to fall to Earth. This question was specifically expressed in the Vedas. *When the great lava fountain formed on the side of priori-Mars facing the Earth, extending perhaps a thousand kilometers toward the Earth, the Greeks got the idea that it was the deity which, situated as it was between the two bodies, supported priori-Mars. This was Atlas, and the 'world' he supported was not ours, but nearby priori-Mars.*

Atlantis

It is said that more books have been published on the subject of Atlantis than any other except possibly chess and the Bible. In a recent one, Zangger points out that Atlantis means 'the daughter of Atlas.'[124] Thus the key to locating Atlantis lies in identifying Atlas, which we have just done. The circular continents which made up the mythical place called Atlantis should sound quite familiar to the reader at this juncture. A very similar geographical description was given in the Hindu myths quoted earlier in which seven circular continents separated by seven circular seas surrounded Mt. Meru on which Brahmā lived. In that instance, the commentator thought that the ancient Hindus were describing the geography which surrounded them on the Earth. By the same token, all of the authors, from Plato to Zangger, have sought Atlantis on the Earth. Today the Earth seems the only possible location, given Plato's detailed description of the dimensions and layout of a great city. However, *no modern scholar has 'discovered' the true Atlantis - because it was not on the Earth. It was nearby however, in a position which provided a birds-eye view. It was because it could be seen from Greece to India, that these descriptions are so similar. Atlantis was on the planet priori-Mars, the nearest point of which was only some 33,000 kilometers above the surface of the Earth during the many 'dance encounters.'*

Hermes

Hermes, the Greek god associated with the planet Mercury, was the solid iron core of priori-Mars, which exited the planet at times of the break off of the dance encounters. He was considered a messenger because he would leave priori-Mars and descend into a lower orbit, closer to the Earth and subsequently return. This gave the impression that he was carrying messages from Zeus to the Earth. Hermes was said to have invented the lyre, which he made out of tortoise shell. The tortoise shell was a description of the surface of the solid iron core of priori-Mars. The strings of the lyre were linear grooves scratched on this surface by the rocks in the lithosphere lining the Valles Marineris, as it exited the rigid outer shell of the planet. These grooves are still visible on the surface of Mercury, and are shown and discussed in the sister volume to this book, entitled *Chaos*, by Angiras.

Prometheus

Another deity in Greek myth that is obviously related to priori-Mars during the 'dance encounters' is Prometheus. He was the most rebellious of gods, tricking Zeus into giving men the best part of the slain ox. Zeus punished mankind by sending Pandora to Earth, who unleashed all forms of misery on mankind. He then punished Prometheus for thwarting the initial effort of Zeus to introduce Pandora into the world and for his role in aiding mortals to control fire, thereby aiding in the development of civilization, which the gods generally discouraged. Prometheus was bound to a rock in the Caucasus Mountains with unbreakable chains, where a vulture tore at his flesh all day, with no rest, no sleep and no respite. Each night his flesh grew back, only to be eaten again the next day. This unusually cruel punishment was actually intended to force Prometheus to reveal a very important secret. Zeus knew that Fate, who brings all things to pass, had decreed that a son would be born to him who would dethrone him, and drive the Olympian gods from their home in heaven. But only Prometheus knew who the mother would be, and refused to tell. Zeus dispatched Hermes, who implored Prometheus to reveal the secret and thus be released:

> Look for no ending of this agony
> Until a god will freely suffer for you,
> Will take on your pain, and in your stead
> Descend to where the sun is turned to darkness,
> The black depths of death.

Pandora may have been a Greek name for the solid core of priori-Mars which scourged the Earth according to Egyptian myth. It might also be a reference to a series of plagues or diseases which descended from priori-Mars along with its dust, water and atmospheric gases during the dance encounters.. The 'fastening' of Prometheus to the Earth suggests he was priori-Mars, since it alone of all heavenly bodies, remained stationary over the Himalayas during the 'dance encounters.' As each day progressed, priori-Mars waxed and waned, exhibiting phases like the Moon, as it orbited the Earth, passing in and out of its shadow. As more and more of the planet passed into darkness, the body of Prometheus was seen to gradually disappear each half day, only to be renewed in the next half day. The fact that the largest form visible (Valles Marineris), when the surface of priori-Mars was in sunlight, resembled a large bird with wings extended, is consistent with the notion that it was a vulture which ate away his flesh every day. This same feature was also referred to as Zeus' eagle.

Prometheus never revealed the secret that Zeus sought, but the reign of the Olympian gods did indeed come to an end when the solid core of priori-Mars, which was Hermes, himself, left the planet and moved into the inner solar system. The mother was proto-Venus, because it was the primary influence on the separation and capture of priori-Mars in orbit around the Earth. The prophecy delivered by Hermes - that another god would have to sacrifice himself in order that Prometheus be freed, was indeed true. It was Ares, that was sent out into the black depths of the solar system, where the Sun barely shines. As prophecied, Zeus, the column of smoke and fire, suffered the same fate, while Hermes was sent into the interior of the solar system, where it is scorching hot.

Cyclops

These were three monsters that appeared in several Greek myths - giants with a single huge eye in the middle of their heads. The Cyclops had nasty dispositions, but praised for their unique ability to work with iron. Their association with iron meant that the Greeks, like the Egyptians, knew that liquid iron was extruded from the outer core of priori-Mars when the solid core became exposed through the Valles Marineris. The evidence was the subsequent falling of pure iron meteorites. Thus the Cyclops was inspired by the same events as the Egyptian hieroglyph 'the eye of Horus.'

Griffin

The liquid iron flowed out into space and formed bizarre shapes, resulting in the form that the Greeks called a Griffin. A drawing of this creature is shown in Figure 11.1. The name comes from the Greek word *gryps*, meaning 'hooked' (as in a beak,) and is also related to the Persian word *griften*, 'to grip or seize.' This implies its correspondence with Horus, a common Egyptian hieroglyph associated with priori-Mars, selected because of the resemblance of the distended Tharsis Bulge to a beak.

The griffins eyes were said to be 'fiery,' it had the form of a lion and the wings of a bird. It was said to guard 'gold deposits' from a one-eyed race called the Arimaspeans. The eyes were 'fiery' because they were two continuously erupting volcanos now called Arsia and Ascreaus Mons. The form of a lion refers to the overall shape of the red planet, which resembled the *head* of a lion, and the one eyed race was the Cyclopes.

Because of the lure of gold, some have tried to associate the griffin with a prehistoric beast, the skeletal remains of which have been found in the Tarim Basin in western China. We believe that such attempts are futile. The idea of 'gold' in the myth probably came from glowing lava flows, which covered a large area in the vicinity of the Tharsis bulge.

Most representations of the griffin have the erect column protruding upward from the head in the same direction as its long ears, as shown in Figure 11.1. It is interesting to note the similarity of these features to the ceremonial hood and stylized 'beard' worn by the pharaohs and the sphinx. It is only necessary to turn the griffin head upside-down in order to recognize the Egyptian interpretation. The column on the head of the griffin becomes the pharaoh's 'beard,' and the long ears become the 'hood' that extends downward on both sides of the pharaoh's head.

Figure 11.1 The head of a griffin, 7th century B.C.

In Timaeus and Critias, most famous for its description of Atlantis, Plato gives a detailed description of the 'universe.' Readers of the classic period up to the present assume that references to 'the world' or 'the earth' describe the Earth. However, these translations were made after the thread of knowledge of the dance encounters was lost, and the authors did not realize that other worlds were a part of the daily experience during the reign of the Olympic gods. As stated previously, priori-Mars was also referred to as 'heavens.' The following is an example of such a statement.

> He decided that it should have as many forms of life as intelligence discerns in the perfect Living Creature. There are four of these: the gods in heaven, birds in the air, animals that live in water and animals that go on the dry land. The divine form he made mostly of fire so that it should be as bright and beautiful to look at as possible; and he made it spherical like the universe and set it to follow the movement of the highest intelligence, distributing it round the circle of heaven to be a kind of universal cosmic embroidery. And he gave each divine being two motions, one uniform in the same place, as each always thinks the same thoughts about the same things, the other forward, as each is subject to the movement of the Same and uniform; but he kept them unaffected by the other five kinds of motion, that each might be as perfect as possible. This is the origin of the fixed stars, which are living beings divine and eternal and remain always rotating in the same place and the same sense; the origin of the planets and their variations, of course we have already described. And the earth our foster-mother, [winding] as she does about the axis of the universe[1], he devised to be the guardian and maker of night and day, and first and oldest of the gods born within the heaven. It would be useless without a visible model to talk about the figures of the dance of these gods, their juxtapositions and the relative counter-revolutions and advances of their orbits, or to describe their conjunctions or oppositions, and, how they periodically hide each other from us, *disappear and then reappear, causing fear and anxious conjecture about the future to those not able to calculate their movement*:
>
> Footnote:1. The meaning of these words, and in particular the word here translated 'winding,' have been much disputed: ... Plato has so far accounted for the apparent movements of the heavenly bodies on the assumption that the earth is at rest at the centre of the universe. And any interpretation of these words must allow for this. Cornford suggests that in order to resist the motion of the Same [the constant rotation of the fixed stars], which it would otherwise share, ... the earth must have a counter-motion of its own, which will exactly cancel it out, and keep the earth at rest without rotation at the centre; and this is what Plato means to convey by the rather odd word 'winding.' [125]

In light of the V/A scenario this passage makes it very clear that the "foster-mother" was *not* the earth, but the nearby 'world' we call priori-Mars. Because it remained fixed in the heavens over the Himalayas it was said to "resist the motion of the Same (the constant rotation of the fixed stars.)" Priori-Mars could be seen rotating about its axis with a slow, wobbling motion while fixed in the sky over the Himalayas. Therefore the motion that the author of the footnote is struggling with should be translated 'wobbling' or 'precessing,' as opposed to 'winding.' This was a description of priori-Mars - not our mother Earth, but our step-mother.

Greek mythology fits the V/A scenario. All the crucial aspects are present and there is a surprising parallel between the Greek, Egyptian and Hindu deities.

12
MYTHOLOGY OF OTHER CULTURES

Science is too important to be left to the scientists.
Isaac Asimov
(paraphrasing Clemenceau)

This chapter discusses some myths from the Middle East, China, Tibet, and Indian tribes from North America, which are consistent with the V/A scenario. Recurrent themes are the great flood, the 'home' god who rushes off to do battle and returns, the prominent bird beak, the celestial circle or mandela, the duality of the Martian surface, and the apparent halting or slowing of the Earths rotation on certain occasions.

Proto-Venus Description from Mesopotamia

The great destructions caused by the two close passes of proto-Venus to the Earth were the result of scorching of the Earth and flooding, due respectively to the incandescent state of proto-Venus and the overturning of the Earth. The following is an Akkadian version of one such encounter which emphasizes the flood and devastation that resulted:

> As soon as a gleam of dawn shown in the sky, came a black cloud from the foundation of heaven. Inside it Adad thundered. Adad's rage reached to the heavens: turning all light to darkness. Six days and nights raged the wind, the flood, the cyclone devastated the land. When the seventh day came ... the battle was over . . .
>
> The sea became calm and the cyclone died away, the flood ceased. And all mankind had turned to clay. The ground was flat like a roof.

Two encounters with proto-Venus reoriented the spin axis of the lithosphere by approximately 180 degrees in a matter of hours, causing the oceans to leave their beds and sweep across the land masses. The myth relates how clay from the ocean bottom settled out on the surface of the Earth, as the waters subsided, filling in the depressions and making the local surface flat.

Venus and priori-Mars Deities

Although proto-Venus' eccentricity was reduced very quickly so that its orbit moved inside that of the Earth, it was still an imposing sight during the Vedic period, recognizable as a 'world' at close approach and exhibiting an ever changing, comet-like tail. Even at this stage, its eccentricity made its path almost impossible for the ancient astronomers to predict. The proximity of its inferior conjunctions varied due to its eccentricity but, every 14 or 15 years, much closer ones occurred. The myths tell that it was at these special times that proto-Venus influenced the capture and release of priori-Mars in orbit around the Earth.

As a result both priori-Mars and proto-Venus were worshiped in all cultures. One of the earliest indications of the parallel worship of two deities comes from Early Bronze I ruins at Megiddo, a former chalcolithic site dated around 3200 B.C.(radiocarbon date), where the only public building was a double temple composed of two broad rooms each with a raised pedestal for the statue of a deity located opposite the entrance. Although the number of deity names multiplied over the many years in which these planets interacted with Earth, the visible features on them, and the effects caused by them, were the source of all the deities which we now refer to as 'pagan.'

The status of major deity was not conferred on the more distant proto-Venus by ancient people simply because of its striking appearance. proto-Venus was worshiped in many cultures for millennia because the ancient people *experienced* the fact that it still affected Earth in three ways: (1) Its gravitational field still had a powerful influence on the tides and quaking of the Earth during inferior conjunctions; (2) When priori-Mars and Earth became aligned with proto-Venus at inferior conjunction, it caused unusually violent eruptions on priori-Mars which expelled all manner of material; (3) proto-Venus influenced the comings and goings of priori-Mars, that is, its entry into and its breaking off of the 'dance encounters' between priori-Mars and the Earth.

In order to get an idea of how the out gassing proto-Venus might have appeared in the second part of the Vedic period in which priori-Mars was the dominant deity, we refer to images of large conventional comets. These images exhibit a great variety of shapes which change even from day to day. A common characteristic is that they exhibit multiple tails. The broader one, which can take

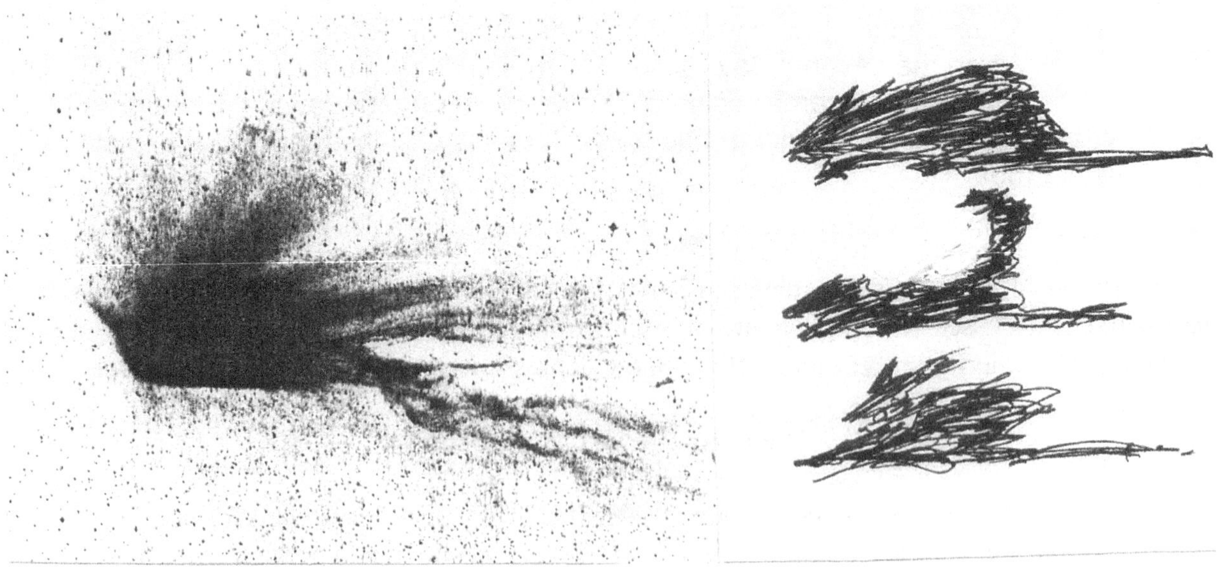

Figure 12.1. Halley's comet is shown in A. B, C, and D are sketches showing how slight variations of this bizarre form taken on by proto-Venus during the Vedic period could have been interpreted as a mouse, a scorpion, or a witch on a broom.

on a multitude of shapes, is due to neutral particles whose paths in space were controlled by the pressure of the solar wind. One or more thin tails are also formed due to ionized gas molecules. These are oriented in a different direction than the neutral particle or dust tail because they are influenced by the interplanetary magnetic field as well as the solar wind. An idea of the more bizarre shapes that can be assumed by conventional comets is illustrated by Halley's comet in Figure 12.1. Based on these examples it is not a difficult to imagine proto-Venus taking on forms such as a scorpion, a mouse, or a witch-on-a-broom in myths cited by Velikovsky. The thinner tail was seen as the mouse tail and the broomstick, while the dust tail was the body of all three.

Because of the great size and proximity of proto-Venus, the ash that was out-gassed blocked some of the Sun's rays when at inferior conjunction, and as a result, its tail could be seen during the day. Velikovsky relates an American Indian myth, told by the Menomini Algonquins, of how the Sun was caught and retarded in its motion for hours until a 'mouse' released it by gnawing through the 'ropes' that bound it. *This myth indicates that an anomaly in the apparent motion of the Sun occurred at an inferior conjunction of proto-Venus. We claim that this corresponded to the release or capture of priori-Mars over the Transhimalayas, at which times the rotation of the lithosphere was retarded or advanced. The Menomini could not have seen priori-Mars because it was on the opposite side of the Earth, but they experienced the apparent retardation of the Sun nonetheless. The retardation was a result of the reorientation of the spin axis of the lithosphere that occurred at these times. The fact that they observed the mouse in direct line with the Sun, proves that proto-Venus was at inferior conjunction at the time of the capture or release of priori-Mars, and corroborates our premise, based on mythology, that this was a critical element in that process.*

Ancient Sumerian Myths

The ancient Sumerian *Epic of Creation* (*Enuma Elish*) speaks of the supreme deity, E.A, also known as EN.KI, sometimes translated 'Lord of the Earth', whose home planet was called Nibiru. His name literally translated means: " He whose House is Water". We suggest that E.A was the column of smoke and fire and the home planet was nearby priori-Mars. The water was the tidal sea out of which the column of smoke and fire rose at the beginning of each dance encounter. EN.LIL, his half brother, was probably proto-Venus. Sitchin[126] believes that these were actual beings or god-men and not planets, and that they came to Earth some 434,000 years ago. He believes that a "Mission Control", called DUR.AN.KI, literally "The Bond Heaven Earth," was established on a mountain top in Nippur to coordinate their activities. We claim this was the central mountain on priori-Mars, where E.A was located, and not on Earth. The "bond" was the mysterious force which held the two planets together during the dance encounters. Sitchen claims that the biblical Eden, or E.DIN, "the abode of the righteous ones," in Sumerian, was in Mesopotamia. In the context of the V/A scenario, this suggests that priori-Mars was this abode. The possibility of intelligent life descending to Earth is discussed further in chapter 17.

Because the deities in most sacred myths were anthropomorphized, they are believed by many to refer to living beings. We claim that the earliest gods in all the myths were the planets or their adjuncts. The events characterized as the initial creation of 'man' in ancient myth were often based on the ejection of bodies launched from volcanic vents on the surface of priori-Mars, just as

Brahmā was imagined creating gods, demons, fathers and men in the Hindu myths. Sitchen provides evidence favoring our claim. He points out that the term used for Adam at the beginning of Genesis is actually 'The Adam,' referring apparently to a generalization or conceptualization, as opposed to a real being. It is not until Chapter 5 that Adam is translated as meaning mankind. This notion is bolstered by the fact that Cain and Able are not included in 'the book of the genealogies of Adam,' which skips from 'the Adam' to Seth.[127] This implies that 'the Adam,' Cain and Able were not human beings, but anthropomorphized features on the surface of priori-Mars.

Another connection with the Vedas arises in the number seven. Sitchen relates that the Sumerians had a symbol for one heavenly body, made up of a circle enclosing seven dots. He concludes, with some difficulty, that this was the Earth, as seen from the perspective of extraterrestrials arriving from outer space, to which the Earth would be the seventh planet encountered. Of course, this view does not take into account that, on the average, some of the planets would have been on the opposite side of the Sun as the visitors approached. We claim that this symbol represented priori-Mars, since the Vedas make very clear that seven permanent volcanos were visible on its surface. These seven 'holy rishis' shown brightly and were easily distinguishable from Earth, even in daylight, during every dance encounter.

One thing that clouds our understanding of the myths was the tendency of renown men to give themselves the same names as prominent features on priori-Mars. For instance, Nimrod, a great Sumerian leader, may have taken the same name as the 'column of smoke and fire' deity. Similarly, the tower of Babel was an attempt to build a structure on Earth that resembled that deity.

One Babylonian priori-Mars deity was Nergal. Velikovsky[128] says that the Babylonians called it the unpredictable planet[129] and carefully watched:

> ... its rise and setting, its disappearance and return ... its position in relation to the equator, the change of its illuminating power, its relation to proto-Venus ... Historical inscriptions of the eighth century speak of the *opposition* of the star priori-Mars (Nergal) ...
>
> Prayers were addressed to Nergal with the lifting of the hands toward the star priori-Mars. "'Thou who walkest in the sky ... with splendor and terror ... king of battle, the raging fire-god, god Nergal." Nergal, the fire star comes like a raging storm. He is also called Sharappu, 'the burner,' and 'the light that flames from heaven,' and 'lord of destruction.' priori-Mars was generally regarded by other peoples too, as a 'fire star.' Ying-Huo, or the fire planet, is the name of priori-Mars in Chinese astronomical charts. Sargon, father of Sennacherib, wrote on one occasion: "In the month of Abu, the month of descent of the fire-god." Isaiah said the world was "moved exceedingly" and "became removed from its place." Other hymns to Nergal were "The heaven he makes dark, he moves the earth off its hinges" and "Nergal ... on high stills the heavens ... causes the earth to shudder."

Obviously, Nergal posed a threat to the people. The notion that priori-Mars in its current orbit inspired these descriptions is preposterous. Later, the Canaanites worshiped Baal and Ashtoreth, as did many Jewish sects. The names of these pagan gods were rarely mentioned in the Bible, to avoid promoting them, but the Jews were condemned by Jehovah for building altars and worshiping these pagan idols in 'high places,' that is, at altars built on the tops of hills.

Flood Myths Associated with priori-Mars

We claim that Genesis begins with the Earth immediately *after* its desolation by the two close passes of proto-Venus very early in the Vedic period. The biblical passge, correctly translated, is that " ... the Earth *became* void and without form ... " Genesis does not describe the proto-Venus floods, only their aftermath. In other words, it is only concerned with events which occurred after about 3300 B.C (radiocarbon) which corresponds to about 3900 B.C..

Because of the great time span covered by the Bible, we consider that the earlier, more mystical or cryptic parts in the first few chapters of Genesis, fall into the category of mythology, in the sense that myths of cultures that predated the Jews are incorporated into the Torah. This does not mean that they do not describe 'true events,' just that they may be anthropomorphized descriptions of cosmic events, requiring some analysis to determine their true meaning, just as other sacred myths. The remainder of the Bible we regard as an historical source, and therefore as a means of corroborating our hypotheses concerning cosmic events during the laater stages of the Vedic period.

A flood legend paralleling the biblical story of Noah and the Fish Avatara involving Manu in the *Rig Veda* was discovered in the library of the ancient city of Nineveh, built by King Ashurbanipal in the seventh century B.C. This Epic of Creation describes the building of an 'ark' and the loading of the animals and plants on it.

The biblical story of the flood of Noah was predated by the Mesopotamian version but undoubtedly was inspired by real events. Floods occurred regularly throughout the vedic period, whenever priori-Mars was captured or released, that is, every fourteen years. Moreover, they occurred on priori-Mars and the Earth at the same time, reinforcing their significance in ancient myth. Normally they were not greatly destructive, as evidenced by the biblical description as a 'raash,' translated by Velikovsky as 'a commotion.' But, depending on the details of the engagement or disengagement of the two planets, these periodic floods varied in intensity. Every thousand years or so, some factors combined to cause unusually high flood levels or tidal waves, resulting in major destructions of the prominent cultures. Such was the flood of Noah.

The narrative of the biblical flood is deceptive in several ways. First, it was not, and need not have been, a flood of the entire world. Most of the population, as well as the animal and plant species in the biblical world was localized, therefore a 'high tide' combined with a tidal wave caused by a concurrent earthquake, occurring in that part of the world, could have killed a large fraction of men and animals. There was only so much water available on the Earth - certainly not enough to completely cover the land masses the world around. Second, the water did not come primarily from rain, but as a result of the tidal force of priori-Mars. Third, it is doubtful that the land was covered for as long as the Bible says, that is, forty days and nights. The Mesopotamian version says the flood

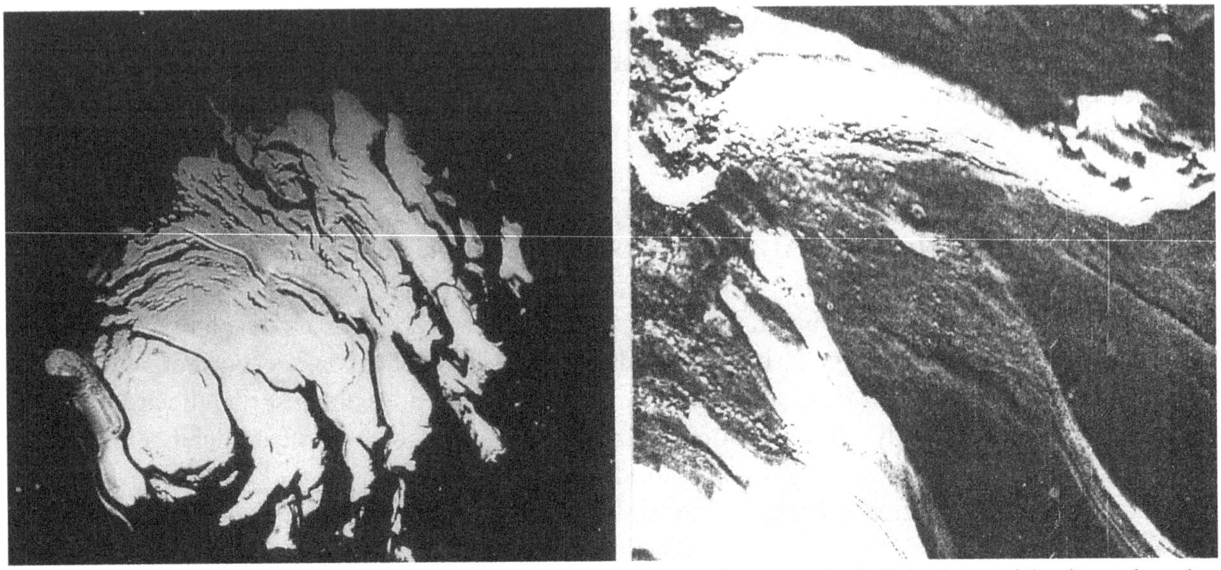

Figure 12.2. Two elephantine formations on the surface of Mars. On left is the residual south polar cap which resembles an elephant with raised trunk, and on right is a the head of another elephant.

lasted one week. Of course, all the land-locked low lying areas would have remained flooded for months, if not years.

Mythical Images Associated With priori-Mars

Most mythical images associated with priori-Mars arose from various surface features - some transient and some permanent. One class is that of natural formations which remain intact today on the surface of the planet, and have been photographed by space probes orbiting priori-Mars. These shapes result from a combination of surface relief, coloration and shadow. They definitely resemble various animals, and there can be little doubt that primitive peoples recognized them as such when priori-Mars was close to the Earth. It is only necessary to look at the NASA images of the red planet with a more child-like attitude in order to become aware of animal forms. The agreement of the features mentioned in the myths and those we find on priori-Mars today is further evidence that priori-Mars was very close to the Earth within proto-historical times since the ancient people had to have observed them with the naked eye.

Not one, but two elephants are recognizable on the surface of priori-Mars (Figure 12.2), and there are two elephants mentioned in the Vedas. The first, Airavata, Indra's elephant, is in all likelihood inspired by the shape of the residual south polar ice cap. This white elephant with trunk raised is also the likely source of the oriental notion that an elephantine form with raised trunk is a sign of good fortune. The second elephant image merely shows the head and trunk. This is probably Ganesa, since the myth involves his head being severed from his body.

The Valles Marineris is easily seen in Figure 12.3 as a great bird in flight with an unusual head or tail. This is the mythical bird Garuda, the 'vahan' or animal which 'carried' Vishnu. It was

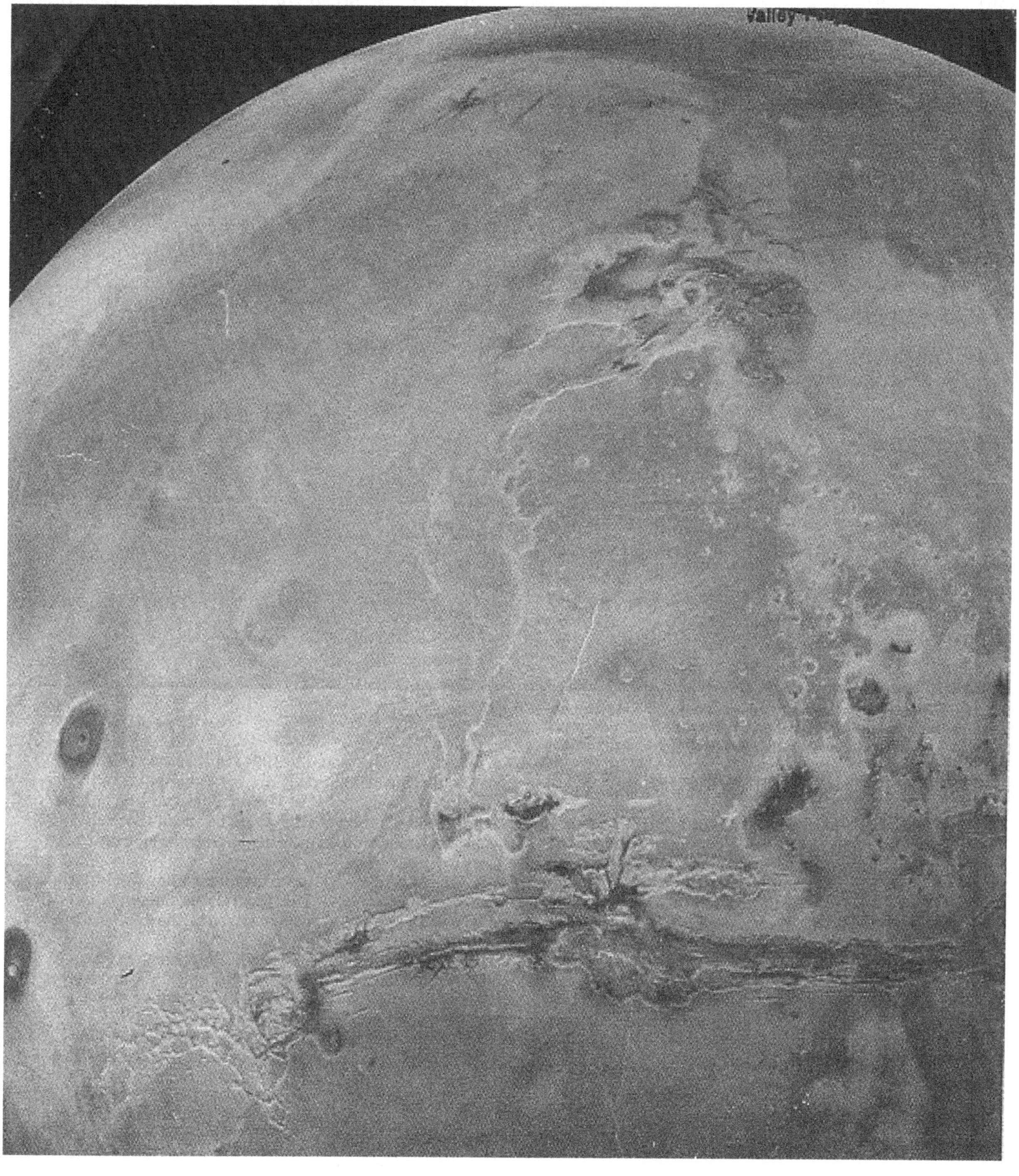

Figure 12.3. Features resembling animals on the surface of Mars. A large bird in flight, a horse or bull head at top, and a monkey at the right. At the lower left, concentric circular faults delineate the great Tharsis Bulge which was extended outward to form a distinctive ‘beak’ or ‘nose during the Vedic Period.

also seen by the Greeks as the vulture which daily tore at the flesh of the imprisoned Prometheus, when it became illuminated by the Sun each day, and was most likely the source of the peacock in Tibetan myth. In the same photo-mosaic, images of a horse or bull (top center) and a monkey (lower right) can also be seen. In Hindu mythology, a bull was said to be the vahan of Siva. The Vedic deity Vayu, (later Pavan), the god of wind and noise that comes propelling the dust and making all things ruddy, is said to have had a son, by a monkey mother.

These same symbols appear in the 'vision' of the biblical prophet Ezekiel, and in the book of Revelations, attributed to the disciple John. Ezekiel saw a creature with four heads. This vision may be a composite of the four headed figure called Brahmā by the Hindus and the features visible today on the surface of priori-Mars. John describes four living creatures around the throne. The first is a lion, the second is an ox, the third has a face like a human being, which may be the figure which we described above as resembling a monkey, and the fourth creature is like a flying eagle. As a result of this and a number of other statements in Revelations, we conclude that the mythical imagery was meant to *reveal* the events observed on priori-Mars by ancient cultures during the Vedic period. This leaves open the question as to exactly how much of this last chapter of the New Testament actually has to do with predictions of the future, as is currently supposed.

At the lower left of Figure 12.3, two of a row of three enormous volcanos are visible at the edge of the mosaic. The two outer volcanos were pictured as the eyes of an enormous face thousands of kilometers across. This is better illustrated in Figure 8.1, Page 103. During the dance encounters, the volcanos erupted continuously and were seen by many ancient peoples as glowing 'eyes' at night and even during the day.

This was the image which inspired the eyes on the sides of the Tibetan Stupas (Figure 12.4), which represents priori-Mars. Note that the eyes are around the sides and not at the apex. This is consistent with the V/A theory that the axis of rotation of priori-Mars, located in the northern plains hemisphere, faced toward the Earth during the dance encounters. Thus the Vedic people would have seen the 'two eyes' rotating around the pole once each day. The eyes are painted on all four sides of the stupa to simulate the rotation of priori-Mars.

Figure 12.4 An elaborate stupa near Kathmandu is the oldest and most significant shrine in Nepal. (From *Himalayas*, design by E.M. Buhrer, Facts on File, NY, 1987)

This image was greatly enhanced by the presence of the great Tharsis Bulge, visible in the photo as several concentric rings, located 'below' and slightly off center relative to these 'eyes.' This image was greatly enhanced by the presence of the Tharsis Bulge, visible in the photo as several concentric rings located 'below' and slightly off-center relative to the 'eyes.' During the dance encounters, this bulge was distended into an enormous prominence, which was interpreted as a 'beak' in many cultures because it was cut open by the western end of the Valles Marineris.

This was why the hieroglyph for Horus (priori-Mars) was a standing bird. Egyptians and Greeks were not the only cultures to see a great curved beak on priori-Mars during the dance encounters.

Figure 12.5 Bird's head mask from Bhutan with third eye in the forehead.

The bird's head mask from Bhutan, shown in Figure 12.5, is also characterized by a prominent open beak Confirmation of our interpretation of this as a Mars deity is supplied by the third eye positioned in the forehead, representing Olympus Mons.

The form of a beak also appears in the carved masks and totem poles of the Kwakiutl Hamatsa Indians of western Canada. Figure 12.6 shows a mask representing their ' Crooked Beak of the Sky,' currently on display at the University of British Columbia Museum of Archaeology. It is particularly interesting that the beak is described as 'crooked.' This is consistent with the off-center location of the Tharsis Bulge, the collapsed form of the beak. In addition, the Vallis Marineris, which constituted the 'opening' in the beak, is not symmetrically positioned. If these people had just 'invented' an imaginary sky god, they would certainly not have made it crooked. It was because they were reproducing what they saw on priori-Mars that they made it match as closely as possible. The entire planet resembled a lion head and was the inspiration for the many lion heads in ancient Egypt and Tibet. In fact there is a good chance that the original head of the great sphinx was made in homage to this image, since it faced priori-Mars constantly during the encounters.

Figure 12.6 Kwakiutl Hamatsa mask of Crooked Beak in the Sky.

Chinese Myth

The close interactions between proto-Venus and priori-Mars, were considered as a conflict between two gods in the myths of many cultures. A Chinese myth says that the first god to come down to Earth was 'Fiery Emperor' but that the most powerful god was 'Yellow Emperor.' Early fragments of myths indicate that the Fiery Emperor led a rebellion against the Yellow Emperor but was defeated. They are called half brothers. We claim that the Fiery Emperor was the flaming proto-Venus while the Yellow Emperor was priori-Mars. The fact that Fiery Emperor came down to Earth first is consistent with the V/A scenario. As indicated previously, proto-Venus moved into an orbit inside that of the Earth after a series of interactions with priori-Mars, and priori-Mars interacted with the Earth for thousands of years afterward. Thus the Yellow Emperor, although he came down to Earth later, was considered the victor by the ancient Chinese.

A notable characteristic of the Fiery Emperor was a fiery crest. The great band across the center of proto-Venus, revealed in NASA Magellan radar images (Figure 12.7) was probably this crest.

Just as the Hindus, the Chinese also named the asterisms, and believed they had a powerful influence on the course of Chinese history. Different asterisms were adopted by each state.[130]

> The historian Szu-ma Ch'ien tells of how Tzu Wei, when in the service to Duke Ching of Sung, observed the planet priori-Mars "guarding" his state's stellar almanac, the hsiu Hsin ... The Duke grew alarmed at what he assumed was an impending personal calamity, so Tzu Wei advised him that the omen could be diverted either to one of his ministers or the common folk, or to the harvest. Ching nobly rejected each of these salvations, whereupon his wily astrologer responded: "Heaven is high above, but it listens to what is said below. Your highness has made three wise decisions. priori-Mars will change its position." And so it did, by some three degrees, according to Tzu Wei's later reckoning, leaving Ching healthy and the state of Sung free to enjoy prosperous times.

This is interesting in that it associates priori-Mars with the asterisms - also in that it finds the 'movement' of priori-Mars unusual. In its current orbit priori-Mars, like all the other planets, is seen to move from day to day relative to the stars. Why then was the motion of priori-Mars seen as a special event in the legend?

Chinese and Tibetan mythologies both contain a duality which represents the two distinctly different regions on priori-Mars, which today are called the southern highlands and the northern plains. The Yin and Yang, shown in Figure 12.8, constitute the two portions of the symbol in the Chinese *Tao*. Similarly, the cupping together of the two hands by Tibetan monks as they chant their '*aum*,' the sacred name of Brahma, is also symbolic of this duality. These parallel the vedic (Har-Hari), the Egyptian '*world of the two lands*,' and a number of other ancient dualisms.

The Nepalese temple shown in Figure 12.9 and the Chinese architectural form called the pagoda are stylized representations of the column of smoke and fire on priori-Mars. The 'stacked' roofs correspond to the horizontal lines that decorate the Egyptian Tet or Djed Pillar.

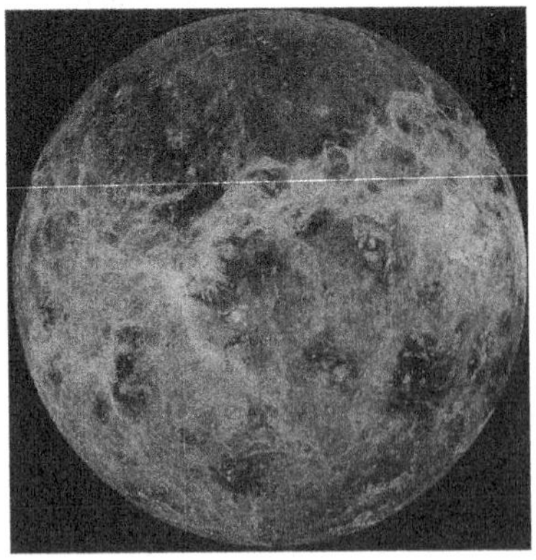

Figure 12.7 Radar image of proto-Venus showing a bright band across the equatorial region on one side of the planet.

The ancient Chinese had a great interest in the heavens and recorded details of many celestial events. But much invaluable data from the Vedic period was lost. The 'burning of the books' by the Ch'in ruler Shih Huang Ti, in 213 B.C. and a purge of astronomical texts by the Sui Dynasty ruler Yang Ti in 610 A. D., resulted in the destruction of virtually all ancient astronomical work.

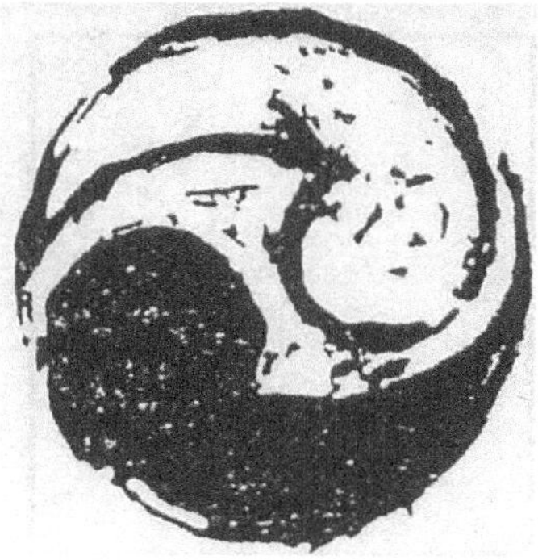

Figure 12.8 The sign of Yin/Yang from the book of Tao.

Some local chronicles covering limited periods of time did survive. One example was the Chunqiu (Spring and Autumn Annals), from the small state of Lu. Among recordings of a number of what are interpreted as solar eclipses, is the oldest reliable record of a meteor shower.[131] The date of this event was March 23, 687 B.C., at which time stars were said 'to fall like rain.' According to Velikovsky, this is the date of the Death of Romulus - the very date we propose that priori-Mars made its final break from the Earth, never to return.

The date of these purges, beginning a few centuries after the end of the Vedic period, suggests the possibility that the astronomical observations collected by the ancient Chinese astronomers during the Vedic period were no longer understood and were destroyed because the post-Vedic astronomers, who had lost the thread of understanding of the Vedic events, advised their rulers that the ancient astronomers 'must have been crazy' - that their observations were nonsense and an embarrassment. Thus, crucial ancient observations were destroyed.

This action paralleled the revision of star positions in ancient Indian astronomy books such as the Surya Siddhanta in the first millennium A.D. Fortunately much of the *text* of the Hindu book was left intact, because there were no observations to confute the descriptive material. Thus considerable material describing the asterisms, meteorite 'falls' at conjunctions have survived, and are discussed in Chapter 15.

Figure 12.9 The Nepalese wooden temple, or Chinese pagoda was a stylized representation of the great solidified lava fountain feature on priori-Mars.

Such destruction of books may seem the unconscionable act of a Stalin or Hitler, but there are parallel examples of such actions much closer to home. The astronomers of the 1950's blackmailed Velikovsky's original publisher, forcing Macmillan to transfer the publication to Doubleday when it was in the top ten list, a*nd would, if they had had their way, have completely destroyed Velikovsky's works.* Fortunately that did not happen, and we have completed his search for the truth.

The best known symbol representing the serpentine tail which extended from proto-Venus, in the earliest part of the vedic period, is the Chinese 'Dragon.' This was the Egyptian Set and the Greek Typhon. The stylized modern version is compared to a sketch of what we believe was its actual appearance, in Figure 12.10.

Figure 12.11 Tibetan painting showing the origin of the four great rivers of Tibet as mythical animals. (From *Himalayas*, Design by E.M. Buhrer, Facts on File, N.Y. 1987)

Water from priori-Mars

Figure 12.11 illustrates the Tibetan monks' belief of how the four great Himalayan rivers, all of which originate in the immediate neighborhood of Mount Kailas had their genesis in four mythical animals. These animals are easily identifiable with priori-Mars. The horse and elephant heads are nearly identical to surface features on priori-Mars, shown in Figures 12.2 and 12.3. The peacock may represent the bird in flight in Figure 12.3. The fact that the artist mistakenly places some of these features on the Earth is attributable to the 'forgetting' that occurred over thousands of years after priori-Mars moved permanently away from the Earth. However, the similarity of the horse and elephant (Ganesa) heads to the actual features on priori-Mars is striking, considering that the painting was made in modern times but long before the Martian surface was photographed by NASA probes. This painting offers further support of our claim that water from priori-Mars gouged out the passages of the four great rivers of Tibet. The lion head represents the overall shape of priori-Mars during the dance encounters.

What is left of priori-Mars has re-formed into a nearly spherical shape since its last 'dance encounter' with Earth, but vestiges of its 'lion head' shape still exist. The so-called dichotomy boundary, the surface discontinuity between the northern plains and the southern highlands, marks the edge of the lion's mane. Although the elevation difference between the two hemispheres is now only four kilometers, it was many times greater during the 'dance encounters,' making the 'mane' much more prominent and a feature that was easily seen from the Earth. The Vedic people imagined this boundary as the edge of the Loka-Loka mountains which encircled the ringed continents. The Tharsis Bulge, which is located toward the edge of the northern plains, is the residual of the 'snout' of the lion.

The water which fell to Earth from priori-Mars collected in large basins in Tibet, constrained on the south by the Himalayas. The following is a Tibetan legend about the formation of the four great Himalayan rivers:

> The sagas tell how " ... a Buddha of that time ... " took pity on the flooded land and with his sword cut a deep swath through the mountains in southeastern Tibet and thus let the water out. Today in this region is one of the deepest and widest gorges in the world, where the Yarlung-Tsangpo cuts through the chain of the 7,776 meter high Namche Barwa and then flows southward as the Brahmaputra into the plains of Assam. [132]

This story implies that great lakes built up in Tibet, and that they were released suddenly and catastrophically, probably through cracks created by violent earthquakes. The force of the enormous volume of water rushing through the original gap cut each of the gorges through the Himalayas in a matter of days. Such a catastrophic process is in direct opposition to that proposed by geologists, which required millions of years.

The Cosmic Mandela

During the dance encounters, the people of the Orient saw priori-Mars above their heads rotating slowly like a great wheel. Because priori-Mars was not symmetrical, this wheel wobbled slightly, but it was still obvious that its axis corresponded roughly with the sacred mountain. This 'wheel in the sky' rotated about once per day, but the ancient peoples were quite aware of its wobbly motion. An image which conveys the off-axis rotation of priori-Mars during the dance encounters can be found carved on the doors of an ancient Tibetan monastery. This Mandela is shown in Figure 12.12.[133] The ellipses represent the paths of the various 'junction stars' (Hindu), which were actually active volcanic vents on the surface of priori-Mars as it rotated throughout the day. They were referred to in ancient Hindu astronomy books as 'circum-polar' stars. 'Constellations' of these points of light were even identified and named in many cultures. These were called 'asterisms' - a separate and distinct word from that used for the fixed stars. They existed on the astral plane - the surface of priori-Mars. It was described as a 'millwheel' being turned by women in Nordic mythology. It was also described as a churn, the Celtic '*quern*,' and as a 'whirlpool' in Homers Odyssey,

Figure 12.12 The 'Cosmic Mandela' - a graphic means of depicting the origins of the universe. From a fresco in Paro Dzong, Bhutan. (From *Himalayas*, Design by E.M.Buhrer, Facts on File, NY, 1987)

Closely related to the cosmic mandela are the myths of many cultures which imply a central 'axis' about which the 'world' (actually priori-Mars) rotated, the *axis mundi*.,'the axis of the worlds.'. This was the great hardened lava column which the Hindus called Brahmā and in some cultures referred to as a 'cosmic tree.' In the Vedas it is the *skambha*, or cosmic pillar. The mast on the stupa in Figure 12.4 represents the holy mountain with Brahmā at the top. The common imagery, expressed in the myths of so many cultures, is unmistakable.

The Sacred Mountain

The Vedic people designated Mt. Kailas, in the Transhimalayas of Tibet, as the earthly home of Indra or Vishnu, that is, the place over which priori-Mars remained stationary during the dance encounters. It is depicted artistically in Figure 12.13. as the 'Mountain of the Gods' showing the stationary form of priori-Mars hovering directly above it. This thanka by Tulku Tsewang, from Tragyam Gompa in Dolpo, Nepal [134], shows pilgrims moving in their clockwise path around the holy mountain. The movement of the worshipers simulates the rotation of the features on priori-Mars, around 'the axis of the world.'

In contrast to this earthly home, however, the Hindu myths designate the *celestial* Mt. Meru as the home of Brahmā. This was a great mountain on priori-Mars on which the lava-tube form of Brahmā arose. Paralleling this description, the Chinese myth referred to in the previous section gives a number of details which agree quite closely with those of the Indian people discussed in the preceding chapters. It refers to the Yellow Emperor, which is consistent with the yellow-red surface color of priori-Mars. After defeating the Fiery Emperor, the Yellow Emperor, who had four faces, built a magnificent palace in the center of the world on top of Mount K'un-lun. K'un-lun ('celestial mountain') was a holy mountain which was unimaginably high. The divine palace was located at the peak on K'un-lun, eleven thousand *li* above the surface. Nine layers of mountains surrounded it, like city walls, and at the bottom lay a deep abyss called the Roushui River. The palace was lit by the light of the 'fire mountains' which bordered the abyss and burnt day and night.[135] This parallels the Indian description in which Brahmā lived on Mount Meru and in which the fire mountains are called the 'seven rishis.' The four faces of the Yellow Emperor correspond to the four heads of Brahmā.

Figure 12.13 A thanka depicting Mt. Kailas, Indra's earthly home, in the Transhimalayas of Tibet. (From *Himalayas*, Design by E.M.Buhrer, Facts on File, NY, 1987)

The Chinese myth even tells that the four heads each spoke with a different voice, as did the four heads of Brahmā. The nine layers of mountains are the Chinese version of the circular continents described in the Vedas and the circular channels surrounding Atlantis, described by Plato.

Practically every ancient eastern culture on Earth describes a sacred mountain which was the home of the greatest deity in their mythology because they all saw the lava-fountain feature on a great mountain on priori-Mars. Most cultures also named the largest earthly mountain range in their vicinity, *that lay in the direction of Mt. Kailas*, where priori-Mars resided, by the same holy name, paralleling the celestial Mountain they saw on priori-Mars. This was probably done after priori-Mars receded from the Earth, and the subsequent generations had lost the real meaning of the myths. Thus Mt. K'un-lun, the Chinese name for a mountain in what is now known as the Hindu Kush, lies to the southwest of China. Similarly, the sacred home of Zeus in Greek mythology, Mt. Olympus, was placed in the Caucasus lying to the east of Greece. This linking of a sacred mountain on priori-Mars and one of the same name on the Earth has led to great confusion among those attempting to understand the myths in modern times.

A Legend of Buddha

The following legend is a Chinese description of events observed on the surface of priori-Mars:

> This is the story of the attainment of the goal of all goals beneath the "tree of awakening," the Bo- or Bodhi-tree (bodhi, awakening) ... coming to the Bodhi-tree, the one who was about to become the Buddha stood on the southern side and faced north. Instantly the southern half of the world sank until it seemed to touch the lowest hell, while the northern rose to the highest heaven.
>
> "Methinks," then said the Buddha-to-be, "this cannot be the place for the attainment of supreme wisdom"; and walking round the tree with his right side toward it, he came to the western side and faced east. Thereupon the western half of the world sank until it seemed to touch the lowest hell, while the eastern half rose to the highest heaven. Indeed, wherever the Blessed One stood, the broad earth rose and fell, as though it were a huge cartwheel lying on its hub and someone were treading on its rim.
>
> "Methinks," then said the Buddha-to-be, "this cannot be the place for the attainment of supreme wisdom"; and walking round the tree with his right side toward it, he came to the northern side and faced south. Thereupon the northern half of the world sank until it seemed to touch the lowest hell, while the southern half rose to the highest heaven.
>
> "Methinks," then said the Buddha-to-be, "this cannot be the place for the attainment of supreme wisdom"; and walking round the

> tree with his right side toward it, he came to the eastern side and faced west ... Now it is on the eastern side of the Bodhi-tree that all the Buddhas have sat down, cross-egged, and that side neither trembles nor quakes ... Then the Great Being saying to himself, "This is the Immoveable Spot on which all the Buddhas have established themselves: this is the place for destroying passion's net," He took hold of a handful of grass by one end and shook it out there. And straightway the blades of grass formed themselves into a seat fourteen cubits long, of such symmetry of shape as not even the most skilled painter or carver could design.[136]

We suggest that this legend arose from watching priori-Mars rotating above the Earth during the dance encounters. Buddha was one of the prominent features on priori-Mars and his apparent motion, both around a central location and up and down, was a description of the precession of priori-Mars as it rotated. The reference to multiple Buddhas .".it is on the eastern side that *all the Buddhas* have sat down ... ," implies that with each new 'dance encounter' a similar form appeared. The rectangular 'seat' was inspired by the same feature that the Greeks interpreted as the plain surrounding the city of Atlantis. Other creatures that are seen to tempt and threaten the Buddha later in this story, described as .' ... demonic forms crystallized, wearing frightening shapes and bearing in their hands bows and arrows, darts, clubs, swords, trees, *and even blazing mountains*. ... ' are lava tube features that erupted over much of the central portion of priori-Mars giving rise to the many heroes and demons in the mythology.

The Death of Romulus, the Founder of Rome

Velikovsky goes into considerable detail in this area. One reason for his attention to this subject, was that the Romans were the first to give the name Mars to priori-Mars. Although he never figured out the truly amazing celestial resonance between priori-Mars, proto-Venus and Earth. His exposition of the 'Romulus' story is most informative, and we paraphrase much of it here.

Rome was founded between 747 and 753 B.C., apparently on the same date as the perturbations of the Earth described in the Bible as the 'raash' of Uzziah (REF). According to persistent Roman legend, the conception of Romulus, the founding of Rome and the death of Romulus, all took place at the times of great commotions, accompanied by celestial motions and perturbations in the motion of the Sun. Calamities of disease, 'a rain of blood,' and continuous earthquakes characterized his reign. Of particular interest was the material that Velikovsky collected in reference to the death of Romulus.

> The death of Romulus occurred when, according to Plutarch, " ... suddenly strange and unaccountable disorders with incredible changes filled the air; the light of the sun failed, and night came down upon them, not with peace and quiet, but with awful peals of thunder and furious blasts ... "

and amidst this storm Romulus disappeared. Ovid's description of the phenomena on the day of Romulus' death is this:

> Both the poles shook, and Atlas shifted the burden of the sky ... The sun vanished and rising clouds obscured the heaven ... The sky was riven by shooting flames. The people fled and the king [Romulus] upon his father's [Mars] steeds soared to the stars. [137]

We claim that the mythical death of Romulus corresponded to the 'death' of priori-Mars, that is, the events surrounding the final release of Mars and its ascent to its current, distant orbit, far from the Earth. The resonant arrangement between priori-Mars, Earth, and proto-Venus, which had been the norm for over three millennia, far longer than the interval from the advent of Jesus Christ to the present, began to come undone in the seventh century B.C. The date at which the Vedic period finally ended was probably 687 B.C. Of particular interest is the statement that " ... Atlas shifted the burden of the sky ... ", since we have explained that previously that Atlas was the 'column of smoke and fire' that was thought to hold priori-Mars in place. This describes priori-Mars as moving from its stationary position over the Himalayas, while at the same time explosions occurred and the Earth was profoundly shaken. "The Sun vanished" could refer to the solid core of priori-Mars leaving the scene. "Rising clouds" probably referred to those on priori-Mars which obscured its surface since it was called "the heaven". This Earth shaking event ended as the body we now called Mars soared to the stars with Romulus, the column of smoke and fire.

Calendars and Henges

Another interesting anomaly occurred around the end of the Vedic period. This concerned the length of the month and the year. Velikovsky devotes several pages to the reporting of variations in the number of days per year in different cultures. [138] The principal numbers, which are found in many cultures throughout the Vedic period are 360 days and 365.5 days. The use of a 360-day calendar year was definitely not the result of observational errors or poor calculations by the ancient people. This was too widely used to be a mistake. An error of this magnitude would become obvious to even the simplest farmer, because it would result in the displacement of the seasons by 200 days in only 40 years. Few realize it, but the 360-day year was the source of the angular measure we use to this day, in which 360 degrees represents an entire circle. At some times during the Vedic period the lunar period was 30 days - a beautifully symmetrical arrangement when it corresponded with the 360-day year.

Apparently the number of days per year, *not necessarily the length of the year*, changed frequently. As a result, there was a continual struggle in all cultures to devise an accurate calendar throughout the Vedic period, due to the planetary chaos characterizing that time. *This was the impetus behind the construction of the ancient henges, and their continual adjustments - to determine the new order of the heavens, after each capture or breakaway of the planet priori-Mars.*

The dilemma of the different number of days per year is easily resolved by the V/A hypothesis. The year comprised 360-days priori-Mars was captured in orbit around the Earth, while it comprised 365.25 days when priori-Mars was in its own orbit around the Sun. The added angular momentum of priori-Mars, when gravitationally bound to the lithosphere, caused the Earth to rotate slightly slower, thereby decreasing the number of days per year.

Because of the difficulty of defining the number of days in the year, some cultures turned to the use of a lunar system, in which the month was the basis. Obviously this was not successful either, because there were periods during which the lunar period was about 36 days. Plutarch wrote that in the time of Romulus, some months were reckoned at 35 and some 36 days in length. The change from a seven to a nine day phase is also found in the traditions of the peoples of Rumania, Lithuania, Sardinia, the Celts, the Mongols, and the tribes of West Africa. Velikovsky concludes:

> Therefore, and in view of the vast material from many peoples, we conclude that at one time during the century of perturbations [the century preceding the death of Romulus], for a period between two catastrophes, the Moon receded to an orbit of thirty-five or thirty-six days duration. It remained on such an orbit for a few decades until, at the next upheaval, it was carried to an orbit of twenty-nine and a half days duration, on which it has proceeded since then. [139]

Thus at the end of the Vedic period, the eccentricity of the lunar orbit was reduced from its (temporarily) larger value at the time, bringing it to a nearly circular orbit that it now occupies. This implies that whatever cosmic interactions that occurred at the time of the final release of Mars from an Earth-crossing orbit involved the Moon in addition to proto-Venus.

13
THE VELIKOVSKY/ANGIRAS SCENARIO

The mere formulation of a problem is far more essential than its solution, which may be merely a matter of mathematical or experimental skill. To raise new questions, new possibilities, to regard old problems from a new angle requires creative imagination and marks real advances in science.

Albert Einstein

The purpose of this chapter is to consolidate what has been gleaned from the myths in the first twelve chapters. To that end we summarize the cosmic events that comprise the V/A scenario and the events which occurred on the Earth as a direct consequence. These are presented without reference to their mythological origins, which have already been related in the previous chapters, and without the validating data from other sources, which are discussed in the subsequent three chapters.

The dates given here are based on corrected radiocarbon 14 dates, since those from the earlier Vedic period are considerably in error. The radiocarbon 14 method is based on the uniformitarian assumption that carbon 14 has continually been produced at the same rate in the upper atmosphere since the occurrence of the dated event. The error accumulated rapidly throughout the Vedic period due to the influx of atmospheric gases from both proto-Venus and priori-Mars proposed in the V/A scenario. Although radiocarbon dating of the beginning of the Vedic period gives 3300 B.C., the corrected dates are close to 4000 B.C., while its end date, 687 B.C., requires little or no correction.

Before the Vedic Period - A Possible Origin of the Moon

Some seven oor eight millennia prior to the Vedic period, two catastrophic events occurred which, we believe, profoundly modified the Earth system. Based on evidence turned up by Velikovsky, this pair of events caused two overturnings of the lithosphere. Our hypothesis concerning these events more distant in time is necessarily more speculative than the events of the Vedic period, but we see them as part of a major metamorphosis of the Earth - resulting in a profoundly different planet. We suggest these were the final encounters that marked the capture of the Moon into its current orbit around the Earth. These two passes of the Moon exerted sufficient torque on the outer, rigid portion of the Earth, the lithosphere, to 'overturn' it. These overturnings resulted from tidal forces on the equatorial bulge and on the Tibetan-Himalayan complex which became uplifted at that time. Each of these close passes resulted in the rotation of the spin axis of the lithosphere, on which we live, by approximately 180 degrees. This meant that, after the first encounter, the Sun rose in the west and set in the east. The second close pass, a millennium later, resulted in a similar event, thus reorienting the lithosphere to its original orientation. These events delineated the Younger Dryas period. It is quite possible that during the Younger Dryas that the

Moon had a number of less traumatic encounters, which, although they did not overturn the Earth, did each expend a significant amount of the lunar orbital energy. The net effect of this sequence of encounters was the capture of the Moon in its current orbit about the Earth.

Although specialists in astrodynamics have never considered this mechanism, we claim that the 'slipping' of the lithosphere relative to the solid core of the Earth provided the 'energy sink' which made the capture possible. These events were extremely destructive, resulting in the extinction of hundreds of animal species, many in North America. But, as a result of this capture, the climate of the Earth was stabilized in a way that has prevented the radical variations of temperature that characterized the climate before that time, i.e. the ice ages and interstadials.

These ancient encounters initiated the up-thrusting of the Himalayan-Tibetan complex, which was already unstable by virtue of the impact of the Indian plate into the Asian plate. Geologists, trained in uniformitarianism, believe that this uplifting took place an inch at a time, and therefore required at least two million years. We claim that it took place as a result of close encounters with the Moon and later with proto-Venus. The uplifting of this massive region had a profound effect on subsequent interactions between priori-Mars and the Earth.

The Pre-Vedic Solar System

Prior to 4000 B.C. there existed only two terrestrial planets, priori-Mars and the Earth. That is, Mercury and proto-Venus were not present in the solar system. Priori-Mars, the more ancient of the two by some 500 million years, was in an orbit similar to that of proto-Venus today. It had a thick atmosphere, lots of surface water, rain, and a myriad of life forms similar to those found on Earth. The entire planet was covered with vegetation.

Jupiter was a calm, bluish, ice planet, similar to Uranus and Neptune as currently shown in NASA Voyager photographs. Saturn was the brightest planet in the heavens, because it already had its extravagant ring system - the result of material splashed into space by an impact a few millennia earlier. Uranus and Neptune also existed but could not be seen by the naked eye.

The Earth was in an orbit similar to its present one, but it was not the hardy planet we know today. It did not have as much oxygen in its atmosphere as it now does. The oceans reached only to the edges of the continental shelves and the Mediterranean comprised three land-locked fresh water lakes. Species of neolithic beings, evolved from lower forms of hominid over millions of years, tended herds, did some farming, built modest shelters, and made tools of Bronze. However, they were profoundly inferior to Homo Sapiens Sapiens.

The Velikovsky/Angiras Scenario

Approximately 6,000 years ago a massive galactic 'traveler' entered our solar system at very high velocity and impacted the calm blue ice planet Jupiter, releasing at least 10^{41} ergs. It bored deep into the icy surface of the planet and vaporized an enormous mass of material which rebounded into space through the atmospheric tunnel created by the impactor. The resulting plasma cloud was thousands of times the volume of Jupiter itself and glowed with a golden color. Much of this mass rebounded with such high velocity that it escaped the gravitational pull of Jupiter and went into an eccentric orbit around the Sun. The plasma cloud quickly contracted due to its mutual gravitational attraction. Its gravitational potential was converted to heat as the converging atoms collided, forming a star-like body or proto-planet, with a temperature above 10,000 Kelvins. This was the birth of proto-Venus.

Its initial perihelion distance was somewhat less than the current distance of proto-Venus from the Sun, but its aphelion was near Jupiter, its point of origin. Its initial orbital period was approximately five years and it crossed the orbits of the Earth and priori-Mars. The great energy of its eccentric orbit was rapidly converted to heat, due primarily to interactions with the Sun at perihelion. Two mechanisms were active in this process. Tidal forces greatly distorted its shape and caused chaotic motions in its fluid interior. Moreover, electromagnetic forces, due to the interaction of the ionized body of the proto-planet with the magnetic field of the Sun converted even more of its orbital energy into internal heat. This resulted in the rapid reduction of its eccentricity. It was reheated again and again at each perihelion passage, repeatedly increasing its temperature above 10,000 degrees for decades. Its high temperature caused great volumes of volatiles, particularly the lighter elements, to be out-gassed and lost to space. This increased its average density from 1, characteristic of the material ejected from Jupiter, to over 5 g/cm^3, characteristic of a terrestrial planet. This is how all terrestrial planets are formed.

It passed close to priori-Mars on numerous occasions because its perihelion was near to the orbital radius of priori-Mars. The net result of these encounters was the ejection of priori-Mars from its ancient inner orbit, to one of larger and larger radii as the two bodies exchanged orbital energy. The combination of the solar and Martian encounters reduced proto-Venus' eccentricity rapidly to the point that it barely reached the vicinity of the Earths orbit in less than fifty years. At the same time the radius of priori-Mars' orbit was moved outward, bringing it closer and closer to the Earth.

During this period the flaming proto-Venus passed extremely close to the Earth on two occasions, with only a few planetary radii separating them, and scorched a significant fraction of the Earth's surface. In addition, its tidal forces on the equatorial bulge and the Himalayan complex exerted an impulsive torque on the lithosphere of the Earth, which, in less than a day resulted in its complete turning over in space, without reducing its rate of spin. The solid core of the Earth, isolated from the lithosphere by several thousand kilometers of liquid iron, continued to spin in the same sense. To the few survivors, the Sun was seen to rise in the west and set in the east. But within a decade proto-Venus returned, torqueing the lithosphere back to its original orientation.

One of these close passes scorched the Earth across North Africa, the Middle East, and northern India, burning everything in its path. The heat was so great along this path that all organic material was obliterated, creating the great deserts that now exist in these areas. All the volcanos of the Earth erupted. The Nile river was completely dried up. These two overturnings of the lithosphere were similar to those earlier ones caused by the final capture of the Moon. They caused kilometer-high tidal waves which swept across entire continents, tumbling great boulders before them, killing most neolithic people and animals and leaving the low lying areas filled with water and clay from the depths of the oceans. Four stages of this process are shown in Figure 13.1. As a result of these catastrophic encounters, which reversed the sense of rotation of the Earth relative to the stars, many surviving cultures recorded in terms of the creation of 'a new sky' or 'a new Sun.'

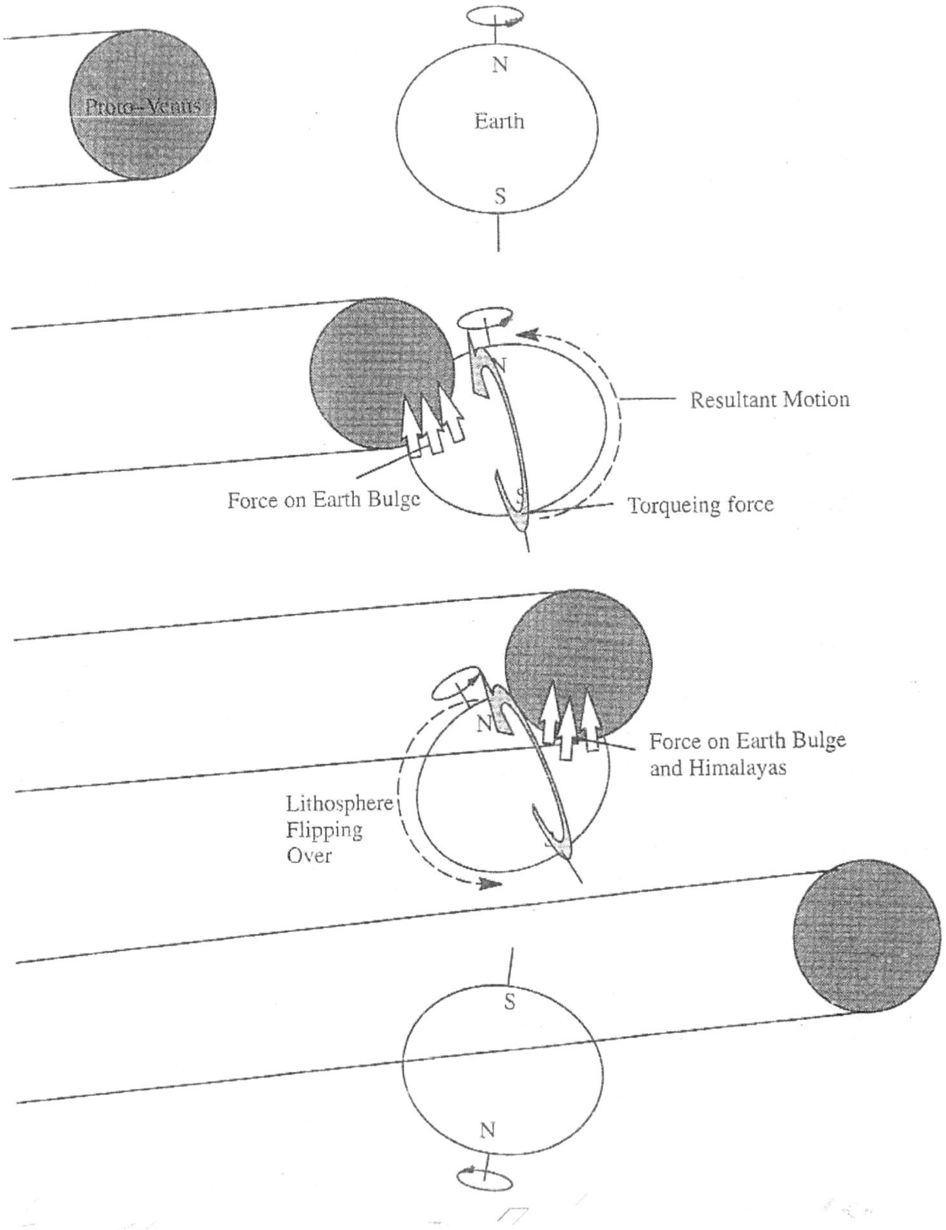

Figure 13.1 Torque by which proto-Venus overturned the Earth in 3300 B.C. was due to the gravitational force on the equatorial bulge of the Earth and the Tibetan-Himalayan complex. Four stages in the process are pictured.

Soon after this, a complex three-planet interaction occurred between priori-Mars, proto-Venus and the Earth. As a result of this encounter sufficient orbital energy was transferred to priori-Mars to send it for the first time into an orbit which crossed that of the Earth. The same interaction further reduced the radius of proto-Venus' orbit, as dictated by the principle of conservation of energy, so that henceforth it remained inside the orbit of the Earth, signaling the end of the most destructive phase of what we term the Vedic period. In this eccentric orbit, proto-Venus still came quite close to the Earth every fourteen of fifteen years, due to the correspondence of its aphelion to its inferior conjunctions (alignments with the Earth and Sun).

The very hot proto-Venus remained in this orbit for millennia and was still an imposing sight, with a beautiful blue atmospheric envelope it had acquired from its numerous close interactions with priori-Mars, and a comet-like tail, the result of out-gassing of hot sulfur through hundreds of thousands of vents in its tenuous crust. It caused continuous earthquakes around the world at its inferior conjunctions. Its powerful gravitational force, at these times, had a profound effect on the capture and release of priori-Mars, events which continued for almost three thousand years. Its eccentric motion created a dilemma for those astronomers assigned to predict its comings and goings.

Priori-Mars then began engaging in a series of encounters with the Earth so amazing that no one in modern times, not even Velikovsky or all the science fiction writers in the world, have guessed. *For some 3000 years priori-Mars engaged repeatedly in 'dances' with the Earth, each of which lasted fourteen or fifteen years, with intervening periods of separation of the same length.* These dances of priori-Mars and the Earth were such that the smaller body, priori-Mars, was captured in a geosynchronous orbit around the Earth - the two planets actually revolving about their common center of mass, like a lop-sided dumbbell, with a period of exactly one day. In this configuration priori-Mars remained over the same place on the Earth - the Himalayan-Tibetan complex, with its north pole oriented toward the Earth.

Each capture of priori-Mars occurred when proto-Venus was at a close inferior conjunction. The subsequent dance encounter lasted about fourteen years, at the end of which time proto- proto-Venus again approached the pair. Its tidal force then caused priori-Mars to break away from the Earth and go into its own orbit around the Sun. That is, the orbits of the three planets were in a unique resonance with one another during the entire Vedic period.

Establishing of the 'dance configuration' required adjustments of the spin axes of both priori-Mars and Earth. As priori-Mars approached, its spin axis was torqued around from its normal orientation, roughly parallel to that of the Earth, until its north pole pointed directly toward the Earth. This was caused by a great anomalous mass near the equator of priori-Mars, which is now known as the Tharsis Bulge. This feature was much more pronounced during the Vedic period, probably due to earlier interactions with proto-Venus. The tidal force on this raised mass resulted in the continual torqueing of the spin axis of priori-Mars in the direction of the center of the Earth throughout every dance encounter. Its motion was manifested to observers as a wobbling or precessing rotation of priori-Mars which also had a period close to one day, at least by the end of the Vedic period.

As priori-Mars approached and its north pole became oriented toward the Earth, the tidal force of the Earth drew all the ground water and lake water in its northern hemisphere toward its north pole. Some of the resulting flows were thousands of times more voluminous than the flow of the greatest rivers on the Earth, and some actually flowed uphill, due to the overwhelming tidal force of the Earth. All water in the northern hemisphere was drawn into a 'tidal sea' at the north pole of priori-Mars. The size of this sea gradually decreased throughout the 2700 year Vedic period, as priori-Mars relinquished more and more of its water to the Earth.

There is a persistently recurring notion in the myths of ancient cultures that the Moon is a great mirror and that the light highlands and dark maria are in actuality reflections of the continents and oceans of the Earth. The idea is sometimes attributed to Aristotle, but there is a hint of the idea in the writings of Homer, dating to the eighth century B.C. In the 19th century Alexander von Humboldt encountered the mirror concept at Isfahan, Persia. Of course this is not true of the Moon, but we believe there was a valid precedent for this idea. It is quite possible that when the tidal sea formed on the side of priori-Mars facing the Earth, it acted as a mirror and therefore the Vedic people might have been able to see *a small image of the Earth in that sea because it formed a convex mirror in space.* The visibility of this image would have depended on the brightness of our sky, whether the side of the Earth toward priori-Mars was illuminated, and whether the surface of priori-Mars was in darkness. A few generations after priori-Mars departed the scene, the true nature of the reflecting body was lost and the best explanation for the heavenly body became the Moon. Thus as amazing as it might seem, the pictures taken by the Apollo astronauts of the beautiful blue Earth suspended in the black void of space may *not* have been the first image of the entire Earth seen by mankind.

In order for priori-Mars to remain stationary in the heavens, and not drift, it was necessary that a 'potential well' exist on the equator of the Earth to which priori-Mars could become gravitationally bound during the dance encounters. The Tibetan-Himalayan complex was the only upraised plateau large enough to serve this purpose. However, under 'normal' circumstances, it was 30 degrees north of the equator. As priori-Mars approached and latched onto this complex, it forced the spin axis of the lithosphere to migrate about thirty degrees, so that the Tibetan-Himalayan complex was on a new temporary equator. As indicated above in the case of the proto-Venus encounters, the lithosphere is free to rotate independently of the solid inner core. As a result of this thirty degree shift, the temporary 'north pole' of the Earth, during each of the dance encounters, was relocated to Hudson Bay, in North America. When priori-Mars broke away every fourteen years, the pole reverted to its current position. Taking the Hindu myths at face value, one hundred such encounters took place during the Vedic period.

The anomalies in the apparent motion of the Sun and stars, chronicled in many cultures, occurred at the transition times, either when priori-Mars was being captured or released. Some of these peoples could not even see priori-Mars, because they were located on the other side of the Earth. These apparent accelerations or retardations of the Sun were due to the reorienting of the spin axis. This process involved relatively little energy dissipation, because the rate of spin of the Earth, did not change greatly. However, the length of the day did change slightly because the tidal bond between the two planets increased the moment of inertia of the system. This was manifested by the fact that there were exactly 360 days per year during the dance encounters and 365.25 during the period when priori-Mars was 'on its own.'

During the 'dance encounters,' observers in the hemisphere centered on the Himalayan-Tibetan complex could see the northern hemisphere of the larger priori-Mars, fixed some 35,000 kilometers above the Himalayas. It appeared to those below like a wobbling celestial wheel with a diameter twenty times larger than the Moon, slowly spinning about its axis once per day. In one *Rig Veda* hymn the Earth and priori-Mars are described as being like two wheels at the opposite ends of an invisible axle. The capture and release of priori-Mars occurred again and again, for almost three thousand years and that repetition was why the events became indelibly etched in the minds and souls of one hundred generations of humans as evidenced by the well established themes in the myths of many cultures.

The tidal disruptions on priori-Mars repeated the same sequence during each dance encounter: the orientation of the spin axis toward the Earth; the formation of the tidal sea facing the Earth; the emergence of the 'golden egg,' the first of the melted interior rocks pushed upward by magma from below to reach the surface in the midst of the waters; followed by the full eruption of seven great volcanos and rising of the lava fountain figures; and finally the periodic blasting of bodies of all sizes into orbit around the Sun or across the gap to the Earth along with great amounts of water, atmospheric gases and choking red dust.

These events were repeated to such a uniform degree that the Vedic hymns specifically divided each encounter into four 'kalpas' of gradually decreasing length, giving the progression of events from one to another in great detail. The different kalpas were characterized primarily by the nature of the bodies that were ejected into space from the interior of priori-Mars in each period. It is stated in many instances that the same progression of events was repeated during each encounter. As a result of these convulsions and the continuous tidal force of the Earth which they punctuated, priori-Mars deviated more and more from a spherical shape - forming a shape that resembled the head of a lion and the even more distorted shape like the head of a pharaoh, with its characteristic headdress extending down on either side of the head and the stylized beard. These distortions may have changed the moments of inertia of priori-Mars until the angular momentum of its spin and revolution around the Earth were incompatible, causing the dance encounter to become unstable just about the time of the next 'favorable' conjunction of proto-Venus, and the two planets disengaged. But the primary influence on the separation was the proximity of proto-Venus.

It seems at first consideration that such a massive body as priori-Mars rotating about the Earth once a day would have traumatic effects on the Earth, this was not the case. Although enormous ocean tides were raised on both bodies, their orbits and spins were oriented in such a way that the tides on both bodies remained stationary. Because priori-Mars was in a geosynchronous orbit the tide on the Earth did not sweep around the planet once per day, as do the tides due to the Sun and Moon. Also because the north pole of priori-Mars continually pointed toward the Earth, its tide remained stationary at the pole. This configuration, once established, was one of equilibrium, or minimum energy dissipation, and therefore relatively quiescent. Nature favors such configurations. The greatest trauma for the Earth occurred during the engagement and disengagement of the two planets, under the auspices of the nearby proto-Venus, at which times earthquakes were continuous. But this was not nearly as traumatic as the earlier close encounters of proto-Venus which literally turned the Earth over and destroyed most animal life.

When captured, priori-Mars continued to spin at approximately the same *rate* as before the encounter, so that its spin angular momentum was essentially conserved during the transition. However, once in the 'dance' configuration there was undoubtedly a weak coupling between the spins and orbits of the two planets which tended to make their spin rates equal. Whatever the original spin rate of priori-Mars, it was gradually induced to conform to that of the Earth, as evidenced by their very similar spin rates today, a fact dismissed by planetary scientists today as mere coincidence. We picture this coupling as being due to the attraction of secondary uplifted areas on both bodies for one another, such as mountain ranges and continents, similar to the meshing of two gears. It was not a strong coupling, but over a period consisting of approximately one hundred of these fourteen year long encounters its influence was manifested.

Flooding on Earth

The transitions to and from the dance encounters were accompanied by *continuous* earthquake activity because the tidal forces exerted on different tectonic plates tended to pull or push the plates horizontally. In those days earthquakes were taken for granted. The most notable effect, particularly in the Mediterranean, the Middle East and the Indus Valley (where a large fraction of the ancient population lived), was tidal flooding. priori-Mars' mass was more than one tenth that of the Earth and its center approached to within 45,000 kilometers of the center of the Earth. As a result, it produced an enormous ocean tide, orders of magnitude greater than the Sun and the Moon. However, unlike the tides we experience today which sweep around the globe as it rotates, priori-Mars' tide remained fixed. This radically changed the sea level around the Earth, raising it in areas near to northern India and lowering it in a peripheral, circular band centered on the Himalayas. As a result the northern part of the Indian continent up to the Himalayan foothills was inundated. At the same time the ocean depth was greatly reduced in a band which included the Arctic Sea north of Asia, the Bering Sea, the Western Pacific, Indonesia, northern Australia, the east African Coast and the Mediterranean Sea.

The flow of the waters to their new tidal levels was, of course, constrained by the local topography. At the onset of each dance encounter the water of the Mediterranean was drawn across the northern Sahara and flooded what is the center of Egypt, Cairo and the Giza plateau, as deep as 50 meters. The Mediterranean also exited through the Red Sea, the Negev, Palestine and further north, through the Gesreel Valley to Mesopotamia and down through the Tigris and Euphrates valley to the sea. This adjustment took, at most, a few days, and the new coastlines that resulted from the presence of priori-Mars remained stationary for the duration of each encounter. Of course, any low-lying areas along these courses remained inundated for months or years after each flood.

During the dance encounters, the Mediterranean and Red Seas were almost completely emptied, and the residual water in them evaporated during the fourteen year encounters. At the same time, all of northern India right up to the foothills of the Himalayas was inundated by some 5 kilometers of sea water.

Because of the lower sea levels on the Earth at the beginning of the Vedic period, the Mediterranean Sea was actually three large clear water lakes, which were not linked to the Atlantic, and the Aegean 'sea' was dry land. As a result, the flooding which punctuated the dance encounters was a great benefit to the surrounding cultures. Each flood supplied vital irrigation to arid areas that lasted for years. This was the origin of the 'Fertile Crescent.'

The flooding in our scenario was more complex than the mere shuttling back and forth of the waters due to the arrival and departure of priori-Mars. Throughout the Vedic period, there was a continual increase in sea level on Earth due to the influx of water from priori-Mars.

The priori-Mars-Forming of Earth

One of the most irresponsible notions advanced recently in the popular space literature is that of 'terra forming' priori-Mars, that is, modifying the priori-Mars system to be more earth-like. The very notion that mankind could modify the environment of an entire planet hundreds of millions of kilometers from Earth is ludicrous, considering that we have been unable to modify the great deserts of the Earth, some of which are adjacent to oceans. What these writers, indeed all the scientists on the Earth, do not understand is that the Earth itself was just revitalized, by the infusion, on a planetary scale, of all the volatiles on priori-Mars, during the three millennia of the Vedic period. As a result, priori-Mars is bone dry, in spite of the land forms left by the massive tidal flows that occurred only a few thousand years ago. Scientists, who believe that these massive flows occurred billions of years ago, still ponder the question of where all the water went. Ironically, it makes up a part of every living thing on this earth. Unfortunately, those proposing to terraform priori-Mars are counting on using Martian water, which they believe resides in the ice caps and the permafrost, just below the surface. *The puzzle of "where the water went?" was intensified in late 1998 when the NASA delivered priori-Mars Orbital Laser Altimeter revealed that the northern ice cap contains only one tenth of the calculated minimum volume of water required to carve the great outflow channels on its surface.* [140]

As priori-Mars orbited the Earth once per day, the pair would periodically pass through alignments with the Moon. This could have happened as frequently as every fifteen days. When the alignments were precise, and in particular when they involved both the Sun and the Moon, they produced great convulsions in the interior of priori-Mars, and caused the ejection of large bodies of rock from the interior through numerous volcanic vents and calderas. Often a series of bodies was launched in rapid succession from deeper and deeper in the planet through the same vent at roughly the same speed. These took nearly the same trajectories in space, forming streams or families of asteroids/meteoroids, many of which still orbit the Sun today. The explosive launching of these bodies often carried with them great volumes of dust and rocks from overlying layers on priori-Mars, Depending on the velocity and direction of the ejected material it ended up on Earth, on the Moon, on Mercury or in orbit around the Sun. The amount lost from priori-Mars to space, by all the combined mechanisms, could easily have been one quarter of its original mass.

The total velocities attained by the ejected material were surprisingly high due to several factors. The tidal force of the Earth acted upon the ejected bodies, negating much of the Martian gravity, so that they rose higher and faster than normal. To the speed of ejection from the vents were added components due to the revolution of the Earth-Mars system around the Sun, the revolution of priori-Mars around the Earth, and the spin of priori-Mars. It is also likely that some ejected bodies received an additional 'slingshot' boost by looping completely around the Earth. Bodies up to 20 kilometers in diameter were launched in this way.

These monumental expulsions carried with them atmospheric gases and water. During each event a portion of the tidal ocean on priori-Mars was evaporated and drawn, along with some atmospheric gases, across space to the Earth. The water fell to Earth primarily in Tibet and the Himalayas and, over time, flooded the high mountain valleys. When the water became too deep to be constrained, it broke through the Himalayas in great catastrophic floods, rapidly carving the four great rivers which originate in the vicinity of Mt. Kailas. Today, kilometer thick layers of ocean sediments, salt and possibly fossils from priori-Mars lay in the Tibetan basins, unsuspected by geologists.

The volatiles on priori-Mars became more and more depleted throughout the Vedic period. As the surface pressure decreased and the temperature increased, more and more water evaporated and in turn was carried across the gap to the Earth. The net result was the nearly complete depletion of the Martian volatiles and their transfer to the Earth. We estimate that the ocean level of the Earth was raised by over half a kilometer during the Vedic period, due to the influx of the entire complement of water originally on priori-Mars at the beginning of the Vedic period. Thus the history of the sea levels on the Earth during the Vedic period was much more complex than currently imagined, consisting of a cyclic change with a period of twenty-eight years, combined with an increase of about 500 meters during the same period.

Since priori-Mars was covered with vegetation at the beginning of the Vedic period, we assume that it must have had vast amounts of nitrogen, oxygen and carbon dioxide. Its entire complement of these precious gases was transferred to the Earth, in addition to its water. The influx of such massive amounts of life-giving materials transformed the original sickly planet Earth into the robust and beautiful planet that is our 'mother' today.

But priori-Mars was devastated in other ways as well. The tidal force of the Earth decreased its internal pressure, causing subsurface rocks to melt. The same force exerted a continual upward pull on the liquified rock, first causing the magma to rise to the surface, then through the tidal sea. Periodic convulsive events, caused by alignments of the rapidly revolving Earth-Mars pair with the Moon, accelerated the process, and lava from vents directly facing the Earth shot higher and higher into the air towards the Earth, forming an enormous fountain. This was at the position on priori-Mars at which the gravitational pull of priori-Mars was partially negated by the Earth. The lava shot many kilometers into the air and, the higher it went, the less was the gravitational pull of priori-Mars. Then another factor came into play. The outer portions of the lava streams hardened because they were exposed to the cold of outer space, and formed tubes or pipes through which the hotter lava in the interior continued to flow. Lava on the Earth does the same thing, the only difference being that the pipes are laying on the ground. On priori-Mars they were strong enough to support themselves in the vertical direction because of the weakness of the net gravitational field. This enormous lava fountain rose innumerable times to an altitude we estimate at a thousand kilometers, and was by far the most imposing feature visible on priori-Mars - a massive column of smoke and fire.

At the times of the convulsions due to the Sun and Moon, large bodies, ejected from the interior, crashed right through this framework of lava tubes and shattered it into millions of pieces which fell to the surface. The growth and destruction of the lava tube feature was repeated many times during each encounter. The net result of the repetition of this process was the pulling the planet 'inside out,' that is, the denser rock from the interior was brought to the surface while the lighter surface rock was buried beneath them. The northern plains of priori-Mars are strewn with the pieces of these shattered lava tube forms, as can be seen in closeup photographs of the surface taken by the NASA Viking landers.

The Break off of the Dance Encounters

As a result of the convulsions, deeper and deeper layers of rock were brought to the surface. They were exposed in concentric circles around the central axis. The planet became more and more distorted from its original spherical shape. Eventually, the liquid iron from the outer core worked its way to the surface and actually flowed out into space above the planet, glowing and forming bizarre shapes. *Toward the end of each encounter priori-Mars became even more distorted, a great gash opened up in its side and the glowing solid iron core was exposed. It looked like an eye and the gash was like the eye socket. At the end of some, or perhaps all, of the encounters, the solid core of priori-Mars actually came out of the planet, with the two parts moving into slightly different orbits with their center of gravity remaining in the original orbit. This separation of the planet was probably the result of the approach of proto-Venus, and these events were part of the termination of the dance encounter.* After a few orbits the two parts of priori-Mars rejoined one another as they drifted away from the Earth. The solid core reentered the lithosphere of priori-Mars once it moved outside the gravitational influence of the Earth. During the ensuing fourteen years, priori-Mars reformed into a relatively normal, spherical shape. Thus when the next encounter occurred, it proceeded in the same manner as all previous captures. Although the details are not clear, the distortion of priori-Mars, combined with the drawing out of its core, resulted in the breaking off of the encounters, under the influence of proto- proto-Venus which was nearby at the time.

The Final Breakaway

The end of the Vedic period was signaled by the breakaway of priori-Mars from the last dance encounter. Based on dates of the death of Romulus, we estimate that this occurred in 687 B.C. At that time something different happened that caused priori-Mars' orbit to slide outward some fifty million kilometers from the Earth, never to return to its vicinity. Specialists in astrodynamics will initially voice disbelief at the notion of this kind of orbital change, claiming that priori-Mars' new orbit would still have to intersect that of the Earth. But once they understand the proposed scenario they will undoubtedly acknowledge its validity. Although the core of priori-Mars had rejoined the lithosphere on all previous occasions, something different happened on the final one. The change in the length of the month just at that time, from 36 to 29.5 days, implies that the Moon may have interacted with one part of priori-Mars while it was separated. *Whatever the details, the iron core of priori-Mars was deflected in a completely different direction from its lithosphere. The result was that the lithosphere drifted outward in the solar system and the core drifted inward, in such a way that conservation of energy and angular momentum was maintained. Where did it go? It went into orbit around the Sun. We now call it the planet Mercury.* At the same time the orbit of proto-Venus probably interacted with the core as it moved sunward, because proto-Venus' orbit is now circular, and a resonance between it and Mercury has been documented. Hopefully more sophisticated computer modeling will allow us to determine the physical details given the scenario presented here as a starting point. No models yet exist because no one has ever come close to understanding that these amazing cosmic interactions ever occurred.

The beautiful Earth we know today did not exist twelve thousand years ago. That ancient Earth was a much more hostile and dangerous place. But beginning with the stabilization of the Earth at the time of the Upper Dryas, followed by the cosmic encounters during the Vedic period, the Earth was surely prepared for mankind. The primary creation was obviously started billions of years ago. But the preparation of the Earth for mankind was made only 'yesterday.' What a profound and exciting discovery!

14
EVIDENCE OF THE PROTO-VENUS ENCOUNTERS

The intuitive mind is a sacred gift and the rational mind is a faithful servant.
We have created a society that honors the servant and has forgotten the gift.
Albert Einstein

In this chapter we present physical and historical evidence of four overturnings of the of the earth. We argue that the first two of these were part of the process in which the Moon was captured by the earth, around 10,000 B.C. and the second two involved close encounters with proto-Venus, at the beginning of the Vedic period, around 3300 B.C.(corrected date 4000 B.C.). Because the majority of the archaeological dates are based on the radiocarbon 14 dating method we give those dates in this chapter with the corrected dates in parenthesis. These encounters destroyed most of the animal life on the Earth at the time, causing the extinction of many species. However, the long-term benefits far outweighed the deficits. They prepared the Earth for the great benefits that were to follow during the Vedic period.

Evidence of the Capture of the Moon

Current geology textbooks identify four major glaciations in North America in the last two million years. It has been suggested by Milutin Milankovitch that the ebb and flow of these ice ages was due to long term, rather subtle changes in the orbit and spin axis of the Earth. These changes are such that they could not have had much effect on the total solar energy reaching the Earth, but could have affected the degree of contrast in the seasons. For instance, if the tilt of the spin axis and the eccentricity of the Earth's orbit combine to create mild winters in the mid and northern latitudes, more snow will fall. Similarly, cooler summers result in less melting of snow. These are the conditions under which, when they occur for a number of consecutive years, glaciers form.

Based on the Greenland ice cores, the last great ice age, the Wisconsinan, ended over 12,000 years ago. These cores, collected by both European and American teams, comprise a continuous record going back over 40,000 years. They are hundreds of meters long and allow the counting of individual years at the top of the column, due to variations in the ice from winter to summer. At deeper levels the time resolution degrades slightly due to the compression caused by weight of the overlying ice layers. These cores show details of a number of seven degree Celsius higher temperature periods called interstadials during the last ice age. See Figure 14.1. The inferred warming appears to occur very rapidly while the return to glacial conditions is more gradual. The periods between episodes have now been correlated with layers of material on the ocean bottom thought to have been deposited by icebergs which broke off the continental glaciers during the interstadials. These warming and cooling cycles occur much too frequently and rapidly to be correlated with the Milankovitch season variations, the shortest cycle of which is 23,000 years. Of

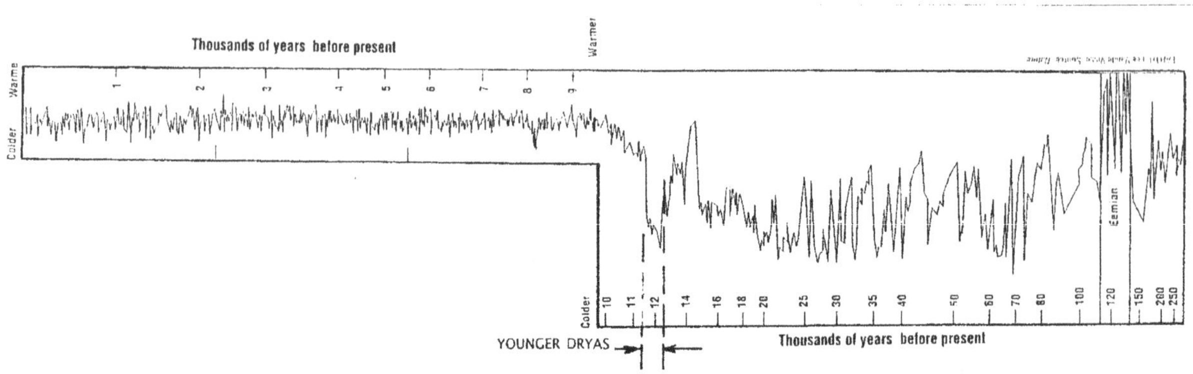

Figure 14.1 Temperature changes from the European Greenland Ice Core Project (GRIP) as a function of time before the present (BP), showing a dramatic change in the climate at the end of the Upper Dryas.

particular interest to our investigation is the fact that the nature of the temperature variations changes completely, after the period labeled the Younger Dryas. The extreme excursions disappear and a smooth gradual warming trend is established which extends to this day. The claim made by some geo-scientists, that we are still in the last ice age and just experiencing another interstadial, seems highly unlikely.

Although not specifically discussed in *Worlds in Collision*, Velikovsky uncovered some interesting clues from very early times which are of renewed interest due to recent data on the Younger Dryas. In discussing the historical evidence related to overturning of the Earth, Velikovsky relates how Egyptian priests told Herodotus around 475 B.C. that their kingdom was 341 generations old. Based on Herodotus' estimate of thirty years per generation, this meant that their culture's 'memory' extended back some 12,770 years before the present. The priests also told him that in the entire history of their kingdom, *the Earth had overturned four times.* Mexican hieroglyphics also describe 'four motions' and 'four prehistoric suns' or 'world ages'

Given the two devastating encounters between proto-Venus and the Earth proposed in the V/A scenario, we infer that another pair of overturnings occurred at a more ancient time within the Egyptian cultural memory. There is convincing evidence from the Greenland ice cores which points to the occurrence of two global traumas within this period. Figure 14.2 focuses specifically on the data from the Younger Dryas, which indicates that two extremely rapid geological changes occurred at that time. This plot shows the approximate date marking the 'memory' of the ancient Egyptians, based on 341 generations of 30 years duration. The correspondence of these dates implies that the two ancient overturnings, referred to by the Egyptians, caused the rapid temperature changes which characterize the younger Dryas.

The only mechanism capable of overturning the lithosphere is the close passage of a massive body. Based on the V/A scenario, the earlier pair of events could not have been caused by proto-Venus because its 'birth' out of Jupiter occurred only about 5,300 years before the present. What then was the causative agent? Again Velikovsky provides some vital information.[141] In an essay published in an assembly of topics concerning his work, he cites Greek references to an ancient people who lived in Arcadia in Greece, before the Hellenes, who claimed that they existed before the Moon circled the Earth. He also cites the oral traditions of the Indians of the Bogota highlands which reminisce about a time before the Moon. Moreover, the Egyptian deity, which we interpret to be the Moon, Isis, literally means '*ancient*,' implying that it is old, but that their ancestors witnessed its birth. Lastly, Velikovsky suggests that the following Scriptures imply that the Moon was not always present in the heavens.

> In Job 25:5 the grandeur of the Lord who "makes peace in the heights" is praised and the time is mentioned "before (there was) a moon and it did not shine." Also in Psalm 72:5 it is said: "Thou wast feared since (the time of) the sun and before (the time of) the moon, a generation of generations. [142]

What is the significance of the Moon as seen by modern science? It is generally acknowledged that the effect of the Moon, in its current orbit, is to give the Earth impunity from the buildup of small perturbations, due to the other planets, that might otherwise cause large spin axis changes over long periods of time. In other words, the Moon stabilizes the Earth against the climatic extremes such as the ice ages that occurred in the past.

The first overturning, which initiated the Younger Dryas, was probably caused by a very close encounter with the Moon. The second one, which marks the termination of the Younger Dryas, marked the capture of the Moon into its current orbit, ending a sequence of chaotic interactions that occurred when it was in an earth crossing orbit. Under this hypothesis, the closely spaced spikes in the atmospheric calcium data, shown in Figure 14.2, which appear to have a period of approximately 67 years, may be indicative of intermediate encounters, which were not so traumatic. The forces on the Earth each time the Moon passed could have caused earthquakes, volcanic eruptions and floods.

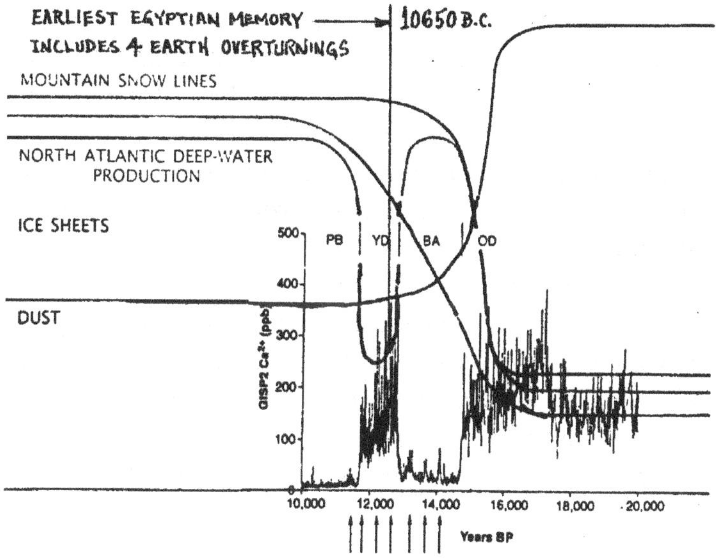

Figure 14.2 Superimposed on the ice core data showing the rapid temperature changes of the Upper Dryas is the date within which the Egyptians 'remembered' four Earth overturnings, assuming 341 generations of 30 years.

The overturnings caused world-wide destruction because the oceans literally swept across continents killing almost all animal life. This is attested to by the many species of animals that became extinct at that time. Velikovsky points out large deposits of millions of broken animal bones of many species, extinct and extant, which were all buried together at that date, as receding flood waters deposited and buried them in silt or clay.

Experts in astrodynamics believe that one body cannot capture another, unless there is a third body involved to carry off the excess orbital energy lost in the capture process. This theory is based on the assumption that the bodies involved have no internal structure. But the two overturnings and the hundreds of 30 degree shifts of the spin axis due to the capture and release of priori-Mars show that the 'slipping' of the lithosphere relative to the core provides an energy 'sink,' making possible capture without the need for a third body.

To summarize, we suggest that the onset and termination of the millennium-long Younger Dryas, was marked by two overturnings of the lithosphere, which resulted from the process by which the Moon was captured in its final orbit around the Earth. The reason for the chaotic climatic changes prior to that time - the ice ages and interstadials - was the lack of the stabilization of the Earth afforded by the Moon.

Evidence of Pre-Vedic Earth

We claim that sea level on the Earth was more than 500 meters below its current level prior to 3300 B.C. (corrected date 4000 B.C.). Strong evidence for lower ocean levels in the past exist around the world. There are numerous instances of intact submerged barrier islands, similar to the ones along current coastlines, located out as far as the edges of the continental shelves. These are particularly significant in that they have little structural integrity and would be swept away by ocean currents in a relatively short time, geologically speaking. Submerged canyons are also found on continental shelves, in many cases extending from the present coastlines to the margin of the shelf. Some examples are Monterey Canyon off the coast of California, the west coast of Florida, and Baltimore Canyon off the middle Atlantic coast of the United States. They have all the characteristics of subaerial (above water) erosion, implying that they were above sea level when the erosion occurred. Exotic mechanisms such as 'turbidity currents' and 'freshwater sapping' have been invented by scientists, desperate to explain these canyons, because they cannot understand how the continental shelves could have been above sea level in the geologically recent past.

Additional evidence for low sea levels five thousand years ago can be found in a completely different discipline. Many living coral reefs have been discovered with bases at depths from hundreds to a thousand meters. Since reefs are formed by living organisms that require sunlight, they cannot grow if their top is deeper than 45 meters. Therefore, these reefs could not have begun growing at the current depths of their bases. Generally there are two explanations offered for these formations. Either the sea level was lower when they first formed and has risen slowly since that time, or the sea bottom on which they are located has gradually subsided since they began growing. The latter is less likely because it requires subsidence of a number of surfaces at different locations around the globe at the same slow rate, whereas a raising of the ocean level would apply to all parts of the world.

Geologists currently have only one explanation of recent sea level increases are currently explained as the result of melting ice. When the last ice age ended and the glaciers melted, the water is thought to have flowed back into the oceans causing a sea level increase. One problem with this explanation is, in order that the resulting rise in sea level not exceed the maximum growth rate of the reefs, a few inches per year, *the melting period is hypothesized to have taken over a thousand years.* This is a highly questionable conclusion.

Even if one accepts this extremely slow melting process, the maximum sea level increase from all the glaciers could not have been larger than 140 meters - far from sufficient to account for the many reefs whose bases lie at depths as great as 500 meters and the submerged canyons which extend to the limits of the continental shelves. Modern scientists cannot conceive of any mechanism whereby the sea level could have been more than 140 meters below the current level in geologically recent times.

We claim that the rise in sea level was due to the influx of water from priori-Mars during the 2800 year Vedic period. This explanation fits the observations particularly well, because the average rate of accumulation would have been (500 meters/3000 years) about 17 centimeters per year, which is just within the rate at which reefs can grow. Any other mechanism such as the subsidence of the sea floor, usually invoked to explain these reefs, would have to honor this rate, the world around - a highly unlikely prospect.

The Himalayan-Tibetan Complex

The vast uplifted area comprising the Himalayas and the Tibetan plateau is shown in Figure 14.3 in relation to other peaks in the world. Not only is this the highest and most massive range, but it is also the youngest. The geology of this region and the surrounding plates in western China (Tarim and Djungar) is extremely complex. There are far more questions than answers for the many puzzles uncovered by geologists.

With the acceptance of continental drift as a valid picture of the evolution of the Earth's surface, it has become a widely accepted idea that the uplifting of the Himalayan complex is tied to the collision of the Indian and Eurasian plates. This collision is estimated to have begun some fifty million years ago. One great paradox in this scenario is that detailed studies of the high Himalayas indicate that the *uplifting* occurred so recently that geologists estimate less than two million years ago. This is 'yesterday' on a geologic scale, and could be a few thousand years ago. The following quote illustrates that even a geologist has misgivings about the currently accepted mountain building process, orogeny, as an explanation for the uplifting of the vast Tibetan-Himalayan complex.

> It is quite astonishing that the last upthrust not only affected the Himalayas, the Transhimalayas and the Karakorum but also the whole of the Tibetan region. With an area of 2.5 million square kilometers (1 million square miles) this region is the highest landmass on Earth, and in the last million years it has risen by between 4,000 and 5,000 meters (13,200 and 16,500 feet), an average of 4 to 5 millimeters ... per year. Today the rate is close to one centimeter per year. [143]

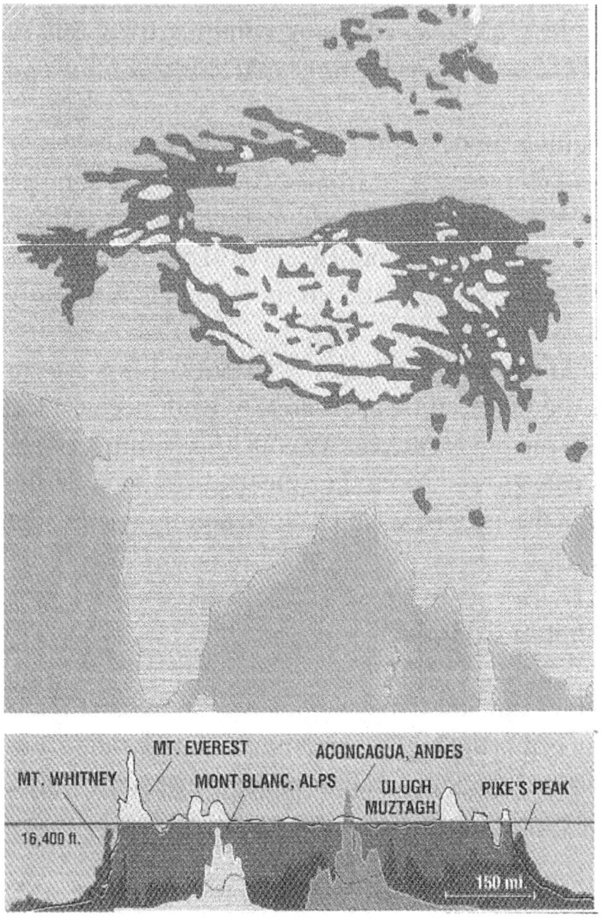

Figure 14.3 Schematic depiction of the enormous height and extent of the Himalayan-Tibetan complex. (Reproduced from Earth magazine, July 1993, p. 29. Source: Peter Molnar)

Geologists estimate the current rate to be about one centimeter per year, apparently driven by the continuing collision of the Indian and Eurasian plates. It apparently is only a guess, however, since recent remeasurements of the altitudes of a number of major peaks indicated apparent errors in the previously accepted heights of the order of ten meters. If the error in the height measurement is this large it is certainly impossible to measure a growth rate of one centimeter per year. The mind set of geologists is such that they cannot conceive the untrusting of any mountain range to occur in less than two million years. This notion is based on the current frequency of earthquakes. They picture a mountain range being uplifted, say, thirty centimeters (one foot) per century, requiring about two million years to attain a height of 7,000 meters. It is on such a basis that they estimate the Himalayas untrusting began two million years ago. It may very well be that they are currently rising at this rate, but that does not preclude the process having been initiated in recent millennia due to catastrophic events, which are not accepted in the geologists 'world view.' Recent studies have detected a layer of partially molten rock perhaps 5 kilometers thick in the crust below southern Tibet. [144] Although the extent of the melting has not yet been determined, it is generally thought that this layer is present throughout the entire Tibetan plateau, because of the common incidence of hot springs throughout the region. Indeed the very reason that it is a plateau may be due to its being decoupled from the underlying geology by the liquid layer. Although geologists and geophysicists believe the partial melting is merely a result of the collision of the Indian and Asian plates they are presently unable to explain it quantitatively without assuming that the crust of Asia was already molten at the onset of the collision - a highly unlikely scenario.

The lack of up-thrusting in the first forty-eight million years is consistent with what has been observed in other parts of the world where continents have collided and no untrusting has resulted. We claim that it was the tidal force of planet-sized bodies passing close to the Earth, that caused the uplifting. We believe this began at the time of the Younger Dryas (about 10000 B.C., corrected date 17000 B.C.?), and not two million years ago. This uplifting was radically increased by the close encounters with proto-Venus at the beginning of the Vedic period. Recalling the *Rig Veda*:

..in those days the earth was unstable.. the mountains had wings.

When thunderbolts were exchanged with proto-Venus, entire mountain ranges disappeared on one hand and sprung up on the opposite hand.

Scorching of the Earth

A few years after its creation, as a result of a great impact on Jupiter, the flaming proto-Venus had two close encounters with the Earth. The destruction caused at these times took two forms. The most immediate was the scorching of a wide swath that extended across Africa, the middle east and the entire Asian continent. We claim that the effect of this scorching was to kill most of the organic material along its trajectory to such an extent that its path is still obvious to this day. Figure 14.4 shows a photometrically accurate rendition of the entire Earth, assembled by the GeoSphere Project, a commercial venture in Santa Monica, California, from thousands of Tiros-N photographs. From this cloud-free vantage it is clear that the scorching passage of proto-Venus progressed across northern Africa, the middle east, and passed north of the Himalayan complex. It created the great Sahara desert near its point of closest approach. Note from the figure how the description of the destruction caused by the legendary Phaëthon matches the path along which the burning took place.

In a recent study of the North African paleoclimate using computer models, the investigators simulated grasslands, a more loamy soil, and increased solar flux due to orbital differences at that date. Even with all of these changes, the model failed to produce the wetter conditions known to have existed there before the beginning of the Bronze Age:

> Fossil, pollen, ancient lake sediments and archaeological evidence from Africa indicate that the Sahel and Sahara regions were considerably wetter than today during the early to middle Holocene period, about 12,000 to 5,000 years ago. Vegetation associated with the modern Sahara/Sahel boundary was about 5 degrees further north, and there were more and larger lakes between 15 and 30 degrees N. There is evidence of [a more loamy soil] from several locations in the Sahara. However these simulations underestimated the consequent monsoon enhancement as inferred from the paleorecords ... None of the simulations produce vegetation in northwest Mali and northeast Sudan as indicated by observations.[145]

Obviously, a more radical influence than those assumed in this model must have occurred at the beginning of the Bronze Age, in order to effect such a major change over an area as large as North Africa. The investigators fail to recognize the magnitude of the traumatic event that was required in order to cause permanent changes in vegetation on half of an entire continent. As we will see below, this penchant for underestimating the cause of profound changes, applies to other

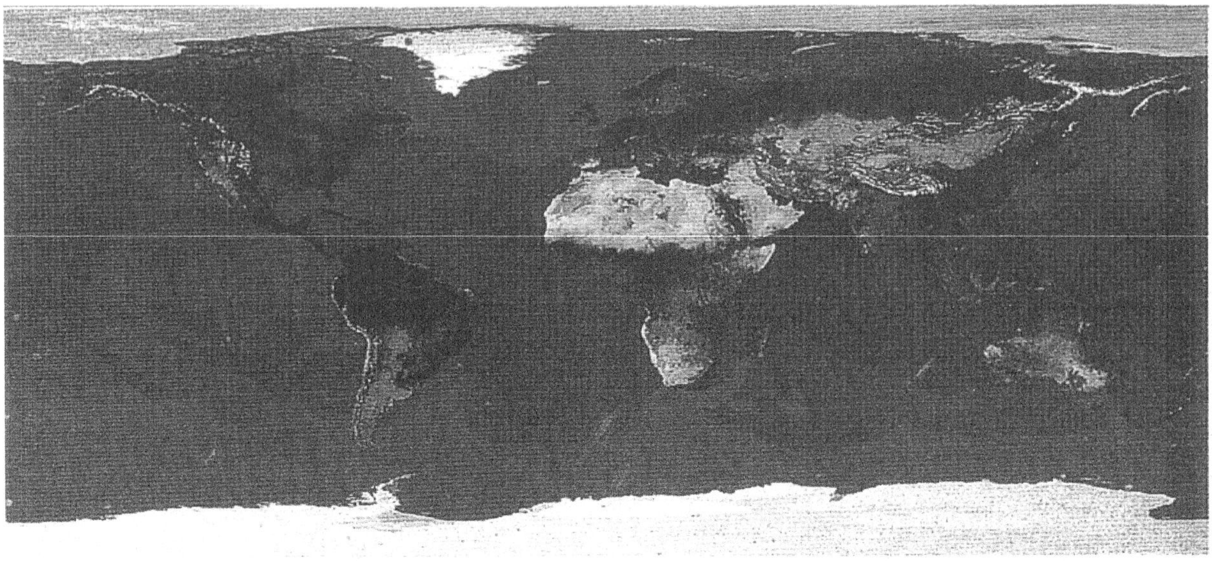

Figure 14.4 A photometrically accurate rendition of the surface of the Earth showing clearly the scorched path followed by the flaming proto-Venus. (Credit The GeoSphere Project.)

disciplines also, and stems from the uniformitarian background on which all the earth and space sciences are based.

Prior to 4000 B.C. northern Africa was a steppe environment inhabited by rhinoceros, elephants, and giraffes. Murals, reliefs, sculptures, and hieroglyphics reflecting an abundance of wildlife are the subject of *An Egyptian Bestiary: Animals in Egyptian Art*, at the Metropolitan Museum of Art in New York City. By the beginning of the Middle Kingdom, all of these species are known to have disappeared. This change is reflected in ancient Egyptian art:

> One of the most striking examples of the change in Egypt's animal life is a late fourth-millennium B.C. elephant-ivory hair comb adorned with a giraffe ... By the beginning of the pharaonic period in the third millennium B.C. neither giraffes nor elephants roamed the Egyptian landscape. [146]

We claim that the delta area and the coast of the southern Mediterranean were scorched by the close passage of the flaming proto-Venus and all organic material including animals, soil and green trees were incinerated, leaving no trace. The dates of the changes in climate and animal species cited above confirm our scenario.

The profound nature and scope of the changes that occurred around 4000 B.C. have already been recognized in a general way by the science of Archaeology, which has designated this date as the end of the Chalcolithic (chalco = copper, lithic = stone) and the beginning of the Bronze Age. This special recognition, the beginning of a new age, is based on a number of major changes in the archaeological record at that date. Three hundred years later began the blossoming of great city-states in the Levant and Mesopotamia, and the beginning of the Pharoanic dynasties of Egypt. This three hundred year period, named the Early Bronze I (EB I), or 'pre-urban' period, is mysteriously lacking archaeological artifacts of any kind. As a result, it is also referred to as the 'silent' period. Unfortunately, archaeologists fail to realize the profound implications of such a long period with no sign of life.

Some two hundred years after the proto-Venus destructions King Narmer, of Upper Egypt, which constituted the portion south of the Nile delta, is thought to have led his armies into the North and assumed control of both provinces. Narmer then established his ruling capital at Memphis, on the border between Upper and Lower Egypt. This period is called the *Predynastic* in Egypt, indicating that it occurred just prior to the beginning of the Pharaonic dynasties. It corresponds to the Pre-urban period discussed above. It is quite possible that King Narmer easily moved into devastated Lower Egypt, and established his capital at Memphis because of its central location on the border between the two Egypts since much of Lower Egypt was destroyed, and no arable land remained there.

In the Levant (the area currently occupied by Syria, Lebanon, Israel, and Palestine) we see the sudden unexplained disappearance of cultures more than a millennia old at this date. The Ghassulian culture (the primary culture of the Chalcolithic Period in the Levant) was characterized by the location of its sites, the large quantities of pottery, and well made copper art objects (436 pieces were found in a single cache in Nahal Mishmar). Most of the sites such as Beer-sheba, Teleilat Ghassul, En-Gedi (Temple) and the Golan Heights are located, not in the more fertile Mediterranean climate zone, but in the (now) semi-arid zones of the northern Negev and the Jordan Valley. These locations are thought to have been only suited for animal herding as opposed to agriculture, although the people obviously grew olive trees for oil in the Golan Heights. Some studies indicate that rainfall may have been greater than at present in these areas. Butter churns are also found at these sites. The development of their technology was sufficiently advanced that they imported arsenic ore from the north in order to facilitate the casting of copper. The end of the Ghassulian culture, which was at least thousands of years old, appears to have happened suddenly as evidenced in the following quote:

> Around 3300 B.C., the Ghassulian culture came to an end under enigmatic circumstances. The most important centers were abandoned, and they remained unoccupied in subsequent times. This is the case at Teleilat Ghassul, at the Beer-sheba sites, in the Judean desert, and on the Golan Heights. The abandonment of the En Gedi temple and the blocking of its entrance, as well as the hiding of the metal treasure in the Nahal Mishmar cave, imply some traumatic event. But no evidence of violence has been found in the last Chalcolithic occupation levels, and the reason for the desertion remains to be determined. [147]

We claim the proto-Venus encounters destroyed this civilization. Archaeologists feel that it was caused by a 'traumatic' event but have no idea that close cosmic encounters even happened.

The Early Bronze Age (EB I) corresponds to the pre-urban or 'silent' period. The settlement pattern of the EB I sites is unusual in several respects. Although there were unfortified agricultural villages as in the Chalcolithic, they were located at lower elevations where the climate was warmer. The sites settled most densely in the earlier age - Golan Heights, the Judean Desert, Teleilat Ghassul and Beer-sheba remained almost totally deserted. Only 30% of the Chalcolithic sites were occupied at all in the EB I period. In the EB I, building outlines became curvilinear, and a unique type of pottery called *Gray Burnished Ware* was introduced which appeared at no other time in the proto-history of the Middle East. Although locally made, the introduction of this foreign material has been interpreted by some as implying the immigration of people from Eastern Anatolia. The only public buildings found were Temples. At Megiddo there are ruins of a temple with two broad rooms each with a raised pedestal for the statue of a deity. The only art works of the EB I are engravings on the walls of the temple courtyard. In a vast cemetery at Bal edh-Dhra', two phases of EB I burials were distinguished. The so-called shaft tombs - caves approached through a vertical shaft, number in the thousands. Since there was no settlement nearby these graves are thought to be those of pastoral nomads (wandering farmers), but this is inconsistent with the large number. In the following EB II period, circular burial structures made out of mud brick were used. These were completely different from the shaft tombs.

There are differing opinions among archaeologists as to the many differences between the Chalcolithic and Early Bronze I period. Changes in settlement patterns, population size and density, agricultural methods, trade relations, and artistic forms are interpreted by some as evidence of a massive immigration from outside into Palestine, while others believe they are a result of more rapid development by indigenous cultures. Comparisons of the shapes of human bones from the two periods show little difference in the makeup of the people and thus favor the latter view.

Although the Sumerian culture is thought to have begun developing in Mesopotamia as early as 3500 B.C. (radiocarbon date), the famous urban center at Ur was not developed until about 3000 B.C. The latter date is consistent with the end of the 'silent' period in the Levant.

We claim that this 'silence' in the archaeological record was a direct result of great destruction caused by the two close encounters of the Earth with proto-Venus. These also the caused of the sudden end of the Ghassulian culture. Hundreds of thousands of people who could not find shelter were incinerated by the radiation from the incandescent proto-Venus and many more lost their lives in the firestorm which resulted when all organic material, living and dead, burst into flame. Those who survived had but a short respite before mile high tidal waves swept over much of the continental areas, drowning most of those still alive. The thousands of unique 'shaft tombs' were the quickest way of burying the dead. Many were in remote sites away from established burying grounds because the dead were buried where the piles of bodies were left by the swirling flow of the subsiding tidal waters. The bones show no anatomical difference between the Chalcolithic and Early Bronze I because the immediate survivors were of the same stock. However, this does rule out the influence of an outside culture in the ensuing centuries.

The few survivors of the Ghassulian culture, which was based on animal herding, are thought to have adopted agriculture and to do this, had to move down from the higher areas into the lower lying flooded areas, which had not been so badly scorched. There were only a very few survivors. This was why most of the densely populated Ghassulian sites were no longer occupied after that date. The building of shelters passed to the survivors who were desperate just to survive. Therefore they abandoned the building style they had used for a thousand years and made the simplest and most expedient shelters they could, with curvilinear walls and no corners.

The profound changes in Middle East climate at that time are just now being discovered by James A. Sauer (Harvard Semitic Museum) and Farouk El-Baz (Boston U.), who have discovered a river bed buried beneath the sand in Saudi Arabia. Based on analyses of pollen in boreholes from the Mediterranean coast of Israel and a dark paleosol layer ten feet thick in Yemen, Sauer concludes *the climate in the area was quite wet until at least 3500 B.C.(radiocarbon date)*

We propose that the unique Gray Burnished Ware, pottery produced only at that time in proto-history, was fabricated from the very clay deposited by the flood waters - the same clay deposit that Woolley found at the bottom of the burial pit at Ur and the same deposit described in the tablets at Nineveh, that left the landscape 'as flat as a roof.' Being uniquely from this period, it is quite possible that this clay contains clues to the close encounters with proto-Venus. It could contain unique chemical signature from the smoke in the 'tail' of proto-Venus which settled to the ground in the months following the encounters. To test this idea, it would be of great interest to do spectrographic analyses of the Gray Burnished Ware clay to see if there are unique trace elements such as sulfur, rare compounds (such as carbon monosulfide, CS), or non-terrestrial isotopic ratios that would indicate an extraterrestrial contamination of this material.

There were few public buildings in the settlements of the Early Bronze I period in the Levant because of shortage of labor. In spite of this the survivors felt it imperative to propitiate these omnipotent deities which suddenly had changed the Earth, and therefore they managed to build a temple. Since priori-Mars became a threat soon after the two destructive encounters, and proto-Venus continued to approach fairly close to the Earth, it makes sense that the people would have worshiped two idols at this time, one for priori-Mars and one for proto-Venus. This is why the temple had two rooms, each with a pedestal for the placement of idols honoring these two 'gods.' This parallels the dual deities Indra and Mitra-Varuna in the *Rig Veda*, and Baal and Ashtoreth of the Canaanites. The lack of art work in EB I was probably due to the killing of the artisans and their supply infrastructure by the great destructions.

Some three hundred years after the proto-Venus encounters, we find the people over a vast area entering into a stage of unprecedented urban development never attained in the Chalcolithic Age. For instance, the year 3113 B.C. (radiocarbon date) is the beginning date of the Mayan 'long count' calendar. We claim that the approximately three hundred year pre-urban period was the time necessary for the rebuilding of the populations. It would have required several hundred years to repopulate from only a few survivors to the tens of thousands of human souls required for the development of the urban centers, for the learning of trades and the organization of the people by the leaders of the time. This process evidently occurred throughout Mesopotamia, the Levant and Egypt. The recovery from the great destruction took at least ten generations. This was a short period archaeologically speaking, considering that the preceding Ghassulian culture had lasted for millennia. Starting with very few people, it is to be expected that the recovery would take at least this long, particularly in light of the dangers of floods and earthquakes, which continued to occur as a result of subsequent interactions with priori-Mars, and the probability of famine and disease against which primitive man had few defenses. *Although 300 years is a short time archaeologically speaking, it is a longer time than the United States has existed, and therefore emphasizes the large scope of the destruction which preceded it.*

Venus Flooding Evidence

Widespread devastation occurred as a result of the Earth being overturned by the two close passages of proto-Venus. These caused tidal waves kilometers high to sweep across entire continents, destroying most of the people and animals in their path. The second encounter, which probably occurred within a few tens of years, reversed this overturning, thereby restoring the orientation to the pre-Venus state, but killing many of the survivors of the first catastrophe. Thus the world population 'passed through the eye of a needle' at that time.

In March 1929, the British archaeologist C. Leonard Woolley was excavating a large burial pit at the site of the ancient Mesopotamian city Ur, called 'Ur of the Chaldees' in the Bible. At a depth of 40 feet the workers announced that they had reached virgin soil and there were no more artifacts. Woolley, believing there must be artifacts from earlier cultures, had his men continue digging. After digging through a layer of eight feet thickness, artifacts of a completely different, pre-Sumerian culture were found. This 'silent' layer was tentatively interpreted by Woolley as evidence of the great flood in the Bible, at which time the layer of clay separating the artifacts of the two cultures, had been deposited. *He dated this flood at 3200 B.C. (radiocarbon date), This provides an important corroboration of the V/A scenario.* However, the same layer of clay was not found everywhere in the area, and where present it varied in thickness. As a result he concluded that it was not from the universal flood associated with the story of Noah in the Bible.

The Mesopotamian flood myth quoted in Chapter 12 fits our scenario of the proto-Venus induced floods. A black cloud came from the foundations of heaven and the storm lasted a week. When the sea became calm, and the flood ceased all mankind had turned to clay. The ground was flat like a roof. The clay carried the ocean waters filled the local depressions, making the earth flat and therefore was not of the same thickness in all places - consistent with the findings at Ur.

We claim that this was evidence of an even more ancient flood, one which resulted from the overturning of the Earth by proto-Venus. This flood is not mentioned in the Bible. However, it set the stage on which the biblical narrative began. The desolation of the Earth in the wake of the proto-Venus encounters is described in the second verse of the first chapter of Genesis:

> 1. In the beginning God created the heaven and the Earth.
> 2. *And the Earth became without form, and void; and darkness was upon the face of the deep.*

The desolation was due to the enormous tidal waves that resulted from the overturning of the lithosphere by proto-Venus. These massive, transient floods did serve a purpose similar to that implied in the biblical story, in that they wiped the slate of Earth clean. What 'people' inherited the world is the subject of Chapter 17.

15

EVIDENCE OF THE PRIORI-MARS ENCOUNTERS - PART 1

Space and Time! now I see it is true, what I guess'd at,
What I guess'd when I loaf'd on the grass,
What I guess'd while I lay alone in my bed,
And again as I walk'd the beach under the paling stars of the morning,
My ties and my ballasts leave me, my elbows rest in sea-gaps,
I skirt sierras, my palms cover continents,
I am afoot with my vision.

Walt Whitman
Leaves of Grass

The evidence for the priori-Mars encounters, or 'dance encounters,' is diverse and voluminous. The clues from around the world are so many that we have broken them up into two parts. This chapter addresses the astronomical evidence and the geological data concerning North American glaciation, and the Himalayas.

The Astronomical Perspective

Scientists interested in the rotation of the Earth have concluded, based on studies of ancient eclipses of the Moon and star occultations, that the length of the day (LOD) was increasing much more rapidly during the last two thousand years, than during the last two hundred years.

> The astronomers who have been concerned with the reduction of observations all agree that the tidal effect during the last 200 years is half what it was during the last 2000 years. Such a change is difficult to interpret from a geophysical point of view. [148]

The V/A scenario provides a rationale for the greater lengthening of the days in past millennia. This was due to the increase in the mass at the surface of the Earth during the Vedic period, caused primarily by the influx of the entire complement of water from priori-Mars. This increased the moment of inertia of the Earth and consequently caused the slowing of its rotation. Although the LOD was increasing more rapidly during the Vedic period, when the water was actually falling, eclipse data from that period would have been invalid because the presence of priori-Mars changed the length of the day every fourteen or fifteen years. The oldest data used in these analyses (deSitter) were dated after the conclusion of the Vedic period (425 B.C.). If any accurate data existed from during the Vedic period, it would have been rejected as spurious because the presence of priori-Mars in orbit around the Earth completely changed the LOD.

Another indication of recent cosmic interactions is the Chandler Wobble. When priori-Mars was captured in orbit around the earth, it became gravitationally bound to the massive uplifted Tibetan-Himalayan complex. The angular momentum of priori-Mars, attached at this point, caused the spin axis of the lithosphere to shift to a point at ninety degrees to Mt. Kailas, where it remained for the duration of each dance encounter. When priori-Mars was released from each encounter, the tidal force of the Moon reoriented the lithosphere approximately to its current location. However, these reorientations introduced transients into the normal rotation of the lithosphere, which would probably take thousands of years to dampen out. We suggest that this unaccounted-for wobbling of the Earth's axis is just such a transient 'left over' from the last encounter with priori-Mars, which was only about 2,700 years ago..

The Earth-Mars Dance Encounters

Although the dance configuration we propose may seem unlikely, it was not unprecedented. *There are actually two astronomical precedents for this type of configuration between heavenly bodies*. The closest to home is Pluto and its moon Charon, which were discovered only recently to be in a similar 'dance' configuration. Figure 15.1 is a photograph of this system, taken by a land-based telescope and processed to remove the effects of atmospheric turbulence. Charon remains some 19,200 miles above the same spot on Pluto's surface. It is roughly 1190 kilometers in diameter, is about half the diameter of Pluto (2350 kilometers) and the ratio of the masses is about 12:1. The Earth-Mars mass ratio was about 6.2:1 at the time of the dance encounters. Scientists have determined that Charon revolves around Pluto at exactly the same rate that Pluto spins, one revolution taking about 6.33 Earth days. Because the resolution of these bodies is so poor, these determinations have been made by matching changes in the reflectivity of very large regions on both bodies. Using conventional reasoning, they infer that Charon's spin axis is roughly perpendicular to its orbital plane and that its keeps the same face toward Pluto by virtue of its spin period being equal to its orbital period, as in the Earth-Moon system. We question whether the reflectivity data is sufficient to determine the spin axis of Charon. If they are correct, then that system is not an exact precedent for the dance encounters, in which the north pole of priori-Mars was kept pointing toward the Earth by tidal torqueing of the

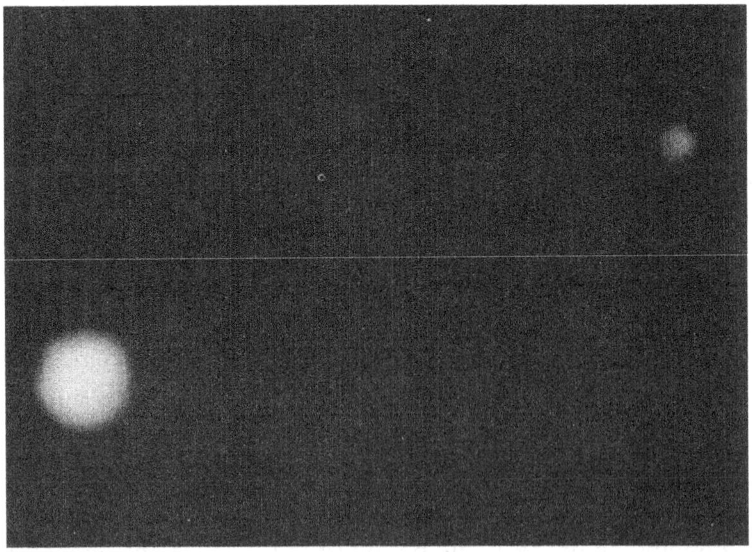

Figure 15.1 Pluto and Charon. A specially processed image.
(Courtesy of G. Weigelt, Physikalishes Institute der Universitat Erlangen-Nurnberg.)

uplifted Tharsis Bulge. The question of the exact configuration of the Pluto-Charon system has yet to be determined.

The second precedent is the common occurrence of binary stars, in which two gravitationally bound stars rotate about a common center of mass. What is particularly interesting about some close binaries is that mass is observed flowing from the less massive to the more massive body. We claim that in exactly the same manner material from priori-Mars flowed to the Earth during the dance encounters. The ancient books clearly indicate that the transfer was not continuous, but occurred at times of alignment of the pair with other heavenly bodies such as proto-Venus, the Moon and the Sun. Whether this is true with stars at much greater distances from one another is a subject for future studies.

The Importance of Hudson Bay

There is acknowledged paleomagnetic evidence that the poles of the Earth have wandered, perhaps hundreds of millions of years ago. Mythology, based on the observations of ancient peoples, not only tells us how such a change can occur, but that *many such changes occurred within the last six thousand years*. Prior to each dance encounter, Mt. Kailas, the point on the Earth over which priori-Mars remained, was at its current latitude, about 30° North. As priori-Mars was captured, the spin axis of the lithosphere shifted by approximately thirty degrees in order that Kailas would fall on the temporary equator. *This shift resulted in the north pole moving to the center of Hudson Bay in Canada.* The Sun and stars shifted to new paths through the heavens, the length of the day changed slightly, and transient flooding occurred as the waters moved to the position that they would occupy for the next fourteen years. These shifts are described in the Bible and in many other ancient texts, but modern scientists belittle them because they cannot conceive of a mechanism whereby they could have occurred.

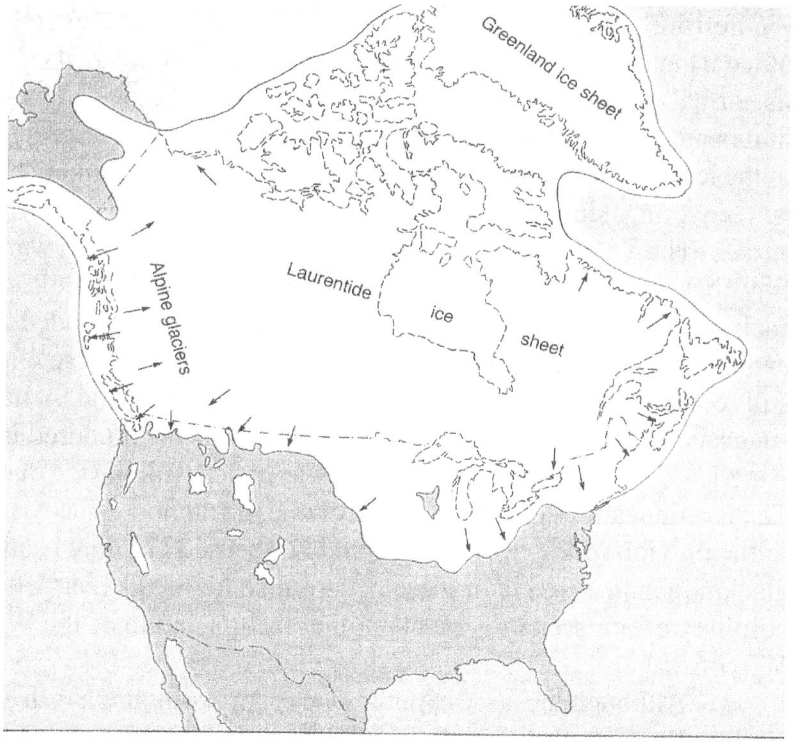

Figure 15.2 The maximum extent of the North American glacier, which was centered in Hudson Bay, not at the north pole. (From *The Earth*, E.J. Tarbuck, & F.K.Lutgens, Merrill Publishing Co.,1984.)

The recognition that the temporary pole was located at

Hudson Bay, during the dance encounters, provides a profound corroboration of our theory. Four Ice Ages have been identified by analysis of ice cores from the permanent glacier on Greenland, the last of which ended some 12,000 years ago. These glaciations are currently thought to coincide with evidence of glacial intrusion into North America that reached as far south as Wisconsin in the USA. In fact, the last of these, the Wisconsinan, was so named because it was thought that it corresponded to the glacial evidence in Wisconsin.

One problem with this connection is the acknowledged fact that the North American glaciation was centered, not at the north pole, but in Hudson Bay. Although the North American glaciers have long since melted, there is ample evidence of their extent and the repeated nature of their intrusions, although it is impossible to determine their number. There are several clues indicating that the glaciation in North Amarica continued much later than the end of the last Ice Age. The glacial deposits look remarkably fresh with only modest signs of weathering on their surface. The landscapes from this period are different from earlier ice ages. Evidence from the bottom of Seneca Lake in northern New York State, in the form of oxygen isotope ratios, indicates that this area remained cold much longer than indicated by the Greenland ice cores.[149] This is currently attributed to the effect of the melt waters from the glacier draining into the lake. The dates of this extended cool period are difficult to determine, but are estimated to extend several thousand years after the Younger Dryas.

Several clues point to the Hudson Bay area in Canada as a focus of glaciation, as shown in Figure 15.2. The bedrock in the Hudson Bay area was apparently so depressed by the overburden of ice which accumulated at the time of the north american glaciers, that it is still rising back toward its normal elevation to this day, as shown in Figure 15.3 [150]. The timing of the relaxation of the crust in this area is not consistent with the end of the last great ice age, 12,000 years ago. If the burden of the ice cap had been removed at that date it is highly unlikely that the crust of the Earth would still be recovering, since the typical isostatic recovery time is about 10,000 years.

The V/A scenario explains why the ice accumulation was centered at the location of Hudson Bay - because this was the location of the temporary North Pole of the Earth (the lithosphere) during each dance encounter. The fact that these encounters only ended some 2,687 years before the present explains the fresh appearance of the glacial evidence. It also explains why the Hudson Bay area is still rebounding today - because this time is well within the isostatic recovery time. Our theory also suggests that there were not four, but as many as one hundred incursions of the glacier into North America. Geologists have no possible way of determining the number because the ice has all melted. The terrain features that remain today belie the number, since each glaciation overlaid the evidence of the previous ones. Research should be focused on more accurate dating of the North American glaciation, since we claim these occurred some 6,000 years after the Wisconsonin ice age. As a corollary of our scenario, we claim that the glaciation of the Wisconsinan Ice Age never reached Wisconsin.

Although the V/A hypothesis explains why the North American glaciers were focused at Hudson Bay, it is not clear why the Greenland glacier, hundreds of meters thick, has remained intact while the enormous mass of the North American glacier has completely melted. One theory is that the sinking of the landmass in the Hudson Bay area brought the ice down to elevations at which the temperature was higher. Greenland was approximately the same distance from both pole positions, and probably remained colder as a result.

Because of Greenland's similar proximity to the current and the temporary north pole at Hudson Bay, the contrast between the ice accumulated during the dance encounters and between them will probably be subtle. This probably accounts for their periodicity not being discovered up to now. However, we believe that a twenty-eight day periodicity will be detected in the Greenland ice cores during the Vedic period, once our scenario is understood. This will not only corroborate the V/A hypothesis, but will also enable us to determine the accuracy of the current corrections to the radiocarbon dating scheme throughout that period.

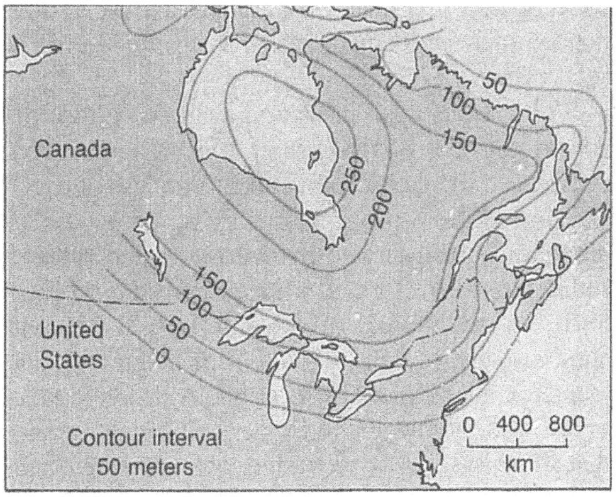

Figure 15.3 Contours showing the amount of rebound (uplift) that has occurred in the Hudson Bay area since the ice sheet disappeared. (From: P.B. King, "Tectonics of Quaternary Time in Middle America," Fig 4, p.836 in *The Quaternary of the United States*, eds. H.E. Wright and D.G. Frey, 1965, Princeton University Press)

The Shifting Heavens

The ancient peoples suffered from a justifiable anxiety complex. As the period between dance encounters wore on, the ancient peoples became more and more concerned about the upcoming trauma that they knew would be unleashed when priori-Mars returned. Velikovsky points out the great importance of Nergal, a Babylonian deity associated with priori-Mars. The movements of this 'unpredictable planet' were closely watched - its rising and setting, its *disappearance and return*, and the various phases of its retrograde motion. Also observed were its brightness and its position relative to the plane of the ecliptic because changes in them were precursors of its approach to the Earth. The brightness measurments were probably a less effective indicator because only a small illuminated crescent contributed to its brightness as it approached the Earth from the Sun side, similar to the appearance of Venus near conjunction today.

After each capture and release of priori-Mars, the problem of determining the new orientation of the stars once again arose. Many ancient structures were built to gauge the ever-changing heavens, to aid in determine the next approach of the threatening planet priori-Mars, and to establish a new calendar. It was particularly important to establish a new calendar, because of its guidance in planting crops. As mentioned in Chapter 13, the number of days per year was 360 during the dance encounters, and 365.25 between encounters. Although priori-Mars and proto-Venus were worshiped as gods, practical consideration reigned for these structures. We claim that this was the reason for the construction of the stone henges or circles such as Stonehenge and Avebury. Of particular significance is the fact that they were rearranged on numerous occasions,

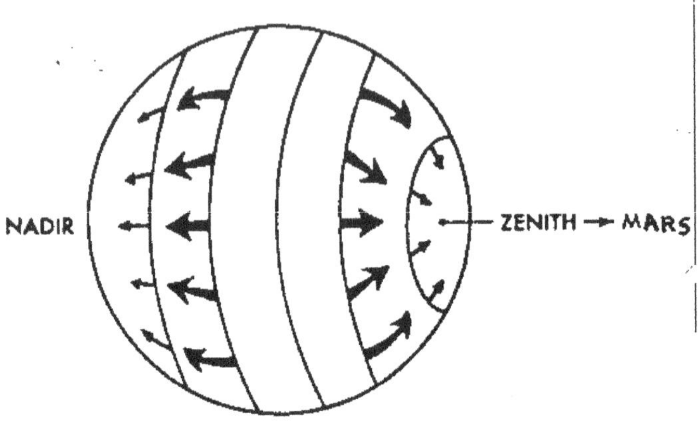

Figure 15.4 Gravitational force of Mars moves ocean waters toward the point beneath it, while pulling them away from the area in the surrounding circular zone.

with the stones being moved, discarded, or new ones being added. We claim that a new alignment had to be made every fourteen years because the orientation of the pole was changed as the Earth and priori-Mars entered into or broke off from a dance encounter. It is because every change in the dance configuration resulted in a somewhat different celestial alignment that no sense can be made from the stones as they now stand.

Other types of structures built for worship were aligned with the latest scheme of the heavens. Velikovsky cites that the Temple of the Lord in Jerusalem was built so that the noonday Sun would shine through several gates directly into the sanctuary at the spring equinox but had to be 'corrected' because the Earth had shifted.

Repeated Flooding of Northern India

The tidal force on the Earth's oceans, due to an external body, is a result of the fact that the gravitational attraction of that body is greater at places on the Earth nearer to it, than it is at places farther away. Thus, although the gravitational force of a body at any place on the Earth is inversely proportional to the square of the distance between the two, the *tidal* force is inversely proportional to the *cube* of the distance between them. As a result, compared to the Moon today, the tidal force due to priori-Mars, centered only 44,400 kilometers above the Earth was some 5,000 times greater. The priori-Mars' tides were enhanced by another factor. The Earth continually rotates relative to the Moon, so the true magnitude of the Moon tide is never realized, because the flow of water in shallow areas never 'catches up.' Because priori-Mars was stationary relative to the surface of the Earth, its tidal effect was fully realized.

The commonly held misconception of tides is that the gravitational force of the attracting body, such as the Moon, lifts the sea directly beneath it upward, causing a high tide. Actually, the vertical force of the Moon is extremely small, only one ten-millionth of the Earth's gravitational force. Thus even the amount of *vertical lifting* of the sea by priori-Mars during the dance encounters, was negligible. The primary effect on the oceans is the gravitational attraction of the extraterrestrial body on the water around the periphery of the spherical zone facing the attracting body, where its tidal force has primarily a horizontal component. This is illustrated in Figure 15.4. The horizontal component of the force pulls the water in that zone horizontally toward the point on the Earth over which the body is situated. The flow of the oceans toward this point causes the water to 'pile up' in

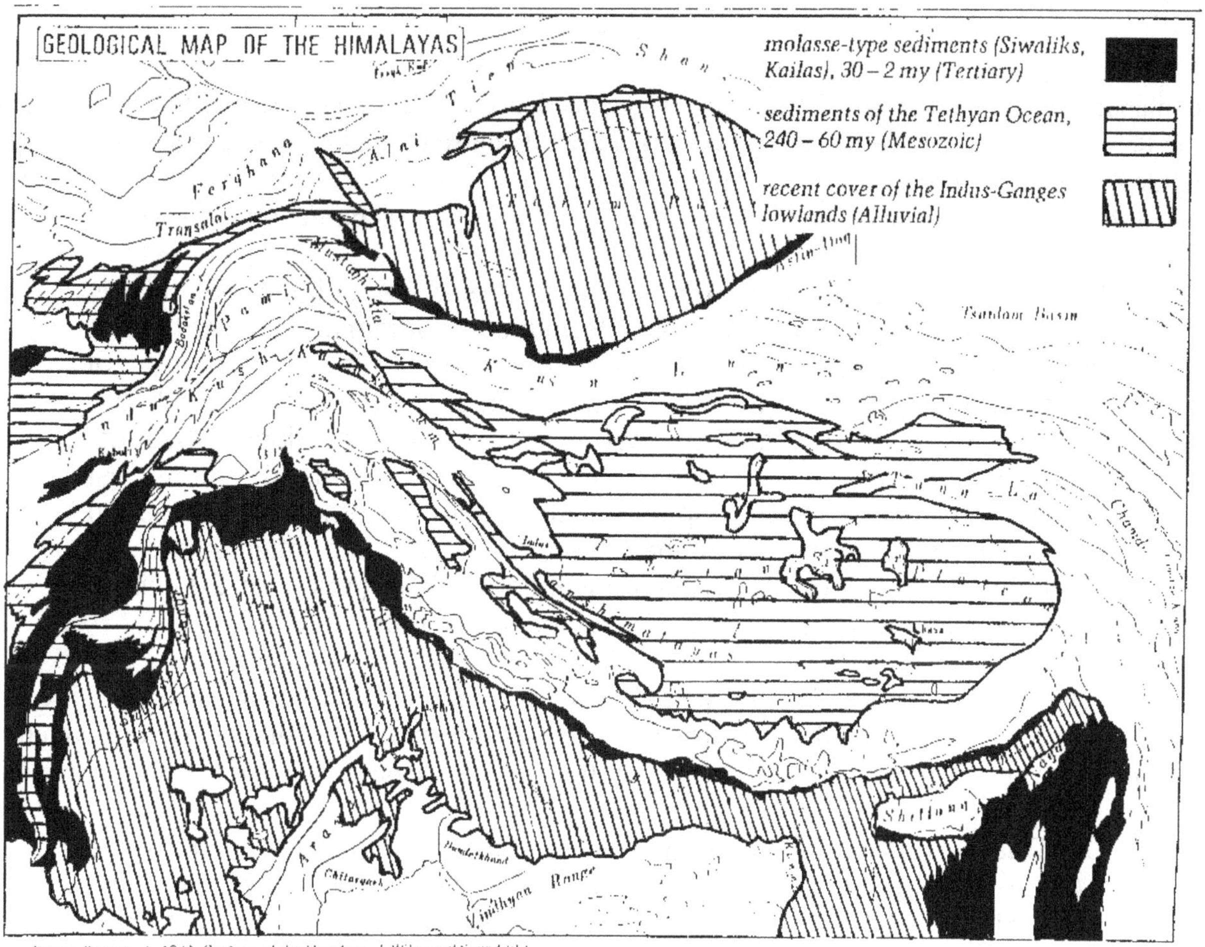

(from: Gansser, A. 1964, Geology of the Himalaya, J. Wiley and Sons Ltd.)

Figure 15.5 Simplified geologic map of northern India and Tibet showing locations of the Siwalik and 'alluvial' deposits. (From: Gansser, A., 1964, Geology of the Himalaya, J. Wiley and Sons Ltd.)

that area. Since priori-Mars was stationary over the Transhimalayas, the result of its tidal force was to pull water toward the southern part of the Asian continent.

During the dance encounters, the low lying areas south of the Himalaya complex were inundated and the coastline of southern Asia subsequently moved inland. The lowest lying portion of India, in the north, was flooded. We claim that this periodic flooding caught millions of animals at sea level, drowned them as it rose, and deposited their carcasses along with sedimentary material along the temporary shoreline at an altitude of some 1200 meters. The resulting Siwalik deposits, shown in black in Figure 15.5, confirm our hypothesis. Their bedding is perfectly horizontal, and contain remains of animals which only live at or near sea level. The region of the priori-Mars high tide flooding is delineated by these deposits in northern India. Although geologists term these 'geologically young,' they mean 'only a few million years old.' We claim that they are less than 6,000 years old. The current view is that the animal bodies were deposited at sea level, and the sedimentary layers in which they were imbedded were subsequently uplifted 1200 meters by thousands of separate earthquakes, inch by inch, while miraculously managing to remain perfectly horizontal over distances of hundreds of kilometers. Even some geologists cannot believe this

explanation, but no other theories have been advanced, because it is incomprehensible to them that the ocean could ever have been 1200 meters above its current level. We consider this a clearly untenable explanation.
The following quote from www.amen.org.uk/vorwelt/megasucc/ is particularly apropo:

> In light of the principle of uniformity, it should be pointed out at this juncture that this kind of rock formation containing its vertebrate fauna cannot be measured according to today's standard of deposition. Their origin was catastrophic. Here lie buried the remains of pigs, mastadons, cattle, deer, giraffes, tigers, hyenas, hippopotamuses, etc. in unbelievable variety. Habitats of the widest variety must have been captured during an extremely short period by the rivers of debris from the rising Himalayan mountain chain. It is absurd to want to claim the conventional time period of 20 million years (middle Miocene to middle Pleistocene) for these events.

The lowland deposits in the flooded area, termed alluvial deposits by geologists, are the sediments left by the tidal flood waters when priori-Mars withdrew. The following quote, which is now over one hundred years old, reinforces the notion of repeated ocean incursion in northern India:

> Dr. Das ... says that there is the clearest mention in the *Rig Veda* (VII. 95.2) that the Saraswati flowed from the Himalaya right into the sea. His natural inference is that in Rig Vedic times there must have been a sea, occupying the desert tract of Rajputana ... Further, he says there is also the clearest mention of the existence of an Eastern Sea ... and as the Sea's contiguity to the Punjab ... is clearly indicated by a number of verses, he naturally infers that it lay to the immediate East of the Punjab ... Surprisingly enough, the existence of a Miocene sea, cutting off the Punjab from Southern India was asserted by some old geologists but this seemed too early and remote for the development of a civilization like that depicted in the *Rig Veda*; he therefore surmises that this sea must have existed down to later geological periods. Mr. D. N. Wadia in his Geology of India mentions the existence of a Pleistocene Sea covering the Gangetic trough at the foot of the Himalayan Range, - a sea that undoubtedly lasted for several thousand years till it was gradually filled in by the waste of the high lands and the alluvium brought down by the Himalayan rivers ... As regards the existence of a sea in Rajputana in the Pleistocene period, the evidence is still very meager, but Mr. Wadia writes of "marine transgressions" which caused portions of Rajputana to be invaded occasionally by the encroachments of the Arabian Sea, such encroachments lasting

> probably for thousands of years at a time. He says such invasions of the sea on land are of comparatively short duration and flood only low level areas, converting them for the time being into "epi-continental seas." The duration of these encroachments was 'short,' only geologically speaking. It must be computed by at least thousands of years. [151]

The documents quoted indicate that the low-lying levels in northern India were definitely flooded or 'invaded,' and the implication is that these encroachments were of short duration and happened repeatedly. The author of the above quote was obviously at a loss to explain the cause of what were reported to be short incursions. Like many other scientists, he could not conceive of a mechanism by which the sea, once it covered the area, could be quickly removed. Thus he hypothesizes that the subsequent lowering of the sea level must have occurred due to sediments from the highlands gradually raising the bottom level. He believes, as any geologist would, that this would have taken a minimum of thousands of years, and repeats this several times.

Our hypothesis of the dance encounters between priori-Mars and the Earth give an explanation of these incursions quite consistent with the events described - their short durations, great depths, their repetitious nature - so many that the number has not been determined from studying the deposits alone.

Although the tidal force of priori-Mars resulted in the raising of the sea level in the area directly below it, it had the opposite effect on the sea levels in areas which fell in the clear circular zones shown in Figure 15.4. In these areas the sea levels were lowered and land normally under water was exposed, because priori-Mars' tidal force pulled the water away from these areas toward the Himalayas. This occurred in a circular band which passed through the Mediterranean, Arctic Ocean north of Asia, Japan, New Guinea, Malaysia, Northern Australia, and the east coast of Africa. The sea level on the far side of the Earth, in the western hemisphere, centered around Easter Island, also rose during these encounters but to a lesser degree than in the Indian Ocean. Given this hypothesis, we refer first to the ancient records to see if there is corroborating evidence of such coastline changes in any of these areas of the globe during the Vedic period.

Amazing support for our hypothesis can be found in a little-known book entitled *The Outline of History*, by the famous H.G. Wells. The author includes the map of Eurasia, shown in Figure 15.6, with the following caption.

> This map represents the present state of our knowledge of the geography of Europe and Western Asia at a period which we guess to be about 50,000 years ago, the Neanderthaler Age. [152]

The exciting thing about this map is that the anomalous water level distributions predicted by our hypothesis are corroborated, not just in the area around the Himalayas but in the Mediterranean and Red Seas as well. This is not just a case of changes in the coastlines due to the lowering or raising of the sea level over the whole Earth. What is actually shown is that the sea level has been raised in the vicinity of India and lowered in the Red Sea and the Mediterranean Sea at the

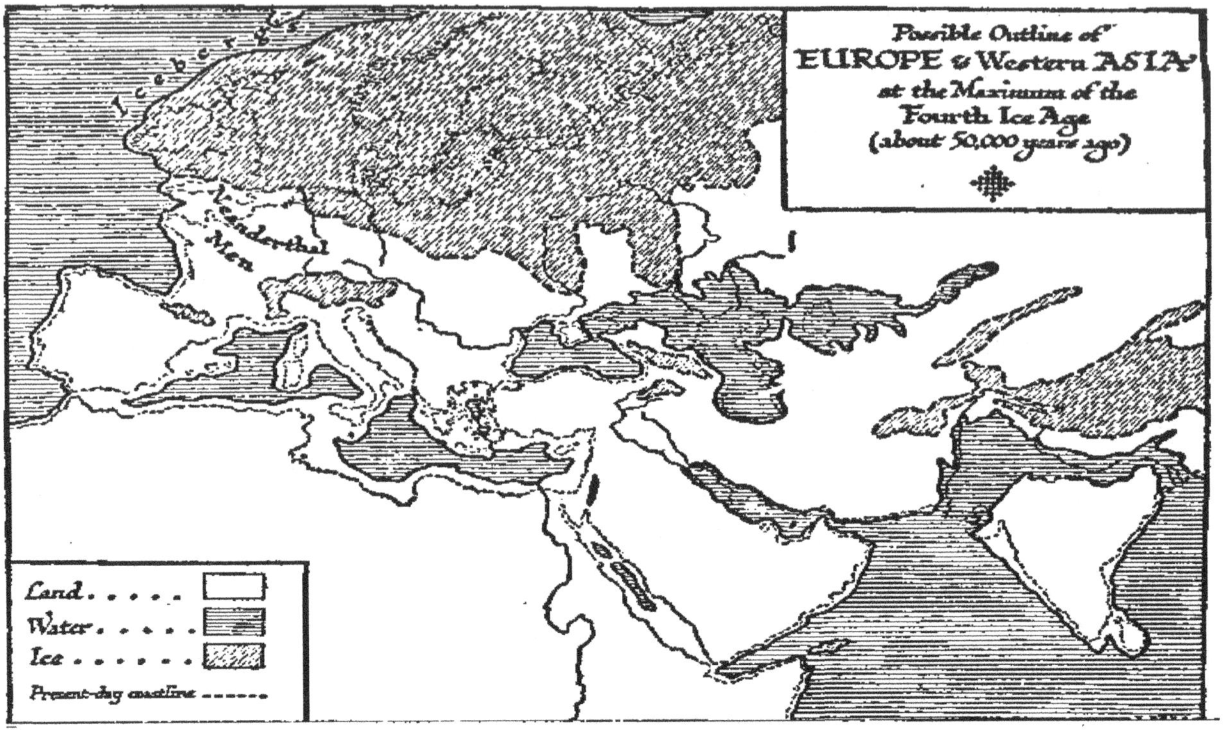

Figure 15.6 Figure on page 77 of *The Outline of History* by H.G. Wells. Note that all of northern India is flooded while the Mediterranean level is so low that it consists of two lakes cut off from the Atlantic Ocean.

same time. How Wells came by this map is not explained. Apparently he considered the source authoritative, since he did not question the obvious ocean level differences. His estimate of the age depicted does not agree with our hypothesis, but he admits this is only a guess on his part. He probably attributed it to a more distant period in the past for the same reason that Dr. Das estimated the sea encroachments would take thousands of years - because the textbooks tell us that large scale geologic changes require long periods of time. We claim that this view is incorrect because there have been many interactions with large bodies in recent times, which were capable of making such changes instantaneously. Specifically, we claim that during the later dance encounters the sea levels conformed essentially to those shown in the map published by H.G. Wells. Therefore we estimate that the map shows coastlines during one of the 'dance encounters'.

The Indus Valley Culture

The Indus river valley, in what is now Pakistan, is the most likely location of the ancient Aryan race that initiated the *Rig Veda*. Based on the research of Jonathan Mark Kenoyer, an archaeologist at the University of Wisconsin at Madison, the major city, Harappa, was founded at the beginning of the Vedic period. It was a commercial city at the crossroads of several trading routes. He characterizes it as a towering brick city surrounded by sturdy walls with imposing gateways, reminiscent of the medieval forts in Delhi and Lahore, but without the moats and sudden turns that would have made them effective for defense. He believes that the buildings were periodically elevated to avoid the sewage runoff from surrounding homes and that the purpose of the walls was to control the flow of goods in and out of the city, because the main gate was only nine feet across. There were no great monuments, but there remain elaborately carved stone seals with inscriptions that no one has been able to decipher. Physical evidence is interpreted to mean that the Indus river shifted, flooding many settlements and disrupting agriculture.[153]

We suggest a completely different explanation of the characteristics of the city. It was built up high so the occupants could escape the periodic floods associated with the capture and release of priori-Mars from the dance encounters. The walls were dams to restrain the flood waters, and the gates were narrow so that they could be quickly blocked when the floods were imminent. In this application moats and sharp turns would have served no purpose. The undeciphered symbols may have been Aryan, since we believe they were the first culture to acquire a spoken language..

There are more realistic views of the nature and the magnitude of the problems that beset the Indus civilizations:[154]

> The second explanation for the demise of the Indus civilization has more geological and archaeological support, but it too leaves questions un answered. It has long been recognized that there were environmental pressures on the Indus cities. The stamp seals at Mohenjo-Daro indicate conditions which do not exist now, for animals (such as tigers, rhinoceroses, crocodiles and elephants) depicted on the seals are no longer indigenous to the area. Humans themselves undoubtedly altered the environment through overgrazing, over cultivation, and deforestation. These factors may have led to a gradual decrease in agricultural production, a breakup of central control, and eventual pressure to locate in new and more suitable areas.
>
> Beyond this, there may also have been environmental changes of a more drastic nature. Geomorphologists and hydrologists have argued recently that the cities along the Indus river, in the province of Sind, Mohenjo-Daro included, were inundated by prolonged flooding in the first half of the second millennium, requiring the construction of large-scale retaining walls to hold back the encroaching waters. This convincing theory receives support from archaeologists who see progressive deterioration in the quality of building and city planning at Mohenjo-Daro.

Geology of the Himalayas

Geological measurements indicate that the Tibetan plateau has slid toward the east some 500 kilometers along two major faults, the Altyn Tagh and Red River. It is thought that this is a result of the colliding of the Indian and Eurasian plates, a process called 'continental escape.' Others contend that mountain untrusting and crustal thickening, which go hand in hand, are the primary result of the collision. We believe that the sliding is one means by which the enormous mass of rock, uplifted by extraterrestrial bodies in the last 12,000 years, is moving back toward sea level. Although the date is at issue, untrusting or uplifting (which we prefer) has certainly occurred and seismic measurements indicate that the crust of the Earth under the Tibetan-Himalayan complex is about twice as thick, some 60 to 85 kilometers, compared with the thickness under the rest of the continents. The question remains as to the source of the additional crustal rocks. We suggest that they developed from mantle rock when the pressure was reduced by the presence of priori-Mars.

Recent studies have detected a layer of partially molten rock perhaps 5 kilometers thick in the crust below southern Tibet.[155] Although the extent of the melting has not yet been determined, it is generally thought that the layer is present throughout the entire Tibetan plateau, because of the common incidence of hot springs throughout the region. This layer provides the decoupling from the underlying geology that allows the sliding mentioned above. Although geologists and geophysicists believe the partial melting is merely a result of the collision of the Indian and Asian plates, their mathematical models are presently unable to show it quantitatively, without assuming that the crust of Asia was already molten at the onset of the collision - a highly unlikely situation.

Tibetan Lakes

There is evidence of the presence of large lakes in the Transhimalayas in the recent past. Geologists have dated these at the termination of the next to last ice age, some 25,000 years ago. At that time, they say, the melting of glaciers created great lakes high in the Himalayas and Tibet. We claim that they formed from vast amounts of water which fell from priori-Mars between 3300 and 687 B.C. They breached their containment, probably as a result of earthquake initiated fissures, and their explosive release cut through the Himalayan range along several different paths, forming five great rivers. Small lakes still form in the high Himalayas, due to the damming effect of moraines from earlier glaciers. These natural dams occasionally break, releasing the accumulated water into high mountain gorges, sometimes carrying millions of large blocks of rock along with them. These provide a small example of the forces unleashed in more ancient times when a vastly larger amount of water finally broke out of its containments.

The small lakes that remain in Tibet are considered sacred by the Buddhists who feel that demons reside in them. Among the Tibetans there are a number of tales of great lakes which existed in much more recent, historical times than those proposed by the geologists. Specifically an ancient monastery in the Ladakh-Himalayas, Lamayuru is said to stand on the site of an ancient lake:

> In the olden times [historically 2500 years ago] when Shakyamuni-Buddha inhabited this world, southern Tibet and the northern

> Himalayas were totally uninhabited. In the valley of Lamaynou there was at that time [geologically more than 25,000 years ago] a clear lake, home of the evil Naga spirits.[156]
>
> In the oldest monastery in Ladakh, Alchi Gompa, which is famous for its splendid Buddhist frescoes, I discovered among the many Bodhisattvas a depiction of two fully manned boats, carrying prayer flags. This unlikely picture in Alchi, where the climate had been desert-like before the monastery's foundation, was perhaps a 'memory' of the time 20,000 to 30,000 years ago when that area had been a region of great lakes.[157]

Tibetans accept the above statement and the fresco which is described, at face value, and we agree with them. These documents place the time of the great lakes in the high Himalayas as recently as 2,500 years ago, around the time of the last last dance encounter. Furthermore, we believe that the inspiration for Shakyamuni Buddha was the same prominent lava fountain visible on the surface of priori-Mars. *The Tibetan claim that Shakyamuni was still present up to 2,500 years ago coincides very closely to the date at which we claim the Vedic period ended. At that time much water remained in the Tibetan lakes, including the last dregs of water from priori-Mars. The date of priori-Mars' final disengagement, some 2,700 years ago, was the first time at which it was safe to go into that area because the onslaught of rocks and thunderbolts from priori-Mars finally ceased.* The Manjushri Mulatantra states that one hundred years after the death of Buddha the lakes in the Land of Snows had shrunk and a forest of sal trees was growing.

It is difficult to ignore the fact that *all the major rivers of the Himalayan and Transhimalayan originate in the vicinity of Mount Kailas*, which lies in southern Tibet, just north of the Indus-Yarlung suture, the line along which the surface rocks of the colliding Indian and Asian plates meet. This is the sacred mountain said to be the 'home of Indra on Earth.' These rivers are shown in Figure 15.7. They are the Yarlung Tsangpo (Brahmaputra), the Indus, the Sutlej, the Ganges, and the Karnali. Because these rivers all cut across the Himalayan ranges, *it is generally believed that they existed before the untrusting began.* This is based on the same dogma of geology - the assumption that mountain building only took place in tiny steps, a few centimeters at a time, over multi-million year periods. The reasoning being that

Figure 15.7 Five major rivers of Tibet and India originate in the area around Mt. Kailas. It is considered sacred to all the major religions of Asia, except Islam. (Courtesy of *Himalayas*, E.M. Buhrer, Facts on File, N.Y.,p. 38.)

because the uplifting took place at such a slow rate, erosion by the existing rivers could keep pace with upward movement, allowing the river to remain in place while the land rose around it. A major problem with this hypothesis is that two of the above rivers, the Indus and the Yarlung-Tsangpo (the Tibetan name) were obviously forced to flow to the west and east respectively, some 1,000 kilometers, presumably due to the presence of the Himalayas, before penetrating the range and finding their way to the sea. *Therefore the mountains must have existed before the rivers.* Even more surprising is the fact that they finally broke through at the points of greatest uplift. This is also true of the Arun river in eastern Nepal. A Tibetan legend tells of the river formation: [158]

> The sagas tell how "a Buddha of that time" took pity on the flooded land and with his sword cut a deep swath through the mountains in southeastern Tibet and thus let the water out. Today in this region is one of the deepest and widest gorges in the world, where the Yarlung-Tsangpo cuts through the chain of the 7,776 meter high Namche Barwa and then flows southward as the Brahmaputra into the plains of Assam

This story implies that great lakes built up in Tibet, and that they were released suddenly and catastrophically, at which time the force of the enormous volume of water cut the gorges through the Himalayas in a matter of days. This mechanism is just the opposite of that proposed by the geologists, which requires millions of years.

Fossil Finds in the Tibetan-Himalayan Area

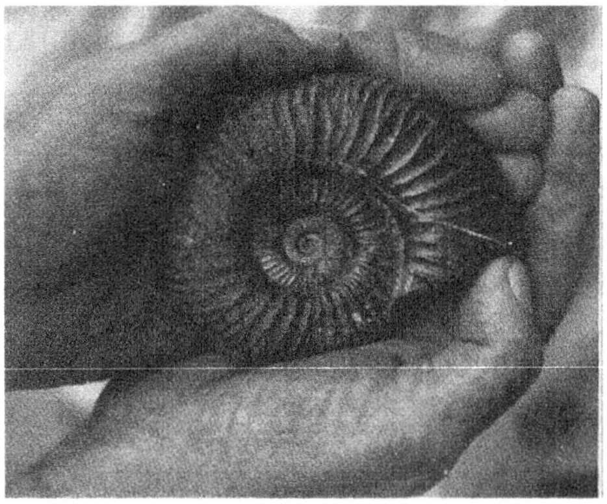

Figure 15.8 Ammonite from the Kali Gandaki valley in Nepal. (Courtesy of *Himalayas*, Design by E.M. Buhrer, Facts on File, N.Y. 1987)

North of the Tibetan plateau in western China lie the Tarim and Djungar basins. These are thought to have been lifted to high altitudes due to the collision with the Indian plate. These basins are separated by the northern Tian Shan, a great range whose name means 'celestial mountains.' So-called 'piggyback basins,' formed by the thrusting and folding of these mountains, are filled with more than 100 meters of Quaternary sediment dated at less than 1.6 million years old. Also seismic reflection records indicate a depth of about 3,000 meters of 'young' sediments across most of the Tarim Basin. The presence of marine fossils in these mountain basins is an enigma to modern day geologists. In actuality the ages indicated by some of the fossil record are much younger than 1.6 million years, so much so that it is played down in most papers because the younger they are, the more difficult the explanation becomes. Fossils include foraminiferans, sea lilies, corals and spiral shelled ammonites encased in black rocks, as shown in Figure 15.8. Some ammonites have been found in the highest passes of the Himalaya - in places where some geologists say they are 'not supposed to be,' that is, in concretions in a 130 million year old layer of black slate on the Jurassic/Cretaceous border. Some say that this is due to the Tibetans carrying these highly prized tokens on their climbs and depositing them in high places as sacrifices to the mountain gods.

An Indian geologist, Vishwa Jit Gupta of Panjab University in Chandigarh, India has devoted his life to the study of the fossil record in the Himalayas. Between 1973 and 1977 he published five volumes on Indian stratigraphy, dealing with fossils from different geologic periods.[159] These became sources on which many researchers relied in order to date the surrounding rock. In 1989 an Australian geologist, John Talent, leveled serious charges against Gupta and his work - essentially charging that his work was fraudulent.

> He's distorted the whole geologic picture there ... It's as if Gupta was claiming to have uncovered rhinos in Rio or kangaroos in Kashmir[160]

Gupta denies the charges and, in his rebuttal, alleges that many of his findings have, or could be, corroborated by researchers and that there are scientific explanations for the data that Talent claims are inexplicable.

> Geology (especially Himalayan geology) is a developing science ... where anomalies should provide food for thought.

Gupta claims that he has unearthed tiny tooth-shaped fossils called conodants in the Himalayas. Talent believes that these came from the only place they are known to be found - a limestone bed in upstate New York. The last word in this controversy has yet to be heard.

In 1984, a Chinese geologist, Hou Xianguang, discovered an amazing concentration of fossils on Maotian Mountain near Cheng Jiang in southwest China, just east of the Himalayas. These are dated as Cambrian period, from 550 to 510 million years old. Several things are unique about this find: the large volume of rocks in the area containing the fossils of these ocean creatures; an amazing variety of fossil creatures never before seen - some 'monsters' as big as a meter long; and, most notable, the degree to which the soft tissue impressions have remained preserved. Is it merely a coincidence that the controversial stratigraphy of Gupta and these unusually well preserved fossils are found in the vicinity of the Himalayas?

The V/A scenario suggests a unique origin for the sedimentary deposits covering much of Tibet and the Tarim Basin. The former are thought to be a relic of an ancient Tethys Sea which existed between India and Eurasia prior to their collision and untrusting. Most of these rocks are found north of the Indus-Yarlung suture, the junction at which the rocks of the two continents are fused together. Just south of this zone are the High Himalayas, which do not contain these marine sediments. Further south, similar marine sediments are found in 15 to 18 kilometer thick layers in the Low Himalayas. We believe that these sediments fell to Earth from priori-Mars along with the entire Martian complement of water. Because they were intermixed with the Martian water, they washed down both sides of the Himalayan range, ending up to the north and south. The majority fell in Tibet, and were immersed in great lakes. Most of these sediments precipitated and were left behind when the lakes breached their boundaries, cutting the major Himalayan rivers.

It is likely that the sedimentary rocks of the Tarim and Djungar basin most likely have the same origin. These are also the result of the tidal ocean from priori-Mars pouring down onto this area - the mythical 'deluge from Heaven' claimed by Plato in *Timeaus and Critius*. Although some of the sediments washed into these mountain valleys are due to material eroded from the bordering mountain ranges as the deluge flowed down over them, we believe that much of these sediments are from priori-Mars - dust and rocks that were blasted across the short space between the planets during hundreds of thousands of convulsive events. The intriguing possibility is that fossils from priori-Mars might be found in these sediments.

Assuming life on priori-Mars followed pretty much the same evolutionary course on Earth, Martian fossils *may have already been* found in the Himalayas and been mistaken for Earth life forms. Such fossils would likely appear to be 'normal' but out of their 'proper' place in the geologic section. Therefore, the controversial fossils reported by Vishwa Jit Gupta from the Himalayas and those discovered by the Chinese paleontologist Hou Xianguang on Maotian Mountain in Cheng Jiang should be reevaluated. Both the fossils and the sedimentary rocks in which they lie should be examined anew with the possibility of an extraterrestrial origin in mind. They should be analyzed for a large range of elements, looking for unusual combinations of elements, isotopic ratios, and even amino acids. Finally, if DNA is associated with any fossils, it might be possible to determine if the life forms are extraterrestrial or conversely to prove that the basic design of life is the same on at least two different planets in our system.

It is difficult to justify spending more billions of dollars in missions to priori-Mars before first examining the great secrets about priori-Mars that may reside in the Himalayan-Tibetan-Chinese area. We propose that geologist-astronauts-in-training from all countries join together in an exciting adventure, practicing their investigative skills along side experienced (but hopefully open-minded) geologists and paleontologists in this vast, cold, unexplored, forbidding ‘world’ on our own planet which is so much like priori-Mars, before venturing that great step to the planet so recently divorced from our own, but now many millions of kilometers distant.

Confirmation of the Dance Encounters in a Hindu Astronomy Book

Although many bodies ejected from priori-Mars during the dance encounters went into orbit around the Sun, these were in the minority because their initial velocity and direction had to be aligned with the direction of priori-Mars motion around the Earth at the time. The great majority were not launched in that direction, and some fell back on priori-Mars. A great number were caught on the stronger gravitational field of Earth. Most of these impacted in the vicinity of the Himalayas, causing the deaths of many people. As a result, the prediction of these events was a subject of great concern, and the ancient astronomers expended much effort trying to predict the occurrence of the next 'fall.' This work is actually documented in one chapter of an ancient Hindu astronomy text called the *Surya Siddhanta*.

Applying the V/A scenario to this work, we find that the discussion of the 'falls,' which have remained obscure for millennia, suddenly becomes clear. What was being explained was the cosmic alignments that controlled the severity of the falls. From the Hindu myths we now understand that the mechanism which caused the repeated expulsion of numerous large bodies from priori-Mars was the alignment of the Sun and Moon with priori-Mars and the Earth as the latter two danced in a tight circle about one another. From the modern point of view these events would be called eclipses of the Sun or Moon by priori-Mars. But these were not the harmless eclipses we now experience, in which the Moon passes in front of the Sun with no danger to human beings. These boded great danger for the Vedic people.

The *Surya Siddhanta* devotes an entire chapter to methods of predicting the falls. It explains that the severity of the eruptions was much greater when both the Moon and the Sun were eclipsed, and more particularly the closer they both were to the center of the disc (mandela) of priori-Mars. Since this book was not translated from Sanskrit into English until long after the cosmic events had ceased, the translator was unable to understand the fundamental situation being described - that priori-Mars was hanging stationary in the sky above the Vedic people. This was obscure in the original text because, to the author, this was a given and there was no need to explain it. Fortunately, notes by later commentators shed light on the point of view of the translators and also give literal translations of words with which they feel the translator has taken license. The translator calls the chapter of interest "Of Certain Malignant Aspects of the Sun and Moon." A commentator points out that the literal title, "patadhikara" or "Chapter of the patas" means literally "Chapter of the *Falls*." The translator also had trouble understanding the usage of the word "mandala" which literally translated means "disc" or "circle." This referred to the disc of priori-Mars which remained stationary in the heavens. Since the usage of mandala did not make sense, the translator changed phrases that should have been translated "within the disc" to "being a circle" or "being a half-circle," implying that the Sun and Moon fell on a circle around the Earth. The following are the first five versus of Chapter XI of the Surya Siddhanta:

> 1. When the sun and moon are upon the same side of either solstice, and when, the sum of their longitudes being a circle, they are of equal declination, it is styled *vaidhrta*.
>
> 2. When the moon and sun are upon opposite sides of either solstice, and their minutes of declination are the same, it is *vyatipata*, the sum of their longitudes being a half-circle.
>
> 3. Owing to the mingling of the nets of their equal rays, the fire arising from the wrathfulness of their gaze, being driven on by the provector (*pravaha*), is originated unto the calamity of mortals.
>
> 4. Since a fault (pata) at this time often causes the destruction of mortals, it is known as *vyatipata* , or by a difference of title, *vaidhrti*.
>
> 5. Being black, of frightful shape, bloody-eyed, big-bellied, the source of misfortune to all, it is produced again and again

Our interpretation of these versus is as follows:

1. When the moon and sun are on opposite sides of priori-Mars and both on the same side of the 'equator' of that planet and both lie within the disc of priori-Mars, it is a *vaidhrti* conjunction.

2. When the moon and the sun are on opposite sides of the 'equator' of priori-Mars and within half the diameter of the priori-Mars disc, it is termed a *vyatipata* conjunction.

3. The closer the Sun and Moon were aligned with one another and the center of priori-Mars, the greater the tidal force on the interior of priori-Mars, and the greater the resulting fall of bodies occurred, killing many mortals.

4. Since the falls at this time often causes the destruction of mortals, it is known as *vyatipata*, a disastrous fall.

5. priori-Mars being at these times black, of frightful shape, swells up and great hot bodies are blasted from numerous volcanic vents into space one after another.

There is about a five degree angle between the orbital plane of the Moon and that of the Earth. Currently their periods are such that the Sun and the Moon return to the same relative position once every eighteen years, but this was probably varied during the Vedic period. The plane of revolution of the Earth-Mars pair was quite inclined relative to the Sun and Moon, but their alignments with the Moon alone could have occurred as frequently as every fourteen days, when the pair passed through the plane of the Moons orbit. As discussed in previous chapters, ejections occurred at these times. In fact, in Hindu astronomy these two points in the Moon's orbit are personified as demons. For example the demon Rahu represents the ascending node of the Moon's orbit.

The Surya Siddhanta explains that the severity of the resulting 'falls' depended also on the position of the Sun, that is, when all three orbital planes came close to alignment. The tides exerted on the Earth today are a combination of the tidal forces of both the Moon and the Sun. The closer their alignment with the center of the priori-Mars mandela, the more violent the eruptions and more the danger for the Vedic people. Any doubt as to this interpretation is dispelled by verse 5, which describes the appearance of priori-Mars at these times, when it literally swelled up as a prelude to the explosions. The meaning of these somewhat obscure verses becomes quite clear when interpreted in the context of the V/A scenario of the 'dance encounters.' Conversely, the passages offer powerful corroboration of our hypothesis.

In the same document there is even stronger, *quantitative* corroboration:

> 19. When the equality of declinations of the sun and moon takes place in the neighborhood of the equator, the aspect may then again occur a second time: in the contrary case it may fail to occur.
>
> 20. If the sum of the longitudes of the sun and the moon, in minutes, on being divided by the portion (bhoga) of an asterism (bha), yields a quotient between sixteen and seventeen, there is another, a third, *vyatipata*.

In order to understand this we need some background data. We claim that during the 'dance encounters' the Earth and priori-Mars revolved about one another at a distance of about 44,400 kilometers between centers. At that distance above the Earth, the disc of priori-Mars, based on its current diameter, would have subtended an angle of about 9.52 degrees when viewed from the Orient. At its present distance from the Earth, the Moon advances some 11.76 degrees per day relative to the stars, while the Sun advances only about one degree per day. In the 'dance' configuration, the Earth-Mars pair revolved about one another once per day. If on one day the Sun and Moon were close enough together to disappear behind the disc of priori-Mars as it revolved about the Earth, the question is 'What is the maximum number of consecutive days this aspect could occur?' Using the current lunar orbit we find that the slowest rate that the Moon advances in one day is 10.77 degrees relative to the Sun. This means the number of consecutive days that they could be within 10.77 degrees of one another at the same time of day would be three - the Moon to the 'right' of the Sun 10.77 degrees on the first day, the Moon exactly aligned with the Sun one day later, and the Moon to the 'left' of the Sun 10.77 degrees at the same time on the next day.

Our calculations of the angle subtended by Mars during the dance encounters give 10.42 degrees, based on the (present) makeup of Mars assumed in Chapter 3. But the angle is about 10.78 degrees based on the makeup of priori-Mars during the Vedic. This is explained in Chapter 6. *The latter value agrees exactly with the size of the disc or mandela that would have been necessary in order that the alignment could sometimes last three days. This tells us that the size of the disc of priori-Mars during the 'dance encounters,' which we have determined using the laws of physics was exactly what was observed by the Vedic people, as interpreted from the Surya Siddhanta.* Therefore the V/A theory not only allows the understanding of these heretofore mysterious verses, but also agrees extremely closely with the alignment data given in that ancient document.

The Surya Siddhanta also contains fascinating descriptions of the terrain on priori-Mars, the progression of the Suns illumination of the surface features as it revolves about the Earth, the division of the disc as related to the 'asterisms,' and a device called an 'armillary sphere' that was apparently used by the Hindus to determine position of the various ' circumpolar stars' (really volcanos) within the mandala or disc of priori-Mars. These all provide corroboration if the V/A hypothesis from ancient science.

Pollution of the Earth by Martian Organisms

The 'nay-sayers' of the scientific world have recently surfaced the horrible notion that organisms returned to Earth by space missions to priori-Mars could 'contaminate' our planet and start epidemics. Pressure from environmentalist organizations could force NASA into adding billions of dollars to the cost of a manned priori-Mars mission. The V/A hypothesis holds that massive amounts of water, air, dust, bacteria, viruses and, perhaps, even animal life fell from priori-Mars for thousands of years during the Vedic period. Indeed, a number of ancient documents lament the disease and pestilence that resulted from these falls. But after a few millennia these have diffused throughout the world and are literally a part of us all. Therefore we can safely dispense with the expensive and time consuming decontamination and quarantine of returning astronauts and their samples as was done with the Apollo lunar missions. The returning astronauts and cosmonauts will merely be returning from a planet which is already a part of every one of us.

Global Warming

The evidence in favor of recent catastrophic cosmic encounters with proto-Venus and priori-Mars prompts us to present our view on the currently popular subject of 'global warming.' The notion has been advanced by some planetary scientists, that the Earth is a delicately balanced ecosystem that could go 'sour' due to the pollution of our air by industrially generated carbon dioxide. This is an extension of the 'runaway greenhouse' mentality, the hypothesis invented to explain the high temperature of proto-Venus. It should be clear to any reader at this juncture that the V/A scenario precludes this 'effect' as an explanation of the high temperature of proto-Venus. It follows that the application of the atmospheric models, created to verify the RGE, to the Earth's global warming, is completely invalid. This point of view is vindicated by the failure of these models to track historical data. It is obvious that the mathematical models, even with accurate data on concentrations of carbon dioxide and other gases, do not come close to predicting what happens in the real world. There are many more complex factors in the real world than can be put in a computer program at our present state of ignorance.

The atmosphere of priori-Mars and practically all of its water was drawn over to the Earth in the last 6,000 years. This included vast amounts of oxygen, nitrogen, and carbon dioxide. Along with this were carried innumerable species of microbes, viruses, plant and animal life. To this, add massive rock bodies, as large as 20 kilometers in diameter, blizzards consisting of millions of tons of smaller rocks, and sufficient amounts of choking red dust that made the rivers of the world run red. Mix in thunderbolts, continuous earthquakes far beyond the Richter scale, wide spread flooding, numerous volcanic eruptions and the internal trauma of the pole shifts. In spite of these massive perturbations occurring in such a short time, and so recently, the Earth has survived quite nicely. In light of this it is difficult to believe that the small amount of carbon dioxide generated by mankind's little factories and automobiles will cause the Earth to 'hiccup.'

The V/A theory provides another argument against the application of the proto-Venus atmospheric model to the Earth. We claim that priori-Mars was in a proto-Venus-like orbit, probably for billions of years prior to the birth of proto-Venus. When priori-Mars first approached the Earth it was still a living planet. Therefore, a planet at that closer distance from the Sun is not doomed to a superheated state, as is currently hypothesized in the case of Venus. It is hot by virtue of its explosive birth and incandescent youth and not due to the sunlight it currently receives, almost none of which reaches the surface.

The fact that the predicted degree of global warming has not occurred, does not excuse our poor stewardship of this beautiful Earth, so recently prepared for us. The population is growing too rapidly, and we are consuming and destroying the bounty of the oceans faster than it can reproduce. Pollution by compounds that, unlike carbon dioxide, were never before introduced into the system, pose a much more dangerous threat.

16

EVIDENCE OF THE PRIORI-MARS ENCOUNTERS - PART 2

The principle of maximum diversity operates both at the physical and mental level. It says that the laws of nature and the initial conditions are such as to make the universe as interesting as possible. As a result, life is possible, but no too easy. Always when things are dull, something new turns up to challenge us and stop us from settling into a rut. Examples of things that make life difficult are all around us: comet impacts, ice ages, weapons, plagues, nuclear fission, computers, sex, sin and death. Not all challenges can be overcome, and so we have tragedy. Maximum diversity often leads to maximum stress. In the end we survive, but only by the skin of our teeth.

Freeman Dyson
Infinite in All Directions

When priori-Mars approached its position over the Himalayas, its tidal force pulled the waters in the clear zone in Figure 15.4 toward northern India. Waters in the open ocean shifted almost unnoticeably because there was nothing to oppose their motion, but the waters in inland seas, such as the Mediterranean, could not move in that direction since there were no immediate channels for their flow. The waters of such seas were first drawn up onto the shore in the direction of the Himalayas, until they reached one or more channels that allowed their flow to the ocean. The channels by which the Mediterranean waters flowed toward northern India were: (1) across the Nile Delta and along the path of the current Suez Canal, west of the Sinai peninsula, into the Red Sea; (2) eastward across the land in the Jezreel valley to the Jordan rift valley and then south to the Red Sea on the east side of the Sinai peninsula; and (3) from the northeastern corner of the Mediterranean across (now) Syria to the Tigris and Euphrates Valleys and down through the Persian Gulf to the Sea. Once the waters found these paths to the sea, the wide flow channels allowed the water level to adjust in a matter of days. Thus the transitional flooding was short-lived, but was many meters deep and was capable of completely destroying any ancient settlements without some form of protection. After each flooding incident, water was trapped in local depressions, in some cases for years. Initially this denied the use of large tracts of land for agriculture. But this process of repeated inundations also had a favorable aspect.

In order to clarify the effects of the dance encounter on the Mediterranean, we must reiterate that sea level around the world was some 500 meters below its current levels at the beginning of the Vedic period. The deepest portion of the passage through the straits of Gibraltar is now about 406 meters deep and in pre-Vedic times may have been considerably shallower. Since the ocean level was below this, the implication is that the Mediterranean was fed only by fresh water runoff. As a result the *pre-Vedic Mediterranean was a group of fresh water lakes*, as shown in Figure 16.1. Corroboration of this hypothesis comes from deep soundings collected for Ehrenberg in 1857, which exhibited the shells of many fresh water creatures with their fleshy parts still in them. This indicates

Figure 16.1 The Mediterranean as it existed at the beginning of the Vedic period, based on the 1000 meter depth contours. It comprised three (fresh water) basins isolated from the Atlantic Ocean while the Aegean and the Adriatic are above water.

that the transient flooding in the Early Bronze Age was not only less severe, but also much more conducive to the irrigation and fertilization of millions of square kilometers of scorched land in the early part of the Vedic period. *In fact, this is what created the area known historically as the 'fertile crescent.'*

In direct opposition to the presence of these recently deposited fresh water creatures is the fact that the current salinity of the Mediterranean Sea, as well as the Red Sea and the Dead Sea is much higher than that of the Atlantic, Pacific, and Indian Oceans. It is currently thought that the high salinity of these bodies is due to the warm climate, which promotes evaporation, combined with the limited exchange of water between them and the major oceans due to the narrowness of the straits which connect them. But this evaporation would have to be rapid enough to overcome all the direct rainfall plus the fresh water runoff from the all the rivers and seas that empty into the Mediterranean. This is a very unlikely scenario.

What is currently translated in the bible as the Tigris and Euphrates Valley, is literally 'the flood':

> Your fathers dwelt on the other side of the flood in old time, even Terah the father of Abraham and the father of Nahor. [161]

At the site of the ancient kingdom of Mari, which is located on the Euphrates plain between Damascus and Mosul, now called Tell Hariri, 24,000 cuneiform documents were unearthed in 1933. These included numerous orders for the construction of canals, locks, dams, and embankments. This is currently interpreted to imply that the prosperity of the country largely depended on the widespread conservation of the Tigris and Euphrates waters for irrigation, under the supervision of government engineers. We claim that the reason for such a great effort was to control the floods that tore through Mesopotamia at the times of the capture and release of priori-Mars, and to reservoir as much water as possible, for use during the fourteen or twenty-eight year periods between them.

In Egypt, a great 'barrier wall' was built by the Pharaohs at the northern boundary of their kingdom, paralleling the route of the Suez Canal today. This was described in Egyptian texts dated at 2650 B.C. and was probably in existence much earlier. It is currently thought that this wall was built to keep the Semites or 'sanddwellers' out of Egypt. This interpretation of its use is questionable, since it would have to have been patrolled along its entire length, hundreds of kilometers, in order to have been effective. We suggest that this wall was essentially a dike, meant to control the flow of Mediterranean water at the times of capture and release of priori-Mars.

Also related to flooding in the Vedic period is the Faiyum depression, a low lying area which lies west of the Nile in northern Egypt. This area was apparently the subject of considerable development during the Twelfth Dynasty (1991-1797 B.C.) - perhaps to revive an agricultural base that had existed for over a thousand years before that time. Herodotus, who lived in the fifth century B.C. describes a lake much larger than that which exists today and surmised that this body of water was not fed directly from submerged springs, but through a canal which brought water in from the Nile via the Bahr Yusef. But *monuments discovered by Petrie and other evidence indicates that the lake was maintained at 16 to 23 meters above sea level during the Middle Kingdom for some 1,500 years down to the Ptolemaic period (first half of the third century B. C.)*, at which time it dropped to two meters below sea level. There is considerable doubt as to how the influx from the Nile, during the short time it was high, could possibly have maintained the lake at its largest extent for such a long period of time, against the losses due to evaporation. In modern times, the lake has been deliberately lowered to make more fertile land available, and the lake itself has become primarily a smelly sump for excess flood waters.

We suggest that the Faiyum depression was maintained at this high level by the transient flooding associated with priori-Mars during the dance encounters. Thus, its waters were replenished every fourteen or twenty-eight years. The influx became saline part way through the Vedic period. This had a negative effect on agriculture, which explains the particular attention paid to this area during the Middle Kingdom. However, each filling of this large lake with sea water brought millions of fish and crustaceans. The fact that the level had dropped to a much lower level at the time of Herodotus and continued dropping to its current low state is consistent with the V/A hypothesis in that after 687 B.C. the floods which had filled it during the Vedic period, no longer occurred.

Ancient Ships Stranded in the Desert

There is more evidence for severe periodic floods in Egypt but, at the insistence of geologists, archaeologists have been forced to accept dates for these that stretch back to the end of the last ice age. This is not consistent with high water marks, some two thirds of the way up the great pyramid at Giza. Further corroboration of the massive flooding in Egypt is documented in the following quote:

> In the wildest part of the desert ... it is possible to wander over plains strewn with shells of innumerable oysters ... and to see in vision a surge of tumbling waters where now all is silent, rockbound, and utterly devoid of movement and life. [162]

Another mystery from Egypt was first reported in the London Guardian, on 21 December 1991, as follows:

> A fleet of 5000-year-old royal ships has been found buried eight miles from the Nile. American and Egyptian archaeologists discovered the 12 large wooden boats at Abydos ... Experts said the boats - which are 50 to 60 feet long - about 5000 years old ... were probably meant for burial so the souls of the pharaohs could be transported in them.

Each boat had been enclosed in a boat-shaped brick 'grave,' and according to the 1991 season research report from David O'Connor, curator of the Egyptian Section of the University of Pennsylvania:

> Each 'grave' had originally been thickly coated with mud plaster and whitewash so the impression would have been that of twelve huge 'boats' moored out in the desert ... The notion of their being moored was taken so seriously that an irregularly shaped small boulder was found placed near the 'prow' or 'stern' of several boat graves ... as 'anchors' intended to help 'moor' the boats.

It is possible that an eccentric Pharaoh might want a 'boat grave,' but an entire fleet seems unnecessarily extravagant, considering the effort required and the value of the boats. The amount of wood available to build boats was severely limited. It all had to be imported from Lebanon. Also it is illogical that these would be placed so remote any body of water.

Our hypothesis gives a rationale whereby this entire fleet of 5000-year-old ships was 'enshrined' some eight miles from the nearest water. These boats were swept into the desert, where they now stand, as a result of one of the floods associated with a capture or escape of priori-Mars. They became stranded there in a local depression when the flood subsided. As long as water remained in this basin, they could still sail within it. So the stone anchors were not just symbolic, they were actually used for the mooring of the boats. The water eventually evaporated and left them useless and rotting. It was at that time that they were enshrined.

Ancient Ships 'Left' on the Bottom of the Mediterranean

Because of the emptying of the Mediterranean Sea, upon each capture of priori-Mars, many ships were stranded when the water was suddenly 'pulled out from under them.' One clue to recognizing the resulting ship 'wrecks' is that they and their cargos would have remained intact but been abandoned on the exposed sea bottom as the waters withdrew. Their sheer weight would have precluded the ships or their cargoes from being moved any significant distance. Since the water vacated for fourteen years, they would have suffered from weathering and rotting to the degree that they would not have floated when the waters returned.

Although boats were not very large in those times and were constructed primarily of wood, a number of wrecks dating from the Bronze Age have been discovered, primarily by sponge divers. Within recent years at least four Bronze and Iron Age ships have been studied and as many as seventy others have been located along the coast of Turkey but not yet investigated. Well documented wrecks include the Late Bronze Age ships at Cape Gelidonya (circa 1200 B. C. Phoenician or Cypriote), and at Ulu Burun near Kas (circa 1300 B. C. Canaanite or Mycenaean). Although most of the timbers have disappeared, 'oxhide' shaped ingots of copper and tin have been found still stacked in orderly fashion and a number of pithoi, large ceramic vessels used to carry all manner of smaller cargo, have been found intact. The fact that these cargos were not scattered on the sea floor implies that the boats did not founder as a result of storms. Two wrecks dated somewhat later also have been found intact on the bottom of the eastern Mediterranean.[163] These were discovered near Kyrenia Castle, Cyprus and off the coast of Israel (Kibbutz Ma'agen Micha'el), twenty miles south of Haifa. The Ma'agen Micha'el ship is estimated to have met its fate in about 400 B.C. and the Kyrenia ship about 75 years later - that is, after the Vedic period. We believe these were probably lost during the Vedic period because the dates given are based primarily on the identification of patterns on the earthen vessels used for storage on-board the ships, which have considerable margin for error.

The following quote from the article on the Ma'agen Micha'el ship ends with a question which is apropos to our hypothesis:

> The pottery remains, certainly one of the main clues by which we attempt to date our ship and follow its sailing route, presently point us to Cyprus and Greece. Yet so many questions remain. Why did the ship land as it did, perpendicular to the coastline? It was abandoned, but was still accessible so close to shore.

The ancient ships found off the coast of Turkey and Israel are good potential candidates for vessels that were stranded by the withdrawal of the waters at the onset of one of the dance encounters, because they were found in areas that were left high and dry at these times. Present day archaeologists interested in these wrecks do not even know that this possibility exists.

Did the Assyrians Hire Phoenician Ships to Attack Cyprus?

Another archaeological discovery is of interest in connection with the proposed V/A scenario. This is summarized in an article concerning the Phoenicians.[164]

> After the Assyrians destroyed Dor in 733 B.C.E. (and presumably deported many of its inhabitants in accordance with Assyrian custom), the city was soon rebuilt. It seems that the Assyrians, who had no navy or maritime expertise, needed the cooperation of the Phoenicians to develop their trade with the maritime states of the Mediterranean. At one point, according to an Assyrian stela of Sargon II found at Kition in Cyprus, the Assyrians also ruled Cyprus. The Assyrian literary evidence is confirmed by excavations on Cyprus. The Assyrians no doubt reached the island only with active Phoenician assistance, and their continued rule of the island was heavily dependent on the Phoenician fleet.

The fact that the Assyrians ruled the island of Cyprus for a period of time around 700 B.C. does not necessarily imply that they needed the cooperation of the Phoenicians. The V/A hypothesis is that the waters of the Mediterranean still repeatedly receded at this date and remained absent for fourteen years at a time. As a result, the Assyrian army could easily have marched to Cyprus across a land bridge to the northeast (Figure 16.1) which is now beneath the waters of the Mediterranean Sea. It would have been exposed if the waters were 800 meters below the current level. We speculate that the Assyrian rule of Cyprus may very well have ended when priori-Mars disengaged from the Earth and the sea returned, since this would have isolated their garrison and allowed the Cypriots to take back their land.

Evidence of commerce or military marches on what is now submerged land may still be detectable by sonar imaging of the bottom of the Mediterranean Sea in the form of roads, disabled vehicles and even buildings constructed during the times when the sea subsided. There are many known ruins along the northern Mediterranean coast which are submerged today as well as evidence of submerged settlements off the coast of Israel. However, actual roads across the bottom, which may now be covered with silt from the Nile, have never been seriously sought. Perhaps some intrepid submariners know these only too well but cannot speak about them for 'security' reasons.

Clues to ocean level changes on the Earth in past ages abound. Higher levels are indicated in a number of currently landlocked reservoirs such as the Jordan River Valley where many ancient water marks are easily visible on the walls of the cliffs that bound the river. Although fed now only by fresh water flowing down from Mount Hermon, the Dead Sea was once full of salt water, as evidenced by the thousands of salt formations which were left behind as the water evaporated.

It is an established fact that the salinity of the Mediterranean Sea and the Red Sea are considerably higher than the oceans.[165] As a result, surface currents of low salinity continually flow into these seas from the ocean at Gibraltar and Aden respectively, while bottom currents of denser, higher salinity water continually flow out. These inward flowing surface currents prevented many ancient sailing vessels from exiting these bodies of water. The Phoenicians demonstrated their seafaring knowledge by lowering weighted 'sails' into the deeper currents, catching the outward flowing bottom currents. This enabled them to fight the inward flowing surface currents, and exit the Mediterranean at will.

The highest salinity ever measured was found in the Discovery Basin in the Eastern Mediterranean. These brines are the result of the dissolution of bischofite ($MgCl_2.6H_2O$). This is the last solid which precipitates when sea water evaporates. Its presence indicates that the water evaporated completely.[166]

> The Mediterranean was drilled at 15 sites during leg 13 . . .
>
> The results of this leg were particularly exciting, because they yielded evidence of drying up of the deep basins, followed by sudden incursions of water, presumably from the Atlantic. Large deformations including overthrust faults were indicated ...

The evidence could not be much clearer. *The salt beds are thousands of feet thick and drilled cores of have yielded evidence that this drying up of the deep basins was repeated on numerous occasions, each followed by a sudden incursion of water, presumably from the Atlantic.* Geologists currently believe that these deposits formed when the drift of Spain into Africa cut off the Mediterranean from the Atlantic Ocean some fifteen million years ago. Then six million years ago the 'wall' holding out the Atlantic was breached and the ocean flowed back, taking a century to fill the Mediterranean to its present level.[167] In their theory the Mediterranean water was trapped and evaporated faster than the influx of clear water from rain and rivers. This theory fails to explain the multiple layers which comprise the deposits. One 'load' of seawater could not possibly provide the amount of salt in these beds. It also fails to account for the shells of freshwater creatures with their fleshy parts intact, in the Ehrenberg drilling samples.

The Sub-Surface Geology of the Mediterranean Sea

The youngest faults on Earth are the Red Sea-East Africa rift valley and the smaller rift valley that extends from the Gulf of Aqaba to the Dead Sea. They are attributed to the Pleistocene (1.8 to 0.5 Million Years Before the Present). These age estimates are based on the amount of salt and other evaporates that have been deposited on top of the igneous rocks (fissure basalts) that rose up from the earth's interior at the time of the faulting. Geologists argue that the deposition of the salt in bodies such as the Red Sea, which are currently in contact with the ocean, must have occurred at a time when they were isolated from the ocean because of a reduced sea level. The only mechanism that they can conceive of, which would have reduced the sea level, is the wide spread glaciation associated with the ice ages, which terminated more than 12,000 years before the present. At these times a large amount of water was thought to have been trapped as ice on land and therefore was absent from the oceans. This leads oceanographers to believe that the evaporates could not have been deposited more recently than about twelve thousand years ago.

We claim that the salt deposits just below the entire floor of the Mediterranean and Red Sea were formed at the *same* date as the intrusion of the igneous rocks. The tidal force of priori-Mars, during the dance encounters, was the cause of the faulting as well as the removal of the waters from the area. Salt was deposited in these seas during the dance encounters when the level of the water was very low. Therefore the presence of the salt deposits does not mean that the intrusions occurred as long ago as the Pleistocene. When priori-Mars was released, the sea water quickly flowed back in, but the salt that had been deposited did not dissolve. This not only explains the high salinity, but also the repeated deposition of bischofites, followed by the rapid submersion.

There is also evidence of recent tectonic activity on the Mediterranean Sea floor, off the coast of Europe and in the south east where deposits from the outflow of the Nile river, the so-called Nile Abyssal Plain, has been lifted to the top of the Mediterranean Ridge.

> In the Hellenic Trench ... lower Cretaceous limestones were found overlying flat bedded middle Pliocene pelagic ooze. In the Levantine Basin it was revealed that the Mediterranean Ridge was part of the Nile Abyssal Plain as late as the Pleistocene. These Nile sediments were upheaved onto the ridge. [168]

Other manifestations of low sea levels in the Mediterranean are submerged erosion features referred to as channels or canyons. When rivers run down to the sea, the landforms associated with them changes abruptly at the coastline. Whereas rivers meander and erode the soil and rock through which they flow in v-shaped troughs when above water, they do just the opposite when they enter the sea. There they release the material eroded from the land surface into alluvial deposits or fans of material under water. When large amounts are eroded it sometimes evidences itself above the surface as a delta such as exist at the mouths of the Nile and Mississippi.

At a number of locations in the Mediterranean, *erosion features similar to those formed by rivers on land are found on the sea floor*. There are v-shaped troughs and canyons and meandering channels cut in alluvial material during earlier epochs. In fact, there is considerable evidence that the flat abyssal plains, the deepest features, are due to such sub-areal erosion, that is, erosion that took place in the continental shelves when they were above sea level.

The discovery of these channels or canyons has caused considerable controversy among marine geologists, who cannot agree on their cause. It is generally accepted that the oceans of the world may have been lowered as much as 140 meters during the peak of the ice ages. Thus at that time rivers could have cut these features on the bottom that was exposed. However, these canyons go to much greater depths than this. Another theory holds that the canyons are more ancient and are below the sea because the coastlines were subsequently lowered as a result of faulting. However, there is little evidence that such activities could have occurred as recently as indicated by the age of the sediments, and unlikely that the same type of tectonic events occurred along so many coastlines. The explanation is made even more difficult by evidence that these erosion events occurred on repeated occasions interspersed with periods in which they were inundated. We claim that these features were eroded during the dance encounters, when the water level was extremely low, as shown in the H.G. Wells map.

Long Term Increases in Sea Level

Although the cyclic sea level changes in the Mediterranean were very pronounced, there was a more subtle systematic change in the sea levels around the world throughout the Vedic period. The temporary ocean tides due to the dance encounters were superimposed on another ocean change, also due to the proximity of priori-Mars to the Earth. One, in fact, which was even more important in shaping the earth as we know it. This was mentioned earlier in connection with the slowing down of the Earth's spin in ancient times and with the filling of great lakes in Tibet. *The sea levels on the Earth continuously increased throughout the entire period of the dance encounters, some 3,000 years, due to the gradual but systematic capturing of the entire water complement of priori-Mars.* The tidal attraction of the Earth, punctuated by periodic internal convulsions caused by alignments with the Moon, drew the atmosphere and along with it all the evaporated water on priori-Mars to the Earth. *As a result of this long term process, we estimate that the ocean levels on the Earth increased by more than 500 meters during the Vedic period.* Thus the background against which the short term tidal events, associated with each dance encounter, took place, gradually changed. This is another reason that almost no ancient settlements have been found along the current coastline of the Mediterranean. There were undoubtedly some coastal settlements built throughout the Vedic period, but as the level of the sea increased they were covered by the waters and are now submerged.

Three of the four great Phoenician ports in the Mediterranean, Arvad, Tyre and Tripoli, as well as the ancient Egyptian port, Alexandria, are centered around offshore islets. It is currently thought that the ancient cultures which built them put them on these islands, but this would have involved the carrying of all construction materials from the shore out to the island. Moreover, many of the ancient harbor works now lie under water. Breakwaters were continuously rebuilt at these sites up to early A.D. At Appolonia, Libya, the ancient port of Cyrene, there is a mile long stretch of underwater works, including what seem to have been large shallow pools for the raising and harvesting of fish or shellfish. There are numerous examples of buildings off the European coast of the Mediterranean now under water, the most extensive of which is the complex of structures along the shores of the Bay of Naples, Pozzuoli. Although these structures do not lie at the great depths of the submerged reef bases and canyons, the increase in sea levels required to submerge them cannot be explained by the current theories because they obviously occurred within proto-historical times. It may be that they were built after the end of the Vedic period, and were inundated as water trapped in Tibetan lakes and the North American glacier made its way into the oceans.

The ancient Canaanite city of Ashdod, which was located on a prominence inland from the present Mediterranean coastline to protect it from the periodic flooding, declined rapidly after 600 B.C. and moved to a new "Ashdod on the Coast." That is, once the periodic floods ceased at the end of the Vedic period and coastline stabilized at its current position, protection from the floods was no longer necessary, and the convenience of the coast for trade and commerce became paramount.

The large salt nodules protruding above the surface in the Dead Sea and in other portions of the low lying Jordan rift valley are due to the evaporation of sea water trapped there after each tidal flood. This rift valley was widened by the tidal force of priori-Mars on the landmass, and was filled with salt water by the transient floods which flowed through it every fourteen years. So much sea water, trapped during the later dance encounters, has evaporated that the water level is now the lowest on Earth, standing some 1,400 feet below sea level.

Strong evidence for *lower* ocean levels in the past also exist around the world. There are numerous instances of submerged barrier islands, similar to the ones along current coastlines, located near the edges of the continental shelves. Submerged canyons are also found on continental shelves, in many cases extending from the present coastlines to the margin of the shelf. Some examples are Monterey Canyon off the coast of California, the west coast of Florida, and Baltimore canyon on the middle Atlantic coast of the United States. They have all the characteristics of subaerial (above water) erosion. Again, the submerged eroded canyons imply that the land over which they used to flow was above water. Exotic mechanisms such as 'turbidity currents' and 'freshwater sapping' have been invoked by scientists to explain these canyons, because they cannot explain how they could have been above water in the geologically recent past.

Deep-Seated Reefs

Additional evidence comes in the form of living coral reefs, around the world, with bases at depths of hundreds up to a thousand meters. Since reefs are formed by living organisms that require sunlight, they cannot grow if their top is deeper than 45 meters. Therefore these reefs could not have begun growing at the current depths of their bases. Generally, there are two explanations offered for these formations. Either the sea level was lower when they first formed and has risen slowly since that time, or the sea bottom on which they are located has gradually subsided since they began growing. The latter is less likely because it requires subsidence of a number of surfaces at different locations around the globe during the same period and at the same slow rate, whereas a gradual raising of the ocean level would apply to all parts of the world.

Some rising of sea levels by limited amounts in the past are currently explained as the result of the ice ages. When the last ice age ended this volume of water is thought to have flowed back into the oceans causing a sea level increase. But in order that the resulting rise in sea level not exceed the maximum growth rate of the reefs, a few inches per year, *the melting period is hypothesized to have taken over a thousand years.*

Even if one accepts this extremely slow melting process, the amount of increase is far from sufficient to account for the many reefs whose bases lie at depths as great as 500 meters and the submerged canyons which extend to the limits of the continental shelves. Since modern scientists cannot conceive of any mechanism whereby the sea level could have been lowered more than 140 meters in geologically recent times, they have been forced to reject the hypothesis of a continuously rising ocean level, and therefore attribute these reefs to subsidence of the ocean floor.

The V/A scenario offers a straightforward solution to this mystery. It is consistent with the fact that these mystery reefs occur in different parts of the world. Perhaps *most importantly, the slow rate of the sea level increase that we propose, that is, 500 meters per 3,000 years, or 17 cm per year, is consistent with the average growth rate of the reefs, which is less than 30 cm per year. Any other mechanism, such as the subsidence of the sea floor at many locations around the globe, would also have to honor this rate - a highly unlikely prospect.*

Tells, Ziggurats, and Pyramids

An important reason why archaeology has been so successful in the Middle East is that one civilization after another have built upon the same elevated sites, or 'tells,' thought to have been fortified city-states. These are found throughout the Levant and Mesopotamia. The sites are easily found because they stand above the local terrain and have flat tops. The concentration of many cultures built on top of one another has greatly simplified the determination of their time ordering and dating. Numerous tells are found in the Levant, most of which were occupied in the Bronze and Iron ages - so many in fact that they have not all been excavated to this day. Many had been buried in desert sands for thousands of years, until they were recognized as sites of ancient civilizations, and began to be excavated by the new science of Archaeology in the nineteenth century. Their biblical names had been long forgotten, and they were known by the Bedouins by names such as Tell es-Sultan, the biblical Jericho. Although they were not interested in digging them up, the arabs who named them knew that they were not natural hills, which in their language is 'jebel,' as opposed to 'tell.' Some lie close to or within the boundaries of modern cities and their ancient secrets are sometimes accidently revealed as a result of public works, as in Tel Aviv.

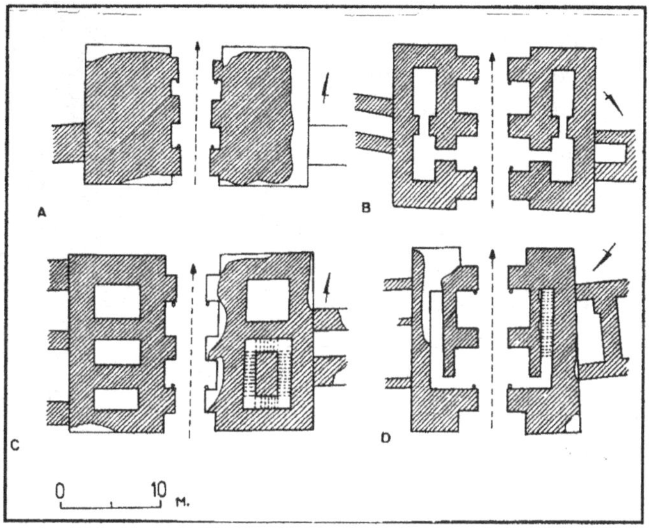

Figure 16.2 Plans of several Middle Bronze II B-C gates: (a) Gezer; (b)Hazor; (c) Yavneh-Yam; (d) Shechem, (after A. Mazar, *Archaeology of the Land of the Bible*, Doubleday, 1992, Fig. 6.12, p.206.)

The 'fortified' cities or towns that occupied the tells were relatively small in size -primarily because of the tremendous manual effort required to build them. Most occupied only a few acres, while the largest were perhaps a few tens of acres. Because of their thick towering

Figure 16.3 Tel Beth-Shean, occupied from the Chalcolithic Age through the Iron age, lies in the Jezreel Valley, in modern day Syria. (after A. Mazar, *Archaeology of the Land of the Bible*, Doubleday, 1992, Fig.1..3,p.10.)

walls, they are thought to have been under continuous threat of attack from nomadic tribes or perhaps the 'armies' of other towns. As the communities grew in numbers, only the privileged lived within the walls while the poorer class lived and worked in the surrounding fields. The fortifications of these cities are quite remarkable, primarily due to their thickness. Typical thicknesses were 3 to 4 meters wide in the Early Bronze II. At many settlements the thickness continued to be increased in the later part of EB II and EB III to 7 or 8 meters. At Yarmuth a second wall was built outside the first and plastered on the outside. The resulting system reached a thickness of approximately 40 meters and surrounded an area of about 40 acres. Jericho was surrounded by two walls 10 or 12 feet apart. The outer wall was some 25 to 30 feet high and the inner wall was more massive - some twelve feet thick. Towers were often associated with the walls. The larger ones attained horizontal dimensions of 8 meters by over twenty meters. Unusually complex gates, examples of which are shown in Figure 16.2, admitted traffic through these massive walls. These are also thought to have improved the ability to defend them against attacking armies.

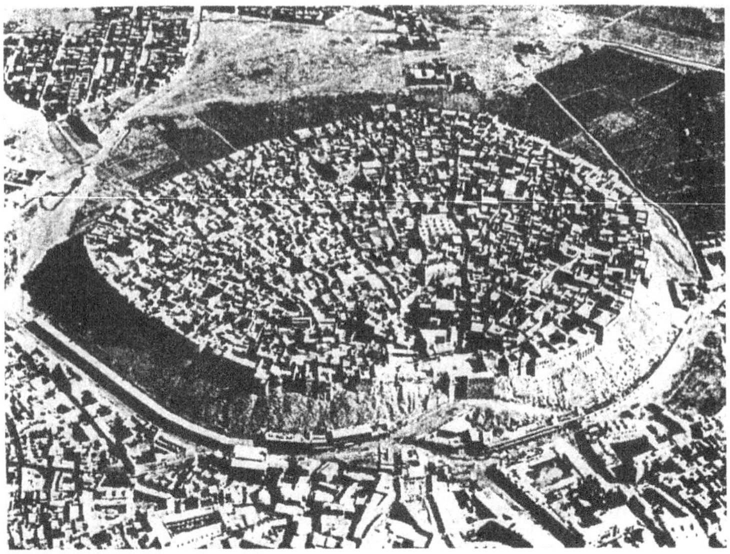

Figure 16.4 Arbela, established by 2850 B.C., the modern city of Erbil, Iraq, gives an idea of how a thriving ancient tel appeared over four thousand years ago.

The famous archaeologist Kathleen M. Kenyon has determined that the walls of Jericho were rebuilt at least seventeen times during the Bronze Age, possibly due to damage from earthquakes or erosion. The early bronze age city walls were sometimes retained by earthwork glacis consisting of layers of earth or crushed lime piled up against their exteriors. These earth ramparts are thought to have been added to reinforce the wall against erosion and prevent an enemy from approaching with siege equipment, but these would only *facilitate* the scaling of the wall.

Although the ancient structures on many tells are in ruins, the tells themselves stand out because of their prominent elevation above the surrounding terrain. Tel Beth-Shean is an example - occupied from the Chalcolithic period through the Iron Age, a period of almost four thousand years, is shown in Figure 16.3. The modern city of Erbil, Iraq, built upon an ancient tell, is shown in Figure 16.4. The close-packed houses and the wall, which is still mostly intact, cause it to appear as did many tells in ancient times.

Tells are currently interpreted as defensive structures, implying that a warlike status existed between Levantine city-states for thousands of years B.C. But it is not clear what adversaries merited such robust defences. The only significant military threat in the area could have come from

Egypt but there is no evidence of Egyptian military activity in these areas in the Early Bronze Age, between the First Dynasty and the Fifth or Sixth Dynasty.

The fortifications became increasingly robust as time went by. In the Middle Bronze II Age they were described as follows:

> During the eighteenth and seventeenth centuries B.C.E. the art of fortification reached a level of unparalleled sophistication. Tremendous efforts were invested by the Middle Bronze Age urban communities to defend their cities utilizing techniques which were by now widespread throughout the entire Levant. The idea was to surround the city with steep artificial slopes which will raise the level of the city wall high above the surrounding area and locate it as far as possible from the foot of the slope so that siege devices such as battering rams, ladders, and tunneling methods would not be effective. ...
>
> ... Earth ramparts were constructed already in MB IIA at a few of the cities on the northern coastal plain. In MB II B they became a common feature at new urban centers which were founded on shallow topography, such as Hazor (the Lower City), Dan, Kabri, Tel Mevorakh, Dor, Shechem, Tel Batash (Timnah), Ashkelon, and Tel Masos (in the northern Negev). The huge earth ramparts were erected by dumping large quantities of earth, sometimes on both sides of a central vertical core which served as a foundation for a freestanding city..The creation of the rampart usually resulted in a high crater, inside of which the city developed. ... Similar earth ramparts are known in Syria ... Some of these ramparts were huge: at Ebla the rampart surrounds an area of almost 150 acres; at Qatna it circumvents 250 acres. Only the site of Hazor with its 200 acres can rival these large Syrian cities ... Ramparts are also known at Tell el Yehudiyeh and Heliopolis in the eastern Delta of Egypt, and they were considered to be a typical Hyksos fortification, originating in the culture of Syria-Palestine. [169]

Because of their similarity to the structures in the Levant, certain stepped pyramids in northern Egypt were thought to have been imported by the 'foreign' rulers, or Hyksos princes which ruled lower Egypt during the Middle Bronze Age. Egypt also falls within the region known as the 'Fertile Crescent.'

Although tells also existed in Mesopotamia, the focus of the great Bronze Age cities were the ziggurats, or towered cities. The famous Ur of the Chaldees was rediscovered near a remote rural train station in southern Iraq when archaeologists investigated a nondescript mound formed by the ruins of a once glorious stepped tower, around which the original city was built. An artists conception of the ziggurat at its prime is shown in Figure 16.5.

In the following quote, Joseph Campbell gives a glorified interpretation of the cultural significance of the ziggurats:

> When the abstract concept of a geometrically organized aesthetic field first appears and the late neolithic market towns begin to elevate temple towers; and now, finally, we have come to an epochal date, c. 3200 B.C., when suddenly, at precisely that geographical point where the rivers Tigris and Euphrates reach the Persian Gulf, the wonderful culture-flower comes to blossom of the hieratic city state.
> ... The whole city now (not simply the temple area) is conceived as an imitation on earth of the celestial order - a sociological middle cosmos, or mesocosm, between the macrocosm of the universe and the microcosm of the individual, making visible their essential form: with the king in the center (either as the sun or the moon, according to the culture) and an organization of the walled city, in the manner of a mandala, about the central sanctum of the palace and the ziggurat. [170]

The timing of the ziggurats appearance corresponds to the beginning of the Bronze Age, that is about 3200 B.C. The ziggurats in Mesopotamia, the pyramids in Egypt, the walled settlements on the tells in the Levant, and the acropoli (fortified hills at the center of each Greek city-state) - all carry through a common theme. Although each culture implemented the idea in its own way, each required a strong will on the part of the inhabitants and a great commitment of labor and organizational skills in order to build and supply them.

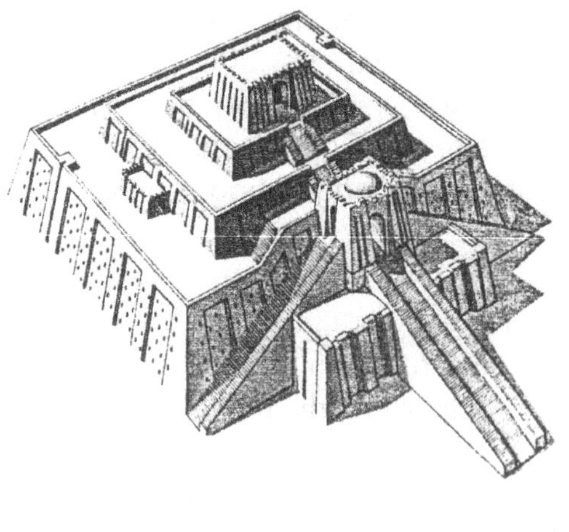

Figure 16.5 Reconstruction of the terraced ziggurat at Ur as it appeared about 2000 B.C.

What evidence do the tells and ziggurats provide that corroborates the V/A scenario? Because of the great time spans compressed in a few tens of meters containing the archaeological evidence on the tells of the Levant, it is very difficult to detect the effects of cycles as short as fourteen years, that might give evidence of the periodic flooding at the onset and termination of each dance encounter. But it turns out that such dissection is not necessary. *The tells in the Levant, the ziggurats in Mesopotamia, the Acropolis in Greece and the pyramids in Egypt are themselves the evidence.* The reason that ancient cultures were concentrated on or near these elevated sites and structures was not because of an enlightened desire to emulate the heavenly sphere or to build extravagant tombs for their rulers. *They were built to allow the populace to survive the ravages of these periodic floods.*

Furthermore, *the ubiquitous thick-walled structures surrounding the tells, interpreted currently as fortifications against attack by enemies, were in fact dams for additional protection against these floods.* The convoluted gates of the cities were not designed to make it difficult for invading armies to pass through them. They were constructed so that earth, stored on top of the wall, could be quickly dumped into them when the flood came and would not wash away. The large towers perched atop the walls were the last refuge of the people when the floods overwhelmed all the other defenses against the rising waters.

An important clue to the true purpose of the 'fortifications' on the tells of the Levant was the piling of earth ramparts or glacis against the walls, sometimes both on the inside and the outside. This method is used to this day in the building of crude dams. The reason for these currently postulated by the archaeologists was to prevent siege equipment from being rolled up to the walls. *But the earth piled up against the wall would facilitate the scaling of the wall, obviating the need for siege devices.* The mixing of lime with the earth, commonly found, was used to stabilize it, to prevent it from being washed away by the flood waters. As noted by Kathleen Kenyon at the site of ancient Jericho, structures were repeatedly repaired to correct the ravages of these floods and the earthquakes which occurred at the same time - both of which were caused by the approach or disengagement of priori-Mars. The ziggurats were the more sophisticated Mesopotamian answer to the same dilemma of periodic flooding, but many tells are also found there.

The answer to the great mystery of why the Egyptians built the pyramids now becomes obvious. *The pyramids were, first and foremost, 'life rafts,' places of refuge for the people living in the area during the periodic floods of the Vedic period.* This interpretation is consistent with their siting on high places, such as the Giza plateau. This was the Egyptian solution to both the flooding and earthquakes - equivalent to the tells and the ziggurats of other cultures in the region. They had essentially no interior space because they were not meant to be lived *in*. Also, they were not primarily tombs, a fact corroborated by the almost complete lack of art and treasures found in their few small rooms. They had steps all around so that thousands of people could rush up simultaneously when the flood waters came. The flood came in a stealthy way, except for the fact that it was accompanied by strong earthquake activity. There was not a long period of rain and a consequent rising of the rivers. *This flood consisted of the entire Mediterranean Sea or Red Sea flowing up over its banks, not just in river and stream beds but everywhere as far as the eye could see. These seas just quietly rose up and flowed across the land as if an immensely powerful hand had tipped the earth.* In our scenario *the construction of larger and larger pyramids during the Bronze and Iron Ages was required to provide safe haven for the continually increasing population.* The pyramid form was chosen to provide quick access from all directions as well as to provide the most stable form to withstand the continual earthquakes and to resist erosion by the flood waters.

This also explains why some of the earlier pyramids and even later ones were roughly constructed, that is, even a roughly constructed one provided refuge. The staged construction of the earlier pyramids was obviously an attempt to accommodate the increasing population in the most expedient way. The great pyramids at Giza must have had the capacity to save tens of thousands of

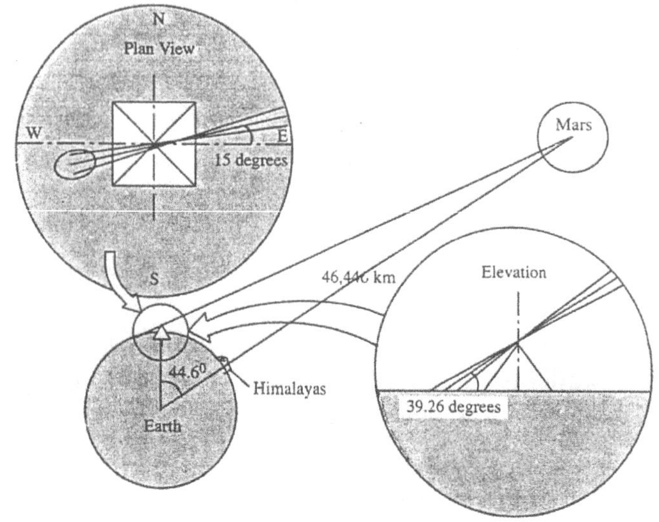

Figure 16.6 Geometrical relationship of Mars with the pyramids at Giza. An observer on the ground within the ellipse in the top view would have seen Mars at the top of the pyramid, no matter the time of day or night.

citizens. Their great size was obviously necessary because there are signs of water levels two thirds of the way up the largest pyramid.

The small 'burial chambers' are located high above ground level to avoid the flood waters and their ventilation shafts, that were such an enormous engineering challenge, slant upward so as to exit the pyramids as high as possible, thereby reducing the chance that the flood waters would enter them.

This implies that survival during these floods was the motivation of the workers who built the pyramids, rather than a greatly overrated devotion toward a great leader who wanted an extravagant 'tomb.' It is an established fact that most Pharaohs were not buried in pyramids. Also, the alignment of the pyramids with the current cardinal directions means they were not aligned with the position of Mars in the sky during the dance encounters. Although Giza and Mt. Kailas are at similar latitudes, an observer at Giza would have had to face some 15 degrees north of east to be directly aligned with priori-Mars during the dance encounters. The Egyptians chose to align these structures with the true north pole and not the temporary pole that existed during the encounters. We infer from this that the primary function of the pyramids was not for worship of Horus or Ra.

The elevation angle of the line of sight from Giza to priori-Mars is also of some interest. From a position on the ground to the west south west of each pyramid priori-Mars could be seen poised at its pinnacle. Since priori-Mars remained stationary over the Himalayas during each dance encounter this viewpoint remained the same regardless of the time of day or night. Taking the position of priori-Mars to be that proposed based on the mass of priori-Mars during the dance encounters, it is possible to calculate this viewpoint. Descriptions in the pyramid texts imply that ancient ceremonies were performed at a point from which the disc appeared at the pinnacle of the pyramid. The Egyptian texts imply Horus(Mars) could be seen poised at the top of the pyramid throughout the climb, as the Pharaoh's body was carried to the top. As can be seen in Figure 16.6, an observer at Giza saw the center of priori-Mars at an elevation angle of about 39 degrees above the eastern horizon during the dance encounters, which does not match the angle of the pyramid structure. But because the angular extent of the disc of priori-Mars was of the order of 10.78 degrees, compared to the Moon which is about 0.5 degrees, and the apparent area of the disc over four hundred times greater, it was not necessary to have an exact alignment for this to be true.

Interestingly, the view of the 'eye of Horus' poised at the top of the pyramid was probably the origin of the mystical Masonic obelisk that appears on the United States one dollar bill. *The eye inside the pinnacle is now called the 'eternal eye of God.'*

There is currently considerable controversy over the age of the Sphinx. As discussed in Chapter 10, the great Sphinx of Egypt faces toward the approximate position occupied by priori-Mars over the Himalayas and its name in Egyptian is 'Horus on the Horizon.' Horus was the primary deity associated with the planet priori-Mars, and was often referred to by a hieroglyphic of a falcon, derived from the beak-like shape of what is now called the Tharsis Bulge. Recently two researchers, Schwaller de Lubicz and Dr. Robert Schoch, have suggested that the Sphinx is much older than the pyramids, that is, the product of a more ancient culture. They base their hypothesis on evidence that the weathering of the Sphinx is characteristic of water. Since they believe that significant rain has not fallen at Giza for some twelve millennia, they suggest that the Sphinx was built that long ago.[171] *The V/A explanation is much simpler. The many periodic priori-Mars-induced floods from the Mediterranean and Red seas are what caused the weathering, and they all occurred during the Vedic period. Thus there is no reason to attribute a much greater age to the Sphinx than to the pyramids.*

The Distribution of City-States During the Bronze Age

The distribution of settlements in the Levant from 3300 to about 687 B.C. provides some interesting clues to the events of that period. In the Early Bronze I period all the settlements are inland, well away from the current coastline of the Mediterranean. The large Chalcolithic centers in the highlands were abandoned and the EB I settlements tended to be in the coastal plane which was better suited to agriculture. The sixty-four dollar question is - *Why are there no sites at all on what is now the coastline of the Mediterranean?* This is very difficult to explain. Even if large ships could not yet be built, there must have been commerce or fishing using smaller ships. No scholars seem to have noticed this discrepancy.

Commerce was carried on via heavily used land routes, particularly the inappropriately named Via Maris, which was used as early as EB II. It stretched from Egypt across the northern Negev and north through Palestine. Many of the ancient Palestinian settlements are along this route. The settlements tended to be on high places near rivers which cross the route on their way toward the Mediterranean Basin but not along the current coastline. During the next eight hundred years (Early Bronze II-III) some of the EB I sites were abandoned. These tended to be the ones closest to the Sea and in the southern part of Palestine.

In this period a relatively large number of fortified cities existed on tells in Palestine. Some of the largest of these seem to be in very remote areas far from roads, water sources and fertile land. The reason for these locations is yet to be explained. Indeed, the reason for the concentration of the population (estimated at 150,000 in EB II and III) in urban centers is a subject of debate.

At the end of the EB III something caused these cities to be abandoned or destroyed when they were at the peak of their development, and replaced by a totally different non-urban pattern which lasted about three hundred years. Theories as to the cause of the abandonment are military attacks by the Egyptian or Ammorites, or possibly a decline in precipitation and a consequent dropping of the water table. The latter could lead to a lack of water for irrigation as well as drinking and eventually to starvation and plagues. Since cisterns were not used, the entire water supply was from springs and collected rainwater. If these stopped producing, the population would have had few choices but to migrate to other areas. The primary argument against this environmental theory is that settlements in Transjordan literally 'on the other side of the river' seemed to have continued unaffected through the period in spite of their greater vulnerability to drought.

For some three hundred years following the collapse of the cities at the end of EB III, Palestine was sparsely populated with pastoral settlements, none of which were on the present coastline of the Mediterranean. This period of decline was paralleled in Egypt. In Syria the (now) coastal cities of Byblos and Ras Shamra suffered a severe destruction. A period of decline paralleling that in Palestine followed in Ras Shamra but in Byblos it appears that commerce was resumed after the destruction.

At the end of the Early Bronze Age and the beginning of the Middle Bronze (EB IV and MB I) pastoral settlements appeared in the Negev. Why they appear in such an arid environment when the area was virtually uninhabited during the previous period of urban development is not known.

Conversely the Transjordan area (Iktanu, Khirbet Iskander, Arock and Ader) north and east of the Dead Sea appears to have maintained its culture virtually unchanged during the period of decline. Pottery clues indicate that people from the entire area of Palestine may have come here to find refuge from whatever caused the devastation.

There were vast cemeteries at this time, the style of which differed from the family graves of the previous period. The shift in custom use to be interpreted as evidence of the invasion of an outside culture but the latest interpretations favor development within the indigenous population as the driver.

The Middle Bronze II period (2000 - 1550 B.C.) was characterized by a revival of urban centers in Palestine. These were more numerous and more highly developed than those of the Early Bronze, some thousand years prior to this time. The culture of this period is referred to as Canaanite and it flourished for about a thousand years and then diminished. The settlement pattern differed from that of the EB II-III period in that it was centered in northern Palestine and a number of cities were located on the coastal plain, whereas this area was hardly settled at all in the earlier period. One of the most remarkable thing about these cities were their massive 'fortifications.'

The end of the Middle Bronze Age is associated with the destruction of most of the great cities in the region. The Hyksos princes, 'foreign rulers' thought to be from Canaan, who ruled lower Egypt for several generations, were thrown out of Egypt and pursued into Palestine. It is believed that the destruction of all these city-states was a result of military turmoil associated with this event and also possibly due to Hittite raids into northern Syria. Unlike the collapse of the cities in EB III, this did not result in the complete destruction of the Canaanite culture.

Some of the major cities in the northern part of the country such as Hazor and Megiddo were soon rebuilt on the same outlines, major buildings such as temples were rehabilitated, and the Canaanite culture, as evidenced by pottery technique, art and craft, did not come to an end.

The distribution of Levantine city states in the Early and Middle Bronze Ages reveal three important patterns. First, their locations were always on elevated terrain. Second, only one or two were on the (current) Mediterranean coast. Third, both ages lasted approximately a thousand years and were terminated by catastrophic events, from which the recovery took some three centuries. The

V/A explanation of all three is based on the tidal flooding which accompanied the capture and release of priori-Mars, plus the inexorable capturing of the entire water complement of priori-Mars throughout the Vedic period. As a result of this gradual increase in sea level, modulated by the large cyclic tidal excursions, which alternately filled and emptied the entire basin, the coastline of the Mediterranean was in a constant state of flux. It would have been foolhardy to locate cities on this existing coastline at any date during the Vedic period. This is not to say that it was not done. But cities built during any of the fourteen year stability intervals, were soon washed away by the advancing waters, and so no trace of them has been found on the bottom of the Mediterranean. The reason one or two ancient cities, such as Byblos, are located on the present coastline is because they were located on elevated places, which fortuitously ended up on the current coastline because of their altitude.

The primary consideration was to locate as high as possible. But this was never sufficient -as evidenced by the more and more extravagant dams constructed and maintained as the ages progressed, and the widespread destructions which punctuated each age. These destructions were not due to the Egyptians, Assyrians and Babylonians. They were due to unique combinations of tidal flooding and earthquakes caused by priori-Mars, which produced tidal waves that overwhelmed all their defenses.

The Mediterranean Cultural Upheaval

The Minoan culture, centered in the Aegean, existed for over a millennium, before its demise in the Middle Bronze Age. The Mycenaean culture, located on the peninsula of what is now Greece, succeeded the Minoan and even surpassed it in cultural accomplishments. Yet it also disappeared under mysterious circumstances.

Some archaeological research in the Levant and Egypt indicates that the 'People of the Sea,' so named in Egyptian documents, who migrated into Canaan and attempted to move into Egypt, were Mycenaean, but this has never been proven. Although they may have come to this area hundreds of years prior to this time, the period beginning around 1400 B.C. (radiocarbon date)is now considered the date of the mass movement of 'The People of the Sea' into the Levant.[172] They comprised a number of tribes, including the Philistines. Their migration is thought to have been the result of a collapse of the Mycenaean culture located on the mainland of Greece and the surrounding islands. The Philistines received some 'bad press' in the Bible because they were adversaries of the Jews, but indications are that this was an advanced culture and one that brought with it a many sophisticated aspects, including political and military leadership, architecture, and artistic forms found on their pottery. This implies that a significant portion of their entire culture had migrated to the Levant.

Reliefs at Medinet Habu, an enormous monument built by Ramases II, depict sea and land battles between the People of the Sea and the Egyptians. They show the ships of the two nations engaged in battle, and illustrate differences not only in the ships but the combat dress of the two armies. An interesting sidelight of the relief depicting a land battle is that it shows what are interpreted as the families and belongings of the People of the Sea being carried in horse drawn wagons.

Figure 16.7 The Acropolis which stands high above the city of Athens was the site of the earliest occupation.

No consensus has been reached among archaeologists as to the reason that these and other peoples decided to move out of the area in which they lived for many centuries and migrate into Palestine. The idea that a natural event, such has drought, is the most popular. We suggest that the ebb and flood of the Mediterranean caused by the advent and retreat of priori-Mars from orbit around the Earth was the immediate cause.

The concentration of second and third millennium B.C. cities on high places with massive fortifications was not unique to the Levant, but was the rule on the Greek peninsula as well:[173]

> Thucydides wrote that prior to King Theseus 'what is now the Acropolis was the city, together with the region at the foot of the Acropolis toward the south. Other literary sources also indicate that the Acropolis was the site of one of these early palaces, and although later construction has eliminated all traces of the building itself, massive fortification walls, a secret fountain, and terrace walls all indicate its position on the citadel.

The importance of high places and fortifications extended also to the Greek peninsula. Figure 16.7 shows how the acropolis towers above the modern city of Athens. Obviously the flooding affected this area also.

Iron Age Settlements

The pattern of settlements located on tells and on natural hilltops surrounded by walls and earth ramparts; and of their periodic destruction and renewal, continued on throughout the iron age. This corresponded to the period in which the Bible becomes the authoritative source of history on the Israelites and all those peoples that came in contact with them. Iron Age I (1200 to 1000 B.C.) corresponds to the period of the Judges, Iron Age IIA (1000 to 925 B.C.) to the United Monarchy (Saul, David and Solomon), and Iron Age II B-C (925 - 586 B.C.) to the Divided Monarchy (Israel in the north and Judah in the south). Some powerful aggressive nations, such as the Assyrians under Sennacherib (around 701 B.C.) and the Babylonians under Nebucanesr (around 527 B.C.) invaded the Levant, and the destruction of many settlements are attributed to them. However, great devastations had been occurring for millennia previous to this time, and these remained the primary destructive force, even throughout the Iron Age.

Major Destructions During the Vedic Period

The period of almost three thousand years, beginning at EB I was one of great accomplishments and great turmoil in Mesopotamia, the Levant, the Mediterranean, and in Egypt. Unprecedented building feats and fortifications were apparently successful in staving off destruction. However, every millennium or so, natural events conspired to overwhelm the best preparations of men. Entire civilizations essentially vanished and others seem to have picked up and migrated thousands of miles. Perhaps most puzzling is the fact that although modern man has uncovered so much information about these sites and deciphered much of the ancient writing, no straightforward explanation of the nature of the turmoil or the destruction has been found.

The V/A scenario explains suggests how the existing cities were practically all destroyed at the end of the Early Bronze III and again at the end of the Middle Bronze Age. As mentioned previously, there was a certain variability in the approach and departure of priori-Mars associated with the dance encounters, with some causing more severe earthquakes than others. During these transitions there were almost continuous earthquakes in the Mediterranean area, especially along the major rifts in the Red Sea and the Jordan River valley. When earthquakes occurred on the bottom of the Mediterranean Sea at the same time as the tidal flooding, the augmented tidal waves could have greatly increased the destruction. It is highly probable that the enormous explosion of the volcano on the island of Santorini, estimated to have taken place around 1600 B.C. (Radiocarbon date) occurred during the capture or release of priori-Mars.

The secret of at least one of the major destructions of cultures in the Bronze Age lies in the long term increase in sea level that took place throughout the Vedic period due to the influx of water from priori-Mars. Unusually large sudden increases in sea level in the Mediterranean area could have been caused by one of several mechanisms.

The most likely event, capable of causing a major devastation in the Mediterranean and adjacent lands, was tectonic in nature. The sill currently separating the Atlantic Ocean from the Mediterranean Sea is only about 400 meters deep. Therefore our assumption that the pre-Vedic sea levels on Earth were more than 500 meters below the current levels, implies that the Mediterranean was a group of fresh water lakes before the beginning of the Vedic period. A possible configuration is shown in Figure 16.1. This means that large areas within the current Mediterranean, such as the Aegean Sea, was normally above water. As a result, the transient flooding in the Early Bronze Age probably consisted of fresh water being drawn up onto the coastal plain and into Mesopotamia. This was more conducive to irrigation and fertilization of the land in the region.

As the Vedic period progressed the sea level rose due to the influx of water from priori-Mars. The Atlantic Ocean finally attained the level of the sill between it and the Mediterranean. At the next release of priori-Mars the Atlantic Ocean waters were high enough to flow into the Mediterranean basin for the first time. This not only increased the flood levels greatly, but began introducing brackish waters into the equation. But the more catastrophic scenario, which would fit the pattern of the destruction of 2250 B.C. is as follows: The sill between the two bodies of water was even higher than now, and as the level of the Atlantic Ocean increased, it exerted greater and greater force on it. This, combined with the earthquakes associated with the subsequent termination of the dance encounter finally ruptured the sill and the sea rushed through the gap with such force that it was quickly widened and deepened, causing a catastrophic influx of sea water into the Mediterranean. This could have initiated a major environmental disaster for the entire area. The flood protection structures that had held up and had been maintained by the hard work of the people between floods, would suddenly have been insufficient. In addition, the salty water would have wiped out much of the agricultural potential in the area.

Simultaneous tidal flooding and earthquakes probably account for the destructions which occurred around 2250 B.C. or 1550 B.C., both of which occurred after approximately a millennium of growth and expansion of great city-states. Both were also followed by a three hundred year recovery period, indicating an almost total annihilation of the population. (It amazes us, that archaeologists treat such 'silent' periods so lightly.) If this was the cause of the destruction across a broad front at the end of the Early Bronze III period (2300 B.C.), it would also explain why the settlements at higher elevations in Transjordan were spared these destructions - because the tidal flood did not reach that altitude. The sudden destructions also explain the large numbers of graves found at certain critical times, and the use of unconventional burial methods. When the floods subsided, a number of bodies were found, all washed together at a remote location. These bodies were then buried at the site by the most expedient means. The abandonment of the family burial plots in the Levant at this time was not due to the introduction of another culture with different traditions, as has been suggested by some archaeologists. This has been corroborated by the comparison of the human bones from before and after the disaster.

Our scenario is also consistent with the recent finding that the Black Sea was also once a freshwater lake, which was suddenly inundated with salt water increasing its level some 500 feet and submerging some 150,000 square km of land. Pitman and Ryan, the geologists who discovered this have dated the event at 7,500 years ago. This is 1,500 years before the date we estimate for the destructive proto-Venus encounters, which caused the seas to sweep across the continents, and therefore could have filled the lake with salt water.

The Fate of the Minoan Culture

There is a long-standing mystery surrounding the demise of the Minoan culture. This culture is known to have occupied a number of islands in the Aegean Sea, which is the shallowest part of the Mediterranean Sea and seems to have been centered on what is now the island of Crete, at the southern boundary of the Aegean Sea. Archaeological evidence suggests that it prospered for a very long time, from 2800 to1600 B.C. (radiocarbon dates), and was characterized as agricultural early on, and later as a seafaring culture. The final disappearance of this civilization has been attributed to the explosion of an enormous volcano which vaporized a large fraction of the island of Thera (Greek) or Santorini (Italian). Evidence suggests that this explosion was greater than that of Krakatoa. However, it appears that this culture had long since passed its prime at the time of this event.

The V/A scenario also offers an explanation for their demise. We claim that at the beginning of the Vedic period the entire Aegean Sea region was above water, as shown in Figure 16.1 and that the Minoan people farmed this entire area. Thus theirs was initially an agrarian culture. Early in the Vedic period, the freshwater flooding may have actually irrigated their land. However, when the

Atlantic Ocean broke through the straights of Gibraltar it caused massive flooding of this area with salt water, killing many and forcing them to abandon their flooded farmland, retreating to the high ground, which is now known collectively as the 'Greek islands.' This forced the survivors to become a seafaring people, and the explosion of Santorini finished them off.

Archaeological clues corroborate the notion of a much lower sea level in the ancient Mediterranean. The following quote, which refers to the period up to about 1000 B.C., is of interest in this regard:[174]

> Not only were the lands of ancient times bound together by water; they also shared a common "Mediterranean" climate that encouraged similar agricultural practices and patterns of human settlement. Still it would be misleading to overemphasize either geographical or cultural uniformities. The Mediterranean was divided into a number of subsidiary seas: the Tyrrhenian or Sardinian west of Italy; the narrow Adriatic down which harsh winds often blew; the island-studded Aegean and others.
>
> ... The movement of physical objects within each basin can be traced back as far back into Neolithic times (down to about 3000 B.C.). Obsidian from the island of Melos even appears on the Greek mainland in the preceding Mesolithic era ... Predynastic Egypt, in the fourth millennium, had established continuous relations with Syria ... By at least the third millennium [3000 to 2000 B.C.] interconnections between basins grew more noticeable ... Across the second millennium, the height of the Bronze Age, physical objects, peoples, and even ideas moved about on a far larger scale, to the point that one can speak of lines of trade.

This indicates exactly the progression we propose - that what are currently recognized as 'basins' within the single Mediterranean Sea, due to their deeper depths, were in ancient times separated by land so that it was not possible for maritime commerce to move from one to the other. This is illustrated in Figure 16.1 which shows the coastlines of the Mediterranean Sea as it would be if the sea level were one thousand meters below the current level. As can be seen, the Adriatic and Aegean Seas would have been above sea level, and quite likely supported entire agricultural societies. The portion that was under water in those days consisted of three large lakes between which there was no maritime connection.

As the sea level rose, evidence of some commerce began to appear between 3000 and 2000 B.C., and by the second millennium, after water from priori-Mars had flowed to the Earth for well over a thousand years, the sea level became high enough to support true commerce throughout the Mediterranean Sea. However, complete dependence upon shipping for commerce was still impossible because of the tidal variations during the dance encounters. These interrupted marine commerce by lowering the sea level in the Mediterranean for the fourteen year periods of each 'dance encounter.' When each one was broken off, the water levels increased for the subsequent fourteen years, allowing more commerce during those periods.

Climatic and Cultural Change in the Pacific Basin

In October of 1998 the 'International Conference on Climate and Culture at 3000 B.C.' was held at the University of Maine. A number of the papers presented indicated that a major climatic change occurred around 5800 years B. P. (before the present)[175], which is equivalent to 5000 years by radiocarbon dating.[176] This was derived from inorganic laminae in alpine lakes in Equador, geoarchaeology in Peru, Andean ice cores, coral records, lake sediments in northern Chile and in the Galapagos. In a number of cases these changes seem to be related to changes to higher cultural achievements involving building and agriculture. In general the climate was warmer *and less variable* for 'several millennia' before 5000 years B.P., than in the immediately following period.

Cores from a lake in Equador were analyzed to determine the periods associated with the depositional changes.[177] One result was evidence of climatic oscillations with a period between fifteen and seventeen years, which began late in the third millennium around B.C. and ceased around 700 B.C. The authors propose that the greater climatic fluctuations after 3000 B.C. (radiocarbon date) were due to the onset of a period of El Nino cycles, but we believe that they were due to much more powerful physical events, involving the capture and release of priori-Mars in orbit around the Earth, which shifted the latitude of Equador some thirty degrees every fourteen years. This data corresponds closely to the proposed interval (15 years) between each capture and release of priori-Mars as well as to the duration of the Vedic period from 4000 B.C. to 687 B.C. B

Biblical 'Evidence'

Believe it or not, the Biblical story of the Exodus of the Jews from Egypt. provides corroboration of the V/A scenario. The 'pillar of smoke and fire' or archangel Michael, was the great lava fountain on priori-Mars. Since this only developed some time into each encounter, the cosmic events at the time of the Exodus must have corresponded to a separation, as opposed to a capture, of priori-Mars from one of the 'dance encounters.' The Red Sea was dry as the Jews approached because it had been empty for some fourteen years. But the Lord instructed Moses to encamp on the shore of the Red Sea until the time was right, in spite of the fact that the Egyptian army was catching up to them. As priori-Mars disengaged from the Himalayas and began drifting westward, the Lord told Moses to cross the Red Sea. The drift of priori-Mars over the middle east is indicated in the Bible which states that the pillar of fire and smoke moved from a position before the Israelites to one between them and the pursuing Egyptian army. Sand that had drifted across the Red Sea during the previous fourteen years may also have slowed the waters, which rose as priori-Mars approached the area. Also the flow into the Red Sea was retarded because the entrance to both the Red sea and the Mediterranean are constricted. The Jews crossed the Red Sea at the last minute, and the pursuing Egyptian army was drowned as the waters closed in.

This scenario answers three questions that any thinking person must have asked themselves. Why did the Lord delay their crossing? To time the crossing of the Egyptian crossing to coincide with the influx of water. How could the Jews be impatient to cross the Red Sea before it was 'parted'? It was already parted. The waters that rose up were the returning waters, lifted by the tidal force of priori-Mars. How could the Jews walk on the muddy bottom of the sea immediately after the waters were withdrawn? The waters had been withdrawn for fourteen years and the bottom was completely dry.

Based on his study of the Torah, Talmud and the Midrashim, Velikovsky discovered a sequence disruptive events, or *raash*, at intervals of 'fourteen to sixteen' years, toward the end of the Vedic period. He attributed them to Mars, although the red planet was not specifically mentioned in the biblical stories. In spite of the fact that he never determined the true nature of the dance encounters, his research does provide considerable corroborative material to the new V/A hypothesis.

Biblical descriptions of these events, involving the reorientation of the axis of the Earth at intervals of about fifteen years, are discussed by Velikovsky in Part 2 of *Worlds in Collision*. Based on his dating of events, we conclude that these were caused by captures and releases of priori-Mars, which occurred in the last sixty-years of the Vedic period. Although he attributes the anomalous motion of the sun and stars to close approaches of Mars, he did not realize that they were due to the *capture and release of priori-Mars*. In fact, the Bible makes no overt reference to an extraterrestrial body.[178] He pieces together the following series of catastrophic events, primarily from the scriptures:

The prophet Amos correctly prophesied a great flood and the early setting of the sun during the reign of King Uzziah in Judah. This was termed a 'raash' or great 'commotion' involving floods and earthquakes. Velikovsky claims that the same flood was described in Egyptian documents during the reign of Osorkon II of the Libyan Dynasty, which he dates at 747 B.C. Isaiah began prophesying immediately after this event, that another catastrophe would occur in fifteen years. This interval was supposedly based on knowledge of a lesser event which afflicted Zebulum and Naphtali fifteen years prior to the raash of Uzziah. He predicted that:

> the stars of heaven ... shall not give their light: the sun shall be darkened in his going forth and the Moon shall not cause her light to shine ... the Earth shall remove out of her place.

The catastrophe occurred as prophesied in the reign of Ahaz: with great earthquakes; the shifting and tilting of the terrestrial axis; the hastening of the sun by several hours; and the 'blacking out' of the Moon and stars at night. Velikovsky claims this event occurred on the day King Ahaz was buried and that it was also noted in the Greek classic Thyestes, by Seneca:

> not yet as day fades into night has the third trump sounded; the ploughman with oxen yet unwearied stands amazed at his supper hour's quick coming. What has driven thee from thy heavenly course? ... *Has Typhoenus (Typhon) thrown off the mountainous mass and set his body free?*

Although it sounds like Seneca might be observing the departure of priori-Mars from its temporary orbit around the earth, he was more likely just describing the abnormal motion of the Earth at the time of a departure. Typhon (Egyptian Set) was the deity name given to the serpentine tail of proto-Venus when it approached the Earth early in the Vedic period, at which time it was seen as a combatant with Pallas-Athene. When this tail dissipated it was thought to have fallen to earth and become imprisoned underground. Thus Seneca was merely referring to a 'mythical' deity whose release, he thought, might have affected the rotation of the Earth. Priori-Mars was certainly visible in Greece on the eastern horizon during the dance encounters. It passed through phases like the Moon every 24 hours as it moved into and out of the shadow of the Earth. The Greeks certainly should have noted its capture and release and correlated these events with anomalies in the rotation rate of the Earth. Clouds or darkness could have prevented their direct observation of these events.

After this catastrophe the heavenly bodies crossed their former paths obliquely, that is, the poles were shifted and the number of days per year changed. Seneca describes the changes in the positions of each of the constellations and indicates that:

> The great bear, which was never bathed in the sea, shall be plunged beneath the all engulfing waves.

He also says that the great bear, or one of its stars never set beneath the horizon, *and thus the polar star was among its stars,* during the prior age.

Isaiah prophesied two more events at intervals of fifteen years. The cynical will say that the prediction of another 'commotion' fifteen years later would have been easy because of the regular intervals between the recurring events. But Isaiah went much further on the last occasion - so much further that no rational person can deny that he had received information from the Lord. At that time the Assyrian general, Sennacherib, had set siege to Jerusalem for the second time, with an army of a hundred and eighty-five thousand men. Isaiah prophesied that the Lord would deliver Judah from the Assyrians and the sign would be that the stars of heaven would be rolled up like a scroll and would not shine. Talmud and Midrashim sources state that the entire Assyrian army was destroyed by a single 'blast' in the middle of the night sent by 'an angel of the Lord, Malach,' 'one who is sent to execute an order,' at the very time this was prophesied.[179] Velikovsky claims that the cause of this destruction was the same as that of the previous *raash* which had occurred at fifteen year intervals, that is, priori-Mars.

> The Talmud and Midrash sources, which are numerous, all agree on the manner in which the Assyrian host was destroyed: a blast fell from the sky on the camp of Sennacherib. It was not a flame, but a consuming blast. Their souls were burnt, though their garments remained intact The phenomena was accompanied by a terrific noise.

Isaiah also prophesied concerning changes in the orientation of the spin axis of the Earth as another sign of Judah's pending deliverance from Sennacherib saying:[180]

> And this shall be a sign unto thee from the Lord ... Behold I will bring again the shadow of the degrees which is gone down on the sundial of Ahaz, ten degrees backward. So the sun returned ten degrees, by which degrees it was gone down.

The Talmud explains that the day had been shortened by some number of 'degrees' back on the day when Ahaz was buried and that part of that time was restored 'by ten degrees' at the time when Sennacherib's army was destroyed, as prophesied. Velikovsky claims that this cosmic event was also noted in several Chinese histories as a time in which the sun went down, then reappeared 'passing through three solar mansions.' He cites Greeks Seneca and Ovid as stating that the sun veered from its normal course, and a Mayan inscription that says a planet brushed close to the Earth.

The Egyptian city, Letapolis, which was also threatened by Sennacherib at the same time, is indicated by the hieroglyphic that means 'thunderbolt.' An additional fact unearthed by Velikovsky is that the 'holy animal' of the Egyptian city Letapolis was the mouse. He relates that Herodotus saw a statue there with a god holding a mouse in his hand, commemorating the annihilation of the army of Sennacherib. To demonstrate the ubiquitous nature of the mouse legend he reveals the myth of the Menomini Indians, an Algonquin tribe, that tells how the Sun was retarded in its motion by 'ropes' until a mouse came along, chewed through the ropes, and freed the Sun. Velikovsky does not give an unequivocal explanation for this puzzle, except to say that the mouse must represent the body that caused the disturbance of the Earth at that time. This correlation of events between the eastern and western hemispheres also makes a strong case for the great age of the American Indian culture.

Velikovsky dates this monumental event at 687 B.C., a date of particular significance, because it signaled the end of the Vedic period. It corresponds to the date of the death of Romulus in Roman myth. It is also noted in ancient Chinese astronomical writings as a year of an intense meteor shower. Additional corroboration of our scenario comes from the fact that King Hezekiah initiated calendar reform of some kind during his reign, which extended beyond 687 B.C., since once priori-Mars disengaged from the Earth, the length of the day changed considerably. He also attributes calendar reform to Numa, the successor of Romulus, who was a contemporary of Hezekiah. The date of the last cataclysm is placed at 660 B.C. by the Japanese, and Diogenes Laertius attributed the earliest determination of the modern calendar to Thales the Milesian in the second half of the seventh century B.C. (650 to 600 B.C.). This coincides with the Babylonian exile of the Jews, and when they returned from captivity they brought with them their present calendar, in which the months are called by Assyro-Babylonian names.

Velikovsky's Biblical Research

The *raash* cited by Velikovsky were incidents for which no specific cause was noted in the Bible or the Talmud. This undoubtedly was a source of frustration to Velikovsky, who made it abundantly clear that he believed these events were caused by Mars, but for which he could find no direct corroboration in the ancient books. In spite of this, Velikovsky's research in this area dovetails nicely with our interpretation of the Hindu myth. It corroborates the deductions we have made concerning the 'dance' configuration - particularly the fact that the orientation of the Earth's spin axis changed and that this occurred at approximately fourteen or fifteen year intervals. It is also consistent with the notion that people living in the middle east might not have been able to see priori-Mars at the time of these transitions, either because it approached from the Sun side or drifted away on the night side in the shadow of the Earth. But it is eminently clear that the ancient Jews observed priori-Mars, in fact their worship of it, as Canaanite priori-Mars deity Aserah, was the cause of the greatest anger on the part of Jehovah, for which he punished them greatly at the hands of the Assyrians and the Babylonians.

The 'ten degrees' lost at the time of King Ahaz's funeral probably corresponds to a shortening of the daylight period. The obvious interpretation in the context of the V/A theory is that this was an occasion on which the north pole shifted so as to move Judah away from the temporary equator toward the present pole. This event was probably one in which priori-Mars was *disengaging* from a dance encounter.

It is doubtful that any ancients understood that priori-Mars could only remain stationary in the heavens if it revolved around the Earth once per day, but they could see the capture and breakaway, assuming that clouds or darkness did not always obscure these events.

The description of the change of paths of the constellations by Seneca points up another unknown. Even if we knew for certain that the pole or spin axis of the lithosphere shifted from its current position to the Hudson Bay area, this does not give its absolute orientation relative to the stars. Seneca's quote tells us that prior to the 'release of Typhon' that the great bear constellation remained above the horizon throughout the year, whereas after the spin axis shifted from Hudson Bay back to a 'more normal' position, part or all of this constellation dipped below the horizon at some point each year. Since the great bear does not now dip into the ocean with the changing seasons as seen from Greece, and since we consider the current configuration to also be the 'normal' state, we conclude that each engagement and disengagement did not result in exactly the same orientation of the Earth relative to the stars. The approximate angular change of the pole orientation required to cause the great bear to pass beneath the ocean, as seen from Greece, is about ten degrees. Also ten degrees was specifically mentioned in connection with the events associated with the destruction of Sennacherib's army. However, because both the location of the spin axis on the surface and its orientation with respect to the stars probably changed it is difficult to reconstruct the actual changes from Seneca's narrative.

It is logical to expect that the successive orientations of the spin axis varied somewhat because they depended on the exact relationship of priori-Mars and the Himalayan-Tibetan complex as it approached and departed. It is also of interest to note that the magnetic north pole of the Earth is currently about eleven degrees from the true north pole. This reflects the net misalignment of the spin axis of the core and lithosphere resulting from the many interactions of the Earth with proto-Venus and priori-Mars.

The 'blast' that decimated Sennacherib's army was a thunderbolt caused by the planet priori-Mars, which, in the process of disengaging, drifted over the middle east at night. As prophesied by Isaiah, the fixed stars were blacked out 'like a scroll being rolled up,' but shooting stars, actually bodies being ejected from volcanic vents of priori-Mars, were seen during this time. They were ejected due to the internal stress associated with the release. When they fell into the Earth's atmosphere at enormous speeds approaching 11 kilometers per second, they shorted out the Earth-ionosphere 'capacitor' and an enormous electrical discharge called a thunderbolt resulted. This was essentially the same as the crack of thunder associated with a nearby stroke of ordinary lightning, except millions of times more powerful. The downward shock caused by this enormous discharge is what killed the Assyrian army. There is nothing in our present day world that can create a blast of this magnitude, except possibly the blast of a nuclear explosion. However the fireball associated with these devices completely incinerates everything in the vicinity so that the effect of the blast wave is not evident. In the case of the thunderbolt there may be a crater but the primary killing mechanism from 'over-pressure.' The effects are even more unique when the blast is directed vertically downward. Everything remains in its place so there is no apparent cause of the death of those unfortunate enough to be present at the site of the strike.

The passage of priori-Mars over this region at night was the reason why the stars and Moon did not appear in the clear sky. Because priori-Mars was still close to the Earth, it was completely in the shadow of the latter, and although it blocked the view of many stars, these were replaced by the light from volcanos and from the rock bodies that were periodically launched from them. These 'shooting stars' were called Maruts in the Vedas, and were quite a familiar and terrifying experience for the Vedic people who lived in the Himalayan foothills. Thousands of these fell in the Himalayas and the thunderbolts which they triggered were pictured as weapons used by Indra against the Vritra, the great serpent-glaciers which confined the precious water in the high mountain valleys. As familiar as they were to the Vedic people, the thunderbolts were not something the people of the middle east had any first hand knowledge of, and the fact that priori-Mars drifted over the middle east in darkness meant that they never saw the source.

Let anyone who tends to dismiss our description of the mass annihilation of Sennacherib's army consider the following: In June of 1993 PBS TV stations in the USA broadcast an historical program about the World War II 'battle of the bulge.' During that show, one veteran recalled an experience which, although it occurred fifty years earlier, was indelibly etched in his mind. After a bombardment of the town in which he was located by German 88 millimeter guns, he came across a house, one entire wall of which had been blown out. Inside of that house an entire family was seated at the dinner table. They were 'frozen' in position and nothing in the room seemed out of order. The entire family had died instantly from a blast but remained exactly as they were at the time of the blast. The soldier said he had to later confirm that he had really seen this, by discussing it with another member of his unit, because it was so bizarre. *This was the result of a downward directed blast, similar to, but many orders of magnitude less energetic than the blast that killed Sennacherib's army of 185,000 men.*

This explanation of the destruction of the Assyrian host does not detract from the miraculous nature of the event which saved the Judean city in the biblical story. It merely implies that the event had an even more miraculous origin - a cosmic one. As with the events at the time of the Exodus, one cannot help but be impressed by the predictions of the prophets that actually detailed upcoming physical events such as the number of degrees that the earth's axis would be reoriented, the passage of priori-Mars over the middle east at night and the consequent blacking out of the sky - let alone the deliverance of the Israelites from the Assyrians by a perfectly directed thunderbolt. The implication of this story is that the Lord had such complete knowledge of natural events which were set in motion millennia beforehand, to make possible such a miracle, must put any thinking person in even more awe of the omnipotence and omniscience of the Lord .

By the time priori-Mars had drifted as far as the western hemisphere it had passed out of the Earth's shadow and was visible. This explains why the Mayans say that a planet had 'brushed close to the Earth' at that time.

Velikovsky quotes from the book of Isaiah as follows:

For as the new heavens and the new earth, which I will [do] make, shall remain before me, saith the Lord, so shall your seed and your name remain.

In this scripture the Lord, through Isaiah, promised that there would be no more cosmic, catastrophic interactions, and the "new heavens" would stay the same from that time on. In other words, the Vedic period was ended, and the people would no longer have to deal with the threat of floods and earthquakes, due to priori-Mars approaching the Earth.

The Firmament

The meaning of *firmament* has been shrouded in mystery for millennia and many scholars of ancient and modern times have attempted to divine what it actually was. The word implies something firm, solid, and strong, and this is consistent with the writings of the most ancient cultures, yet it has now come to mean the sky. How could such a diametrically opposite translation be explained?

One reason for this interpretation was the reference in many ancient writings to the firmament as the location of the fixed or circumpolar stars. The Egyptians conceived the 'heavens' to be an arched iron ceiling containing the stars, from which a ladder was lowered toward the Earth. They called the pure iron, obtainable only from meteorites, 'celestial metal' because it had fallen from 'heaven.' Similarly, the Babylonians saw the firmament as an immense dome forged from the hardest metal by Marduk (Mars). The early Greeks and Romans saw it as a great vault of crystal in which the stars were suspended.

In the first biblical account of creation, God created 'the firmament,' also called 'the heavens' in some translations. It was said 'to separate the waters above it from the waters below.' The Hebrew word which was translated into firmament was *raqia*, which literally means something 'hammered out' as a metal and thus expanded or extended. In Job 37:18 the 'heavens' are referred to as 'most strong, as if made of brass'. The ancient Jews postulated a three storied 'universe', in which the ceiling of Sheol, the underworld, is the surface of the Earth, and the ceiling of this middle story is the firmament with its ocean which is, in turn, the floor of the top story. The firmament separates the waters above from the waters below. It is said that God could open the flood gates of heaven and cause a deluge of the earth below.

These descriptions fit closely the V/A scenario in every respect. As priori-Mars approached the Earth it appeared larger and larger. This gave rise to the notion that it was 'expanded.' It then remained suspended above the Himalayas for fifteen years at a time. It appeared as a vast dome over twenty times the diameter and four hundred times the area of the full Moon. At the beginning of each encounter between the two, an entire ocean of water was drawn to the side facing the Earth. Melted subsurface rock was gradually drawn up through the tidal sea, thus the creation story states the land (the 'dry' in Hebrew) rose up through the waters, separating the waters above from the waters below. Because they are not aware of priori-Mars dominance of the sky at the time, these events are thought to be descriptions of events on Earth. Many volcanos erupted on the side of priori-Mars facing the Earth. These appeared as bright stars since human beings were looking directly down into them. There were seven dominant volcanos which reappeared during each dance encounter. They were the source of the sacred number seven in the myths of many cultures. They combined with many lesser volcanos to form patterns called 'asterisms', which dominated the constellations of true stars during the 'dance encounters.' These were described as 'fixed' or 'circum-polar' stars in many cultures, because they remained in the same place in the sky day and night, rotating with priori-Mars. The fact that it was said to contain the 'fixed stars' is one factor which has led to the misinterpretation of firmament as the sky in modern times, indeed, in all generations since the last dance encounter, around 687 BC.

When alignments of the Earth-priori-Mars pair with the Moon occurred, moreover with the Sun and Moon combined, great convulsions shot gigantic hot rocks, including some of pure iron from the interior of priori-Mars into space. Vast amounts of water and atmospheric gases were carried along and fell to Earth. This was the opening of the flood-gates of heaven.

Of course the primary reason why firmament is no longer understood is because the 'dance encounters' between priori-Mars and the Earth ceased some 2,700 years ago. In only a few generations people lost the thread of understanding what their ancient counterparts had lived with for millennia. It was then that the modern rationalist attitude took hold and began the denigration of the earlier ideas of the heavens, accepting the same order of things that we observe today. Now scientists lead the people to believe that the chaos described in the myths never happened.

As a corollary, they belittle the Christian belief in the intervention of God in our world. This attitude was forecast in the Bible. Quoting 2 Peter 3:

> 3 Knowing this first, that there shall come in the last days scoffers, walking after their own lusts,
> 4 And saying, Where is the promise of his coming? for since the fathers fell asleep, all things continue as they were from the beginning of the creation.
> 5 For this they willingly are ignorant of, that by the word of God the heavens were of old, and the earth standing out of the water and in the water:
> 6 Whereby the world that then was, being overflowed with water, perished:
> 7 But the heavens and the earth, which are now, by the same word are kept in store, reserved unto fire against the day of judgment and perdition of ungodly men.

This describes the 'sophisticated' views of scientists who are the 'scoffers.' The uniformitarian dogma that they embrace deliberately ignores the fact that the 'heaven of old', i.e. the firmament, was put in place by the power of God. In his omniscience, He prophesied this state of affairs in which modern academics, those 'ungodly men' claim that science, in the form of

Darwinian evolution, has replaced the need for God. Their ignorance will be exposed in the 'end times', which are soon to arrive.

Ancient Interest in Eclipses

The question arises as to why the ancient peoples were so interested in eclipses, particularly at times when priori-Mars and proto-Venus both posed much greater threats to the Earth. After all, a few hours dimming of the sun's light alone causes no great harm, and a lunar eclipse could easily go completely undetected.

The V/A scenario easily explains this. The fact that the most violent eruptions of priori-Mars occurred at times of its alignment with the Earth, the Sun and the Moon was the reason for the ancients' continuing concern with solar and lunar eclipses. The solar eclipse meant that the Sun and the Moon were aligned on the same side of the Earth. The lunar eclipse marked the times when the Moon and Sun were exactly aligned on opposite sides of the Earth. If the rapidly revolving Earth-priori-Mars pair passed through an alignment with the Sun and Moon, the most deadly meteorite falls occurred. Although the solar eclipse caused the worst disruption of priori-Mars the lunar eclipse was a close second.

17
ON THE ORIGIN OF MODERN MAN

Long enough have dream'd contemptible dreams,
Now I wash the gum from your eyes,
You must habit yourself to the dazzle of the light and of every moment of your life.

Long have you timidly waded holding a plank by the shore,
Now I will you to be a bold swimmer,
To jump off in the midst of the sea, rise again, nod to me, shout, and laughingly dash with your hair.

Walt Whitman
Song of Myself

The Velikovsky/Angiras theory postulates that recent catastrophism in the solar system was a crucial aspect in the advent of modern man, Homo Sapiens Sapiens, on earth. Anthropologists, like geologists, who are unaware of these global scale events, believe that modern man developed from hominid species that occupied the earth over the last 40,000 years or more. Archaeologists have recognized the sudden, miraculous development of intelligence at the beginning of the Bronze Age, but lacking an explanation, they have collectively come to downplay the magnitude of this revolution. Not only is the archaeological data on the side of the sudden blossoming of intelligence, but the ancient texts, including the Bible, describe how modern man became established on the earth, if people would only listen. Perhaps the one thing more exciting than the uncovering of the cosmic events that changed the Earth in the last 12,000 years, is the realization that these miraculous events led to the establishment of modern man on earth. Believe - because its you, and its true!

The Beginning of Civilization

The V/A theory postulates that massive tidal waves resulting from two close encounters of proto-Venus destroyed almost all animal life, and left the Earth 'flat, like a roof.' Low-lying landlocked areas remained inundated for years or even decades thereafter. In spite of the destruction, the descendants of the survivors appeared more innovative and intelligent as evidenced by the great flourishing of city-states in lands from the Indus Valley, to Mesopotamia and Egypt during the Early Bronze II (E. B. II) period. Possibly these civilizations arose only out of necessity, in order for the people to survive in the aftermath of near annihilation and the repeated onslaughts of priori-Mars. Perhaps this process facilitated a mini-Darwinian evolution in which only the most resourceful peoples survived. Perhaps not.

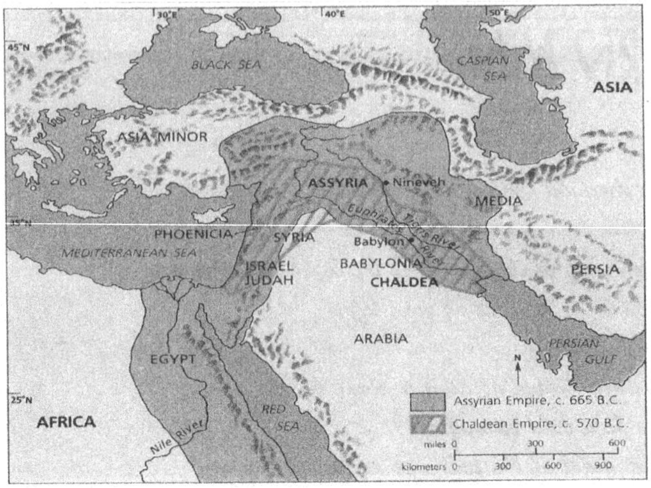

Figure 17.1 The region now known as the 'Fertile Crescent' was repeatedly flooded as a result of the capture and release of Mars in orbit around the Earth.

The priori-Mars-Earth dance encounters, which followed the two great destructions, had a profound effect on the development of civilization on Earth because each one lasted for years and they continued to occur for thousands of years. During that period they produced transient, but not totally destructive flooding at many locations around the world every fourteen years. The prime example of the positive effects of this flooding occurred in the territories to the east and southeast of the Mediterranean. The result of which is now known as the *Fertile Crescent*, shown in Figure 17.1. Other possible physical effects were due to the long term influx of enormous amounts of oxygen and water from priori-Mars. We claim that the increase in sea level by over 500 meters combined with the increased atmospheric oxygen profoundly improved the climate world wide. The higher concentration of oxygen invigorated the people and aided in building the great civilizations of the Early Bronze and later ages.

In addition to physical changes, these events produced major cultural changes. The periodic flooding due to priori-Mars engagements may have forced the people, who might otherwise have lived a tribal, nomadic life, into concentrated population centers for protection from the floods. This in turn could have encouraged specialization of trades, the interchange of ideas, and better organization of the society. On the negative side, sudden local increases in sea level, such as that resulting from the breakthrough of the Atlantic Ocean into the Mediterranean displaced or decimated entire cultures.

Lower ocean levels in the early part of the Vedic period also had a significant cultural influence. There existed land 'bridges' between many islands and continents and greatly reduced the over-sea distances between the rest. This, combined with the tidal shifts due to the periodic presence of priori-Mars, allowed peoples to migrate easily from continent to continent during the Bronze Age. For example, there was certainly a bridge from northeastern Asia to Alaska over which the antecedents of the Amerind peoples could have migrated, perhaps as late as 687 B.C.

The Aryans

Although these physical and cultural factors were real, we believe there were more profound reasons for the great leap in civilization which occurred in B.A. II. The evidence suggests that *the devastation caused by proto-Venus provided the opportunity for a benign but genetically superior race to expand, unchallenged, over the entire world at the beginning of the Bronze Age.*

The *Rig Veda* was apparently begun after the proto-Venus destructions, and was initially composed by a people known as the 'Aryans.' We therefore suggest that they were this 'superior race.' Although this does not reveal their origins, it does give them a name, and indicates they lived in the Indus Valley early in the Vedic period. The Early Bronze I, the three hundred year 'silent' or 'pre-urban' period was the time during which the population of the world was rebuilt. We propose that this people also profoundly influenced the early civilizations in Mesopotamia and Egypt and the land bridges mentioned previously made it possible for them to propagate throughout most of the world. Although three hundred years is thought of as 'the blink of an eye' in geologic time, it was certainly sufficient time for the Aryan race to propagate across the entire Eurasian continent. Consider how the world has changes since the founding of the United States of America only about two hundred and twenty-five years ago.

In spite of their acknowledgment that the Bronze Age was a watershed in proto-history, modern day anthropologists fail to recognize the profound changes that occurred at that date - the destruction of an older species and the propagation of a new, modern man, at that late date.

Linguistic, Archaeological and Genetic Evidence of the Kurgan Expansion

Researchers in these disciplines who study the proto-history of ancient cultures in Eurasia are currently at loggerheads because the evidence of each group is not consistent with that of the others. Archaeologists have discovered many sites in Europe and Asia indicating the presence of cultures that stretch back as far 9,000 years in the past. They estimate the population of European settlements to have reached significant proportions by the *beginning* of the Bronze Age. They reason that each relatively isolated settlement would have developed its own language, because of the limited means of transportation in the ancient world, when horses were not domesticated and the wheel not yet invented. However, linguists find striking similarities in nearly all the modern European languages, leading them to the conclusion that practically all of them derive from the same proto-Indo-European language.

Based on research by the British historian V. Gordon Childe, archaeologist Marija Gimbutas and more recently David W. Anthony, an archaeologist at Hartwick College and linguist Johanna Nichols of the University of California, a proposed origin of this language has been discovered in the western steppes north of the Caspian Sea, in what is now Ukraine, and dated at EB I.

Several theories have been advanced in an attempt to explain how one language, whose origin has been so specifically determined spatially and temporally, could have come to dominate the languages of so many other cultures that predated it, but no clear explanation has been put forth. The use of horses for pulling carts and chariots has been proven at the Ukrainian site, by the discovery of horses and chariots which were buried in the graves of their owners, some four hundred years earlier than these capabilities were thought to exist in Europe. One hypothesis is that their domestication of horses and the use of chariots enabled these 'Kurgan' peoples to dominate other cultures by their military might. Many question the notion that the use of chariots with the crudest wheels could be effective militarily at a time when no roads existed. Another hypothesis is that these people developed more advanced agricultural techniques and spread out to acquire more land, coming to dominate the local languages because of their more sophisticated technology. No

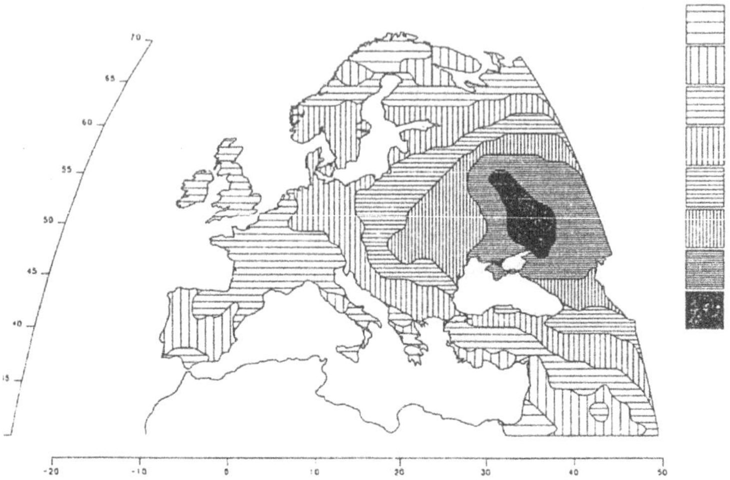

Figure 17.2 Synthetic gene map of Europe and western Asia obtained using the third principal component. (Courtesy of L.L. Cavalli-Sforza and F. *Cavalli-Sforza, The Great Human Diasporas*, Addison-Wesley, 1995, Figure 6.12, p.155.)

explanation has yet gained a consensus and the dominance of proto-Indo-European as the precursor of practically all modern languages throughout Europe and western Asia remains a mystery.

Our hypothesis of an Aryan expansion is greatly strengthened by the fact that these sciences agree on a specific location, language and a date identical to ours, for the origin of a unique people and their expansion from this area, named the 'Kurgan Expansion' by Gimbutas.

This theory finds support in the work of micro-biologists Cavalli-Sforza, Menozzi, and Piazza[181]. They map genetic variations, determined from blood samples of many modern people spread over the world, in such a way that it is possible to trace major expansions of the progeny from a source. Their method of 'principal components' allows the isolation of major expansions in spite of earlier and later expansions which overlap them spatially. The map shown in Figure 17.2 shows their 'third principal component' in Europe and western Asia. This indicates an outward expansion from the area north of the Caspian Sea, thus adding credence to the Kurgan expansion proposed by Gimbutas.

The Dominance of a Single Indo-European Language

In this chapter we suggest how one language could have come to dominate the Eurasian cultures. In the V/A scenario, the cultures for which there exist archaeological sites dated prior to the Vedic period were either completely destroyed at the time of the two proto-Venus encounters, or their populations were reduced to a very few people. The Aryan race was created in the aftermath of the devastating proto-Venus encounters and spread out over the continent in the following centuries populating, or influencing the population of the unoccupied land. In cases where a few survivors of local cultures remained they would have had little choice but to accept the language and the genes of this dominant group. In this way the Aryan genes and language were impressed on all future generations.

It is unfortunate that a stigma has become associated with the name Aryan in our time, because in myth this was always thought of as a venerable race. Ancient myths picture the Aryans as a cultured and benevolent people, not one that killed the indigenous survivors.

Some of these Aryans lived in the Indus valley and the foothills of the Himalayas, as close as they dared to the point above which priori-Mars remained 'stationary' during the 'dance encounters.' They probably combined their superior genes with those of the surviving indigenous people of the area, producing the four distinct 'castes' of India. The combining of two peoples to produce four such distinct offspring may have implications for the 'purity' of one or both races. The intermixing and obfuscation of these castes in India has been prevented by strict rules against their intermarriage. This obviously has not been the case in most other parts of Indo-Europe to which these people also migrated.

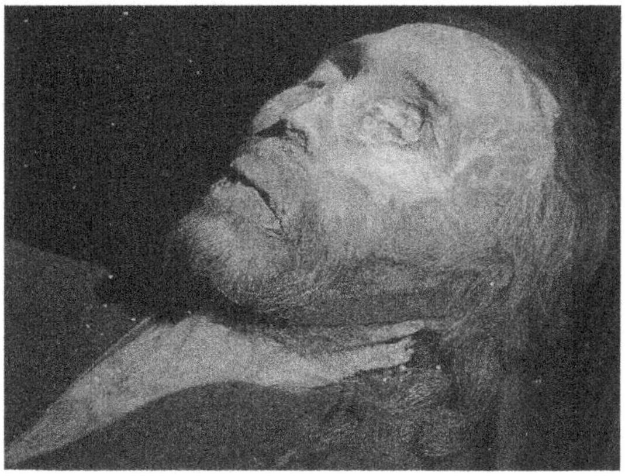

Figure 17.3 A well preserved mummy of a 3000 year old man discovered in the Tarim Basin of western China.

The Aryans initiated the Vedic hymns and formed the priesthood that continued their composition and memorization for thousands of years. The responsibility was passed down to the highest caste of their descendants, the Brahmins, who became the stewards of this sacred body of knowledge and much later were also the authors of the Hindu myths that were based on the same cosmic events.

To the best of our knowledge a pure strain of this Aryan people no longer exists in the Indus Valley or anywhere else in the world. If it does, these people are not presently recognized. However, well preserved mummies of a unique Caucasoid people have been found in the Tarim Basin of western China - north of Tibet, which may be direct descendants of these same Aryans. These are a tall people, with blonde or red hair, and long noses. See Figure 17.3. The mummies of these people have been preserved, not by special preparation of the bodies, but by the cold, extremely dry climate of the area in which they were buried. The soft tissue, skin and even the clothes are extremely well preserved.

These finds have forced a reexamination of historical or legendary Chinese stories of figures with great height, with deep-set blue or green eyes, long noses, and full beards of red or blonde hair. They are known as the Tocharians, but scholars believe they had been known in earlier times as the Yuezhi or Greater Yeuzhi. Stories about these people were suppressed by the imperialistic leadership of China in the late 19th and early 20th centuries. Some scholars that believe the modern Chinese race evolved from an intermixing of this race with local, indigenous peoples in ancient times. Consistent with description of the Aryan race, archaeological evidence indicates that these were a peaceful, egalitarian people, as evidenced by the complete lack of weaponry and few indications of status differences at their burial sites. Even though these mummies are dated at only 1000 B.C., their preserved bodies may be the closest we can come to finding pure Aryan DNA.[182] Unfortunately our biotechnology does not currently understand enough about human DNA to determine how the minds of these people might have differed from those of the pre-Vedic people.

The Man in the Ice

In 1991, the well preserved body of a very ancient man was recovered from a glacier in the Otztaler Alps, providing a wealth of information about neolithic man that could never have been compiled otherwise. Of particular interest to the V/A theory is the age of the body as determined by radio carbon dating. Fifteen samples from the body itself and the grass stuffed in his shoes give a range of dates from 3365 to 2940 B.C., with an average of 3300 B.C.! *This is the uncorrected radiocarbon date for the beginning of the Bronze Age, BA I. This places his death at exactly the time at which our theory claims the devastations due to proto-Venus took place.* X-rays indicate that *Otzi*, as he was named by the examining scientists, had several broken ribs which had little time to heal before his demise. One conclusion, based on the study of other forensic evidence is:[183]

> The most likely explanation, however, is an accident shortly before the Iceman's death.
>
> If this injury is indeed viewed as the result of a violent conflict, this would support a scenario which explains the Iceman's inadequate equipment. His physical impairment due to the pain must have been a serious handicap. Nevertheless, he set off for an inhospitable region, and without sufficient food. All this can be regarded as an indication of 'flight with his last remaining strength'

The suggested scenario leading to his death is not ours and the author cannot relate Otzi's plight to any specific event. But the V/A scenario provides one, exactly at the date of his demise. It is quite possible that this man was a witness to and perhaps, for a short time, a wounded survivor of one of the proto-Venus encounters. Evidence derived from microscopic furrows in a finger nail is interpreted as showing stress, disease, or malnourishment in the four months before his death. We imagine that he somehow survived the encounter, perhaps because he was in a cave, but succumbed to cold or the noxious sulfurous gases from proto-Venus' smoke tail. A *clue in favor of the latter scenario is the fact that the interior of his lungs were blackened by smoke.* Or perhaps he saw the blazing planet approaching and made haste for the mountains with whatever equipment and food he had available. When he saw the tidal wave, kilometers high, approaching, he may have taken shelter between some rocks where the wave hit him, broke his ribs and covered him over. The water left from the passage of the wave might have quickly frozen and preserved his body in the ice until it was discovered by mountain climbers. Perhaps forensic science can tell us more, given these clues. For instance, a spectrographic analysis of the black material coating his lungs might reveal the presence of an unusual element, compound or isotope ratio - similar perhaps to ones found in the Gray Burnished Ware from the same period in the Levant. This could very well have been volcanic ash from the 'tail' of the blazing proto-Venus which engulfed the Earth at the time of the close encounters.

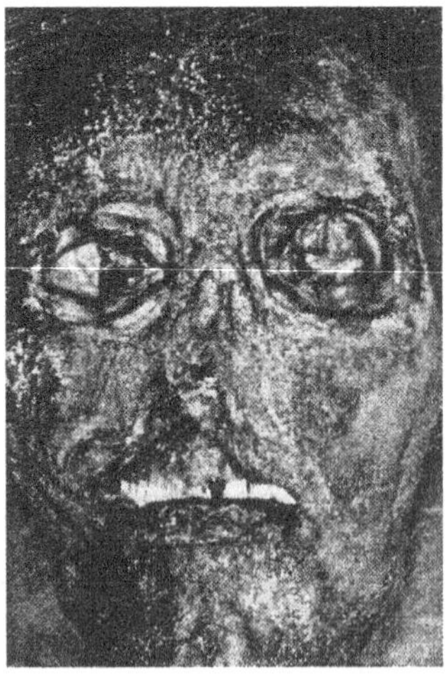

Figure 17.4 The face of Otzi showing indications of a hare lip.

In addition to this anecdotal data, there are some physiological characteristics of Otzi which have a bearing on the question of the advent of modern man on the Earth at the beginning of the Vedic period. The upper jaw of Otzi exhibits a four millimeter *diastema* or space between the two first incisors and an unusually broad *canalis incisivum* which leads to the palate behind the incisors. This is indicative of what is colloquially known as a 'hare lip,' as implied by the photograph in Figure 17.4. This is a very common defect found in at least twenty percent of neolithic skulls. Furthermore, he has only 11 pairs of ribs instead of the usual 12, an abnormally thick bony ridge above the eyes, and shin bones that are thinner and longer than normal. We suggest that these attributes are anatomical evidence of a fundamental genetic difference between the pre-Vedic hominid species and modern man. Even more striking is the contrast between the stately Tocharians and the diminutive neolithic people exemplified by Otzi.

Vach, the Vedic God of Speech

One of the faculties most closely related to intelligence in the modern world is speech, which is indicative of a more profound mental capability of conceptualization. Indeed, from our perspective it is impossible to conceive of one without the other. In keeping with our approach of seeking truth in the ancient myths, we find a startling clue to the advent of language in the *Rig Veda*. The *rishis* offered praise to a rather obscure deity known as '*Vach*.' This is generally thought to refer to *speech*. But what could possibly have caused it to assume such great importance to the Vedic people that they declared it a goddess? Initially we attributed this deity to a sound that originated in the body of the Earth at the times of capture and release of priori-Mars, due to the great tidal stresses, such as the 'trumpets' which sounded when the walls of Jericho fell. However, the *Rig Veda*, literally 'knowledge,' tells us in unequivocal terms of a profound change that took place involving language at the beginning of the Vedic period:

> There are, no doubt, in the hymns some apparent traces of an idea that the authors were inspired; as in the following texts: R.V.i.37,4: "Sing the god-given prayer." In R.V.x.71,3, it is said of the goddess Vach: "By sacrifice they followed the track of Vach: they found her residing in the rishis [priests]."
>
> In R.V. x. 125, 5, again Vach is made to say: Every man whom I love, I make him terrible; [I make] him a priest; [I make] him a rishi; [I make] him intelligent."
>
> In a Valakhilya (or apocryphal hymn) ... the following verse occurs: "Indra and Varuna, I have seen through devotion that which, after it was heard in the beginning, you gave to the poets - wisdom, understanding of speech;
>
> ... the following curious passage of an ancient Brahmana grammar, also, by indicating the crude forms and the affixes, is useful for determining the character of words and their signification. And accordingly it is thus related ... :

'Vach (Speech) spoke confusedly, and without articulation. The gods said to Indra, Make this Vach to become articulate to us ... Indra then, dividing Speech in sunder in the middle, rendered her articulate. Hence she is spoken articulately.' The sense of this quotation, says Sayana, is this: ***Speech, such as the ..(first verse in the Rigveda), was originally confused, i.e. unvaried like the roar of the sea, etc., and undistinguished, i.e. without articulation to denote crude [primitive] forms, inflections, words, and sentences, etc. Then Indra, being solicited by the gods ... divided in the middle speech, which had previously been without division, and introduced everywhere the distinction of crude forms, inflections, etc. In consequence, this Speech, being now distinguished in its parts ... is pronounced by all men."***[184]

As with most of the new ideas comprising the V/A hypothesis, the notion that speech was an important part of the change imparted to humans at the beginning of the Vedic period is not mere idle speculation. Taken literally this implies *that the genetic leap which occurred at the beginning of the Vedic period, involving the Aryan race, gave both intelligence and the very first capability of speech to man.* The *Rig Veda* clearly implies that Indra (priori-Mars), which first came on the scene after the proto-Venus destruction, *gave man the ability to speak for the first time in more than grunts and gestures.* The *sudden introduction of speech at this date, to a people who never before had such an ability, would explain why it was elevated to the position of a goddess.*

It is significant that *Indra* was the first to give speech and the example given by the commentator Sayana uses the very first verse in the *Rig Veda*. The implication is that the refinement of speech was necessary to compose the hymns, and because it was given by Indra, must have occurred at the beginning of the Vedic period, because that was the date of the first Indra. Although credit is given to the deity Indra for the gift of language, it was the pure Aryans who were the first *rishis*, the priests that presided over the composition and maintenance of the *Rig Veda*, and therefore they must have been among the first modern humans who were given language. Were the new race of intelligent beings taught in classes given by the 'elohiym or was it infused as part of the genetic upgrade in a way that is still beyond our ken? We suggest the latter. This is reinforced by another biblical story concerning the 'tower of Babel', in which the power of the Lord over language was dramatically demonstrated.

If this was indeed the case then much more than just a rebuilding of the population was taking place during the pre-urban period. There may be an unintended irony in the naming of this the 'silent period,' because it very well may have been that *this was the period during which the first articulated language was used on the Earth. Indeed, language was probably the single most important manifestation of the increased intelligence because it made possible the definition and consensus on abstract concepts, that made possible the creation of the great civilizations in B.A. II.*

To the best of our knowledge no anthropologists have considered the possibility that neolithic man did not possess the capability of articulated speech prior to the Bronze Age. One reason is undoubtedly due to the fact that these people could weave cloth, smelt copper using arsenic, and carry on trade, as evidenced by the equipment carried by Otzi. *But there is no proof of any link between these capabilities and the use of language.* The fact that the chalcolithic people had a kind of 'technology' *does not necessarily mean they had the ability to articulate speech.* Today we view the situation from a biased point of view since all of our institutions and learning are language based. But these were primitive people and could easily have passed on their experiential knowledge by example and gesture.

In the more traditional view, anthropologists consider language to have developed slowly with 'intelligence' over a period of at least 150,000 years. They measure 'technology' in terms of the quality and variety of spear points and scrapers. It is strictly an assumption that the development of language paralleled these technological achievements. This epoch is considered significant because the cranial cavity had grown almost to its current volume. Considerable debate has been devoted to other anatomical parameters that are thought to be related to articulated speech, but, as most concisely stated by John Shea, of SUNY, Stony Brook:

Speech does not fossilize

Currently, the most radical school of thought is one that favors the 'sudden' appearance of language about 40,000 years ago. This is based in large degree on the sudden appearance of regionally unique stone and bone scrapers and spear heads at that date in Europe, replacing the cruder stone-chipped tools used before that time. Scientists were so impressed by this sudden change that they named this the *Upper Paleolithic explosion*, attributing it to a major increase in intelligence and even the development of articulated speech.

We cannot accept the notion that a new variety of stone and bone scrapers is sufficient justification for claiming the advent of Homo Sapiens intelligence, and certainly not articulated speech. Indeed, we are not convinced that the chalcolithic people, which populated the Levant prior to the Vedic period, needed or were capable of articulated speech. *It is difficult to even compare the significance of changes of stone and bone implements with the rise of major civilizations that appeared for the first time in the Early Bronze II.* The latter certainly has the stronger claim to marking the advent of modern intelligence and with it the development of language.

We have argued earlier that one reason for the dominance of the proto-Indo-European language was because most of the people living at the end of the Chalcolithic period were killed by the proto-Venus encounters - the implication being that any archaic languages died with them. But an even more compelling reason now presents itself. The *Rig Veda* states that the first articulated speech was introduced at the time of the first Indra, that is, the first priori-Mars-Earth encounter. In that case the dominance of this language was *ensured* because at the time of its introduction *there were no competing languages. It was the first language ever spoken on earth.*

A possible clue to the late advent of language in Homo Sapiens Sapiens may lay in the development of each modern human being. It is a known fact that human babies are incapable of real speech at birth - not because of lack of intelligence but because the position of the voice box makes it impossible. It is only after a number of months of physical development that the larynx drops to a position at which articulated speech is physically possible. It is said that the stages of human development in the womb, and even after birth, mimic the process of evolution through ancient life forms, that culminated in man. If this is true, then perhaps the fact that the capability of speech in humans develops *months after birth,* that is, well after intelligence is established, implies that speech was the most recent step in the development of modern man.

The Development of Life on priori-Mars and the Earth

Counter to the currently accepted theory that all the planets were created at the same time the V/A theory postulates that the age of each terrestrial planet is unique. This follows logically from the creation of proto-Venus only 6,000 years ago. This demonstrated the process whereby each terrestrial body was formed. One which is completely unknown to the planetary science community. Because the ages of the oldest Martian meteorites are calculated to be 4.7 billion years, while the oldest Earth rock is only 3.9 billion years, we conclude that *priori-Mars was created hundreds of millions of years before the Earth.* Scientists who believe that all the planets are the same age attribute this difference to plate tectonics having recycled the most ancient terrestrial rock back into the interior. We claim that the odds that every single rock formation was recycled, are small, and that the oldest rocks have been found. Thus the age of the Earth is approximately 3.9 billion years.

Additional evidence comes from the work of two Australian geophysists who published a paper in 1999, in which they use the Niobium/Thorium ratios in many ancient terrestrial rock layers to estimate the amount of crustal rock on Earth as a function of time. They conclude that there was no significant crustal rock prior to 3.8 billion years ago. In their abstract they state:

> Increase in the Niobium/Thorium ratio shows involvement of hydrated lithosphere in differentiation of Earth since approximately 3.8 billion years ago.[185]

It is difficult to believe that the Earth existed for some 800 million years before the crustal rock started to develop, and if so why did this not happen on Mars? Based on this evidence we claim that the age of the Earth is about 3.9 billion years old while that of Mars (actually priori-Mars) is 4.7 billion years old. *This implies that priori-Mars wais 800 million years older than the Earth.*

According to the V/A scenario, priori-Mars' original inner orbit provided plenty of sunlight, and, as clearly shown by NASA images, was blessed with rivers, lakes and rainfall. Moreover, the *Rig Veda* and Egyptian painting of Anubis indicate that the southern highlands were covered with green vegetation in the early part of the Vedic period. Based on this we conclude that priori-Mars harbored abundant life well into the Vedic period. *Its earlier origin implies that life began to evolve there some hundreds of millions of years before life on Earth.* As a result it is probable that there were not only intelligent beings living on priori-Mars at the beginning of the Vedic period, but that they were far more advanced than the neolithic beings on Earth at that time.

Our hypothesis regarding the beings living on the two planets at that time must also consider the effect of the devastating encounters of proto-Venus with both bodies. We postulate that all but a very few neolithic specimens were killed by two such encounters. Many races were completely obliterated. This left the Earth essentially 'a world for the taking.'

Based on the *Rig Veda*, priori-Mars suffered 'innumerable' encounters with proto-Venus. In fact, those encounters were responsible for displacing priori-Mars from its original orbit in the inner solar system, to one that intersected the Earth's orbit. Therefore any beings on priori-Mars at the time would already have been hard pressed, not only because of 'marsquakes' and tidal flooding, but also due to loss of the water and atmosphere to proto-Venus. Once priori-Mars was captured in orbit around the Earth the first time, these beings would have to have realized that further deterioration of their planet, due to the tidal force of the larger Earth during the dance encounters, was inevitable.

Assuming their superior intelligence, it would not have been a difficult task for these advanced Martians to traverse the short distance to the Earth while the two planets orbited one another. The long term proximity of the two planets, the stronger gravitational force of the Earth, and the flow of atmospheric gases to the Earth during the many convulsive periods, would have greatly simplified the transfer problem.

Although it is a hackneyed theme from the realm of science fiction that advanced Martians, realizing their planet was dying, came to colonize the Earth - we suggest that it is actually true. *Our hypothesis of 'dance encounters' adds a new twist to the plot, because in it, priori-Mars was essentially the 'spaceship' that carried them to the vicinity of the Earth and the Earth was both the destroyer of priori-Mars and the benefactor of those encounters..*

It is probable that the Martians had just witnessed the two overturnings of the Earth by proto-Venus and might have realized that these encounters had killed most of the beings alive on the Earth at the time. Assuming their higher intelligence, they also would have realized that the very encounter of their planet with proto-Venus which thrust it toward the Earth, also served the purpose of placing proto-Venus in an orbit that was no longer a threat to the Earth, so that it was a safe place to go. Once the systematic dismemberment of their world by the Earth encounters became a foregone conclusion, it would have been logical for the surviving Martians to cross the short distance between the two planets and establish a colony on the Earth, perhaps in the steppes of central Asia, or the Indus Valley, which were not too far from the Himalayas. It is interesting to note that the Tocharians, who may be most typical of the Aryans, were quite tall. This characteristic is one that might be expected of beings from a small planet, such as priori-Mars, on which the gravity is considerably less than at the surface of the Earth.

It is almost a certainty that these beings would have been considered 'gods' of some sort, because they were strikingly different in appearance from the neolithic people, exemplified by Otzi, and, of course, because they came from the sky in some sort of machines. But the Earth had recently been threatened by more awesome, cosmic gods such as proto-Venus and 'the column of smoke and fire' on priori-Mars, and these were considered 'true' gods. As a result these human-like Martians might have been called 'ordinary gods' (*'elohiym*) or something along those lines.

The Biblical Age of Mankind

We believe that the creation of man, described in Genesis 1 and 2, took place just after the complete destructions of the Earth by the two proto-Venus encounters, and that the darkness and flood from which God recreated the light and the solid land described the aftermath of those destructions. Is there any corroboration of this hypothesis in the Bible?

According to the Bible the ages of the Earth are (a) from Adam to Noah, also known as the antediluvian (before the flood), (b) the present (evil) world, and (c) the world to come. Genesis gives the years from the creation of Adam to the flood of Noah, from Noah to Abraham, and from Abraham to Christ, in terms of the ages of renown men in a continuous blood line. This dates the creation of man at 4064 B.C. The V/A theory postulates that the creation of modern man occurred in the aftermath of the proto-Venus encounters at the same date, i.e. 4000 B.C., which in turn corresponds to the beginning of the EB I, once the correction is applied to the radiocarbon 14 date. This brings these dates into almost exact agreement with the biblical age of man.

Corrections to the Radiocarbon 14 Dating Method

Scientists have come to recognize that there are large systematic errors in the radiocarbon 14 dating of ancient artifacts and remains and that these increase as we go back in time through the Vedic period. They have devised alternate means to estimate the size of the corrections that must be applied, but they have no idea why these corrections are required. The V/A scenario explains why these corrections are required. The fundamental assumption on which this method is based, is that the ratio of radioactive carbon 14 (^{14}C) to normal carbon 12 (^{12}C) in the atmosphere has remained constant over the entire period of interest, i.e. actually from the death of a living object in ancient times to the present. At that time the animal or plant ceased ingesting atmospheric gases including carbon dioxide. The ingested ^{14}C then begins to decay at a known rate while ^{12}C does not, thus the ratio of the two isotopes in the wood or bone of interest allows the age since death to be calculated.

However, the fundamental assumption on which the method is based is that the ratio $^{14}C/^{12}C$ in the atmosphere is the same today as it was in ancient times. Since the ^{14}C in the atmosphere also decays this implies that it has continually been produced in the atmosphere at the same rate throughout the time of interest. The production is thought to be the result of a two-stage process. In the first, high energy particles from the sun split the nuclei of atoms in the upper atmosphere and release neutrons. In the second, some of these neutrons enter the nuclei of Nitrogen (^{14}N) atoms, the most plentiful of the atmospheric gases, and eject a proton, producing ^{14}C, which quickly becomes incorporated into carbon dioxide gas.

Any past events which might have caused the $^{14}C/^{12}C$ ratio in the atmosphere as a whole to change between the death of the sample and today, would result in this method producing erroneous dates. *The change would have to be of global proportions in order to change the ratio in the entire atmosphere of the Earth. We claim that the source of these errors are the very events revealed in the V/A scenario - the interactions of proto-Venus and priori-Mars with the Earth - events of which modern science is blissfully unaware.*

The graph in Figure 17.5 shows unmistakably that the corrections required to bring the theoretical carbon 14 dates into alignment with the other sources changed monotonically during the entire Vedic Period, whereas before and after it the corrections remain relatively constant. It is of no surprise that the magnitude of the cosmic encounters proposed in the V/A scenario could produce just such fluctuations. Indeed, events of such magnitude would be required to change the entire atmosphere of the Earth. For example: (a) the proto-Venus interactions with the Earth, at which times the planet-sized tail of volcanic ash completely enveloped the Earth; (b) the many explosive exchanges of atmosphere from priori-Mars, which, because it was closer to the Sun than the Earth before the birth of proto-Venus, might have had a different $^{14}C/^{12}C$ ratio or a different proportion of Nitrogen in its atmosphere; and (c) the evidence from the Hindu myths that at least on one occasion, a naked fission reaction occurred at the top of the column of smoke and fire, lasting days or weeks, which would have resulted in the Earths atmosphere being bombarded by a massive flux of neutrons. These are all events which had the potential of modifying the $^{14}C/^{12}C$ ratio in our atmosphere, thereby upsetting the fundamental assumptions on which radiocarbon dating is based. For example, if they produced a decrease in the ratio, say by increasing the proportion of oxygen to nitrogen in the atmosphere, the method would now give anomalously younger ages for artifacts that originated before the decreases. This would mean that the actual dating of the beginning of the Bronze Age would have been earlier than given based on the current assumptions, bringing it into closer agreement with the biblical date - *exactly what has been determined by those specializing in the calibration of the method. Moreover, the error increases as the ages of objects increases throughout the Vedic period and is essentially zero after 700 B.C.*

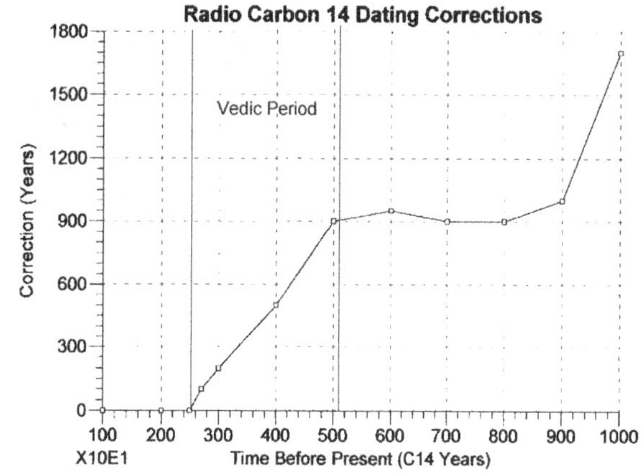

Figure 17.5 Currently accepted RadioCarbon 14 age corrections show a continuous change during the period of cosmic interactions proposed in the V/A scenario. Earlier changes were the result of cosmic interactions during the Upper Dryas.
Source:www.esd.ornl.gov/projects/qen/nerc14C.html

The Creation of Man

The clear implication of the biblical scriptures is that a superior race somehow 'created' modern man. Our interpretation is that they 'upgraded' the genes of the very few survivors of the destructive proto-Venus overturnings of the Earth:

> " 1:26 And the gods ('elohiym) said, Let us create man in our image, after our likeness ... 1:27 So the gods ('elohiym) created man in their *own* image, "in their own image created they him; male and female created they them1:28 And the gods blessed them, and the gods said unto them, Be fruitful, and multiply, and *replenish* the Earth, and subdue it:.."[186]

Where we have taken the liberty of translating 'elohiym in the plural, which is its actual form. Its translation given in Strong's Hebrew-Greek is:

> 430. 'elohiym, el-o-heem'; plur. of H433, gods in the ordinary sense, but spec. used (in the plur. thus, esp. with the art.) of the supreme God, occassionally applied by way of deference to magistrates, and sometimes as a superlative:--angels, X exceeding, God (gods) (-dess, -ly), X (very) great, judges, X mighty.

'Elohiym is a masculine plural that is followed by a feminine singular verb when referring to the Deity, making it a feminine name for God. But El is a masculine name of God. Therefore *'elohiym* is not only plural but is a nesting of the feminine within the masculine. The origin of this form is unknown and there is no absolutely correct interpretation. Many scholars have struggled to rationalize the plural form. For example:

> As long as God is still creating, He does not in fact say "I," He says "We," an absolute, all-inclusive term which does not refer to an I outside of the self but is the plural of an all-encompassing majesty. It is an impersonal, an I that does not face another Thou, that does not reveal anything but lives, like the metaphysical God of procreation, only in itself.[187]

Another interpretation is that the plural form is a reference to the Father, Son and Holy Ghost, the Holy Trinity.

Although the literal translation is 'ordinary gods,' the rabbis circumvented the problem by declaring that the use of the plural in this case really does not mean plural, but instead is a way of indicating 'extraordinary,' that is, the supreme God. It might be argued that the ordinary gods were merely an intelligent race from priori-Mars, but were not the real God community, that is, a people fulfilling the will of the supreme deity. This might be a justifiable interpretation of the Genesis 1 verses, because only the word *'elohiym* is present. But in the Genesis 2 verse below, the term *Yehovah 'elohiym*, indicating the supreme God, is used, apparently in reference to the same creation. We believe that the reconciliation of the concept of a single Godhead with the apparent plurality of the 'Kingdom of God' implied by Jesus Christ when he spoke of his disciples joining him in heaven, are beyond human ability to comprehend because there is no absolute consent and obedience in our world. It is not possible for mere man to really comprehend the mind of God and to even attempt it only exposes our ignorance and arrogance.

The fact that the literal translation of *'elohiym* is "gods in the ordinary sense" is consistent with our hypothesis that *a group of advanced people came to the Earth and either created or genetically modified the indigenous survivors of the proto-Venus encounters.* We believe the translation is a clue to allow us to understand that it was truly a group or community of Gods that transformed or created modern man. The concept of ordinary gods is consistent with the way that the neolithic people would have seen these arrivals from the sky. Although miraculous, they would not be afforded the same veneration reserved for the true gods, like proto-Venus, priori-Mars and the column of smoke and fire. Thus "ordinary gods" seems a most fitting description.

The 'creation' of modern man is described in a slightly different way in Genesis 2:7:

> And the Lord God formed Adam of the dust of the ground, and breathed into his nostrils the breath of life; and Adam became a living soul

Or, as we prefer to interpret it:

> And the 'elohiym took a neolithic 'man' of the earth and breathed into him a spirit; and he became a living soul, called Adam.

In our interpretation, the neolithic man would represent 'the Earth,' that is, the highest level attained by earthly evolution up to that time - but a species that was still profoundly inferior to modern man, like Otzi. The 'elohiym somehow infused the few survivors of that primitive species with a genetic upgrade, and sent the newly transformed beings out to 'replenish' or 'repopulate' the Earth. The notion of replenishing and repopulating the Earth is also consistent with our scenario, because in it the Earth had just been decimated by proto-Venus.

It is also interesting to note that the Hebrew word for man, ***adam***, literally means 'ruddy' or 'having a reddish complexion,' since we know from the high state of their preservation that the tall mummies of the Tarim Basin also had red hair. We suggest that the mummies of the Tarim Basin may be purer descendants of the Aryans, who in turn may have been the first race of man created by the ordinary gods.

The human-like characteristics and the plurality of the gods is further established beginning in Genesis 6:

> " 1. When men began to multiply on the face of the ground, and daughters were born to them, 2. the sons of God ('elohiym) saw that they were fair, and they took to wife such of them as they chose. 3. Then the Lord said , "My spirit shall not abide in man for ever, for he is flesh, but his days shall be a hundred and twenty years. 4. There were giants on the earth in those days, and also afterward, when the sons of God came into the daughters of men, and they bore children to them. These were the mighty men that were of old, the men of renown."

Obviously the "sons of God" (the 'eloyhim?) were similar enough to men to interbreed and lived on the earth for quite a while. Still, the Lord (*Yehovah elohiym* used in this passage) acknowledges the base nature of men and declares that it will not always be so, perhaps referring to the flood of Noah. The change in the nature of man after Noah is obvious from the differences in the ages of the patriarchs when their children were conceived as compared to those before this date.

The Flood of Noah

Just as the 'silent period' is recognized by archaeologists at the beginning of the Bronze Age, another similar break is evident around 2250 to 2000 B.C. (Uncorrected carbon 14 date). In the Levant a number of coastal cities were completely destroyed and many unorthodox graves were dug. In the subsequent centuries major shifts in population took place, particularly into the Negev which was not populated before or after this brief period. Generally the destruction and population loss was not as great at higher elevations such as the Golan heights and in Transjordan. A profound decline took place in Egypt at the same time. Although some scholars have suggested that the destruction at this time was caused by a massive invasion from the East, most believe that it was due to a major environmental disaster, such as a flood. They are at a loss to explain the cause of such a destructive event at that time.

We suggest that this destruction was described as the flood of Noah in the Bible, and that it was caused, not by rain, but by a release of priori-Mars from its orbit around the Earth. During these releases, priori-Mars drifted westward from the Transhimalayas over the middle east, thereby accelerating the return of the waters to the Mediterranean. As discussed previously, it is probable that this event was the one in which the Atlantic Ocean first broke through the sill at Gibraltar and decimated the entire Mediterranean area with salt water. This disaster was likely to have been abetted by tidal waves induced by a major earthquake in the Mediterranean.

The timing of this event matches the biblical scenario quite well. Genesis gives some 1650 years from Adam to Noah. Taking the date of Adam as 4000 B.C., and subtracting 1650 gives 2350 B.C. which compares favorably with the corrected radiocarbon 14 date of this destructive event.

The Tower of Babel

The story of the tower of Babel as told in Genesis is one of the most difficult for Christians to understand, because it is not clear why the Lord reacted as He did. Why should God be upset by people working together to build a great building? Our interpretation is that the powerful leader, Nimrod, who commanded its construction, led his people in the worship of the 'column of smoke and fire' on priori-Mars. Indeed, the tower built by Nimrod was a reflection of the one visible on priori-Mars. The Bible is full of stories concerning the Lord's impatience with the worship of 'idols,' that is, the worship of priori-Mars and proto-Venus deities. We therefore suggest that His wrath was really directed against this in the story of Babel. However, the story is cast in a way that seems more modern, that is, the Lords impatience with the arrogance of man, and particularly scientists, who believe that they do not need God. He brings the building of the tower to a halt in a unique and imaginative way. Genesis, Chapter 11 begins:

> "1. And the whole earth was of one language, and of one speech.
> 2. And it came to pass, as they journeyed from the east, that they found a plain in the land of Shinar [Sumer] and they dwelt there ...
> 5. And the Lord came down to see the city and the tower which the children of man builded. 6. And the Lord said, Behold the people is

> one, and they all have one language; and this they begin to do: and now nothing will be restrained from them, which they wish to do. 7. ... let us go down, and there confound their language, that they may not understand one another's speech. 8. So the Lord scattered them abroad from thence upon the face of all the earth, and they left off to build the city. 9. Therefore is the name of it called Babel, because the Lord did there confound the language of all the earth: and from thence did the Lord scatter them abroad upon the face of the Earth

The very first verse points out that there was only one language at this date. Unfortunately this sheds little light on our hypothesis that the first language on Earth appeared with the advent of modern man, because the story of Babel followed the flood of Noah, and if indeed all men were killed except Noah and his immediate family, that alone would be a very good reason why only one language was spoken in the following centuries. But as mentioned above, the archaeological evidence does not support the notion that all men were killed by the flood of Noah. In the 1650 years between Adam and Noah mankind had already spread over a large portion of the world, and that flood did not affect the entire world.

The story of Babel corroborates evidence of a continuing diaspora at that date, prompted by the confounding of the language spoken at Babylon. Two factors of this story are of significance to our thesis. First, the language that was spoken in Sumeria, proto-Indo European, was closely related to that of the Kurgan peoples and was the precursor of most modern European languages. Second, this story illustrates the great power that the Lord (Yehovah 'elohiym) exercised over speech in men. The capability to confound the languages of the entire population of Babel implies that God also had the power to give language originally. This single historical event could have precipitated the dispersion of these people throughout Eurasia, each to his own designated place.

Myth and the Origin of Modern Man

We believe that the sacred myths are a source of profound knowledge. These books were not kept sacred for thousands of years without a reason. They are not idle shepard's tales. They are narratives of real events. All the evidence leads to a recent preparation of the Earth and a miraculous creation of modern man. The biblical creation story does not apply to some hominid species created *ex nihilo* millions of years ago, but rather to a recent creation in which intervention by a benevolent race produced a leap in the evolutionary process - one that may never have occurred otherwise. This leap included the gift of articulated speech. As a corollary, we claim that Genesis refers only to events that have occurred since EB I.

From the earliest biblical texts we conclude that the 'Elohiym were a community of 'ordinary gods,' that is, with the outward appearance of human beings. The V/A scenario strongly suggests that they came from priori-Mars at the time of its first capture in orbit around the Earth. They transformed the surviving sub-human chalcolithic people into Homo Sapiens Sapiens, or modern man, around 3300 B.C. and instilled in this new man the gift of language and speech for the first time on this Earth. The first generations of man had many of the characteristics of the gods that created them, were 'men of renown' and in mythology they were known as the Aryans. They spread out over the recently devastated Earth and while their genes dominated those of the few surviving indigenous peoples, the process of mixing gradually destroyed the purity of the original beings as told in the Bible. Hesiod, the chronicler of ancient Greek mythology states that the gods made the first race of humans from gold, the second race of silver, the third of bronze, then came the Homeric heroes and finally his own age. Perhaps the Tocharians of the Tarim Basin were from the gold or silver races of men. In spite of our facile sense of modern superiority we are of the last and most base age, about which Hesiod wrote:

> " Now is the race of iron, when never by day they cease from toil and woe, nor by night from being worn."

Creation versus Evolution

We believe the V/A scenario provides a means of resolving the long standing controversy between Evolutionism and Creationism. The former claims that Homo Sapiens Sapiens evolved from the lower animals based on Darwin's theory, the DNA basis of all forms of life on earth, and the gradually increasing complexity of ancient life-forms leading to the various hominid species, as evidence. The creationists interpret the biblical story literally, claiming that man is a separate creature, created 'from the Earth' by God at the beginning of time. Can there be a compromise solution to this debate?

We suggest that evolutionary processes on Earth produced hominid species as early as 2 million years ago, and eventually neolithic man, the highest attainment of which is exemplified by Otzi. However, even as late as EB I, these species were profoundly inferior to modern man. Indeed, it is doubtful that another 800 million years of random mutations introduced by DNA replication errors or cosmic rays, would ever have produced modern man. *Therefore, although the concept of evolution, exemplified by the theory of Darwin, and refined by modern science, may explain the development of species up to the time of the great destructions, it does not explain the advent of modern man and does not prove that we are directly evolved from the ancient hominid species.*

Recent isolation of the mitochondrial DNA of a Neanderthal man, interpreted as showing there is no connection between them and modern man, favors our interpretation, but is not the final word. Although this is useful in tracing the propagation of ancient races, it is not nuclear DNA, the DNA in the nucleus of each cell, which uniquely describes our very essence. All we can say is that this evidence *favors* the Bible story that modern man was indeed 'created,' not from the dust, but as a result of a purposeful intercession of a group of omniscient beings - from priori-Mars.

If one believes that the Bible is divinely inspired, then the conclusion inevitably follows that every passage is present for a specific purpose, even though modern man cannot understand the meanings. Perhaps the most radical expression of this idea comes from the mystical Christian philosopher Immanuel Swedenborg, who postulated that there is a homomorphism, a one to one mapping, of the Bible into the brain of every human being, such that the reading of the entire book *with a childlike acceptance of every word*, can result in a mystical transformation of the person to a higher spiritual level. One example of such a transformation is that of Edgar Casey, a mystical

American healer of the 1930's, who was afforded special powers as a result of devout reading and rereading of the Bible at age twelve.[188] The implications of his empowerment are truly staggering.

Mankind's Destiny?

A devout, world respected interpreter of the Bible, Herbert W. Armstrong, in his book *Mystery of the Ages*, claims that the Lord has promised man dominion over the entire universe, but that this power is as yet only that of the Lord who sustains the entire universe by his limitless divine power. Although men are currently below the angels who administer to us, unseen, only man has the potential to join the community or Kingdom of God. Upon their resurrection, the chosen few will join this select community, have bodies like ours, and live on a planet like the Earth, but comprise a spiritual aspect with the full powers of the Lord. The surprising aspect of his interpretation resides in *the proposed function of this god-community. It is to create new planets and revive decaying ones*:

> These scriptures indicate we shall impart life to billions and billions of dead planets, as life has been imparted to this earth."

This was truly an amazing statement coming from a man who had spent his entire life studying the Bible, and had no knowledge of or interest in science or technology. Dr. Armstrong suggests *that the gods are man-like, with physical bodies imbued with a highly spiritual nature, and have the power to create and modify planets.* The very idea of man-like beings having power so far beyond the capability of modern man, is unthinkable, in spite of naive articles about the 'greening of priori-Mars.' After thousands of years we have not been successful in significantly modifying any of the great deserts on our own planet, even those bordered by the oceans. What kind of power could these gods possibly have, especially ones that look like human beings, that would enable them to create or renew a planet?

The V/A scenario reveals a clue as to how such events might be set in motion. It requires not so much power, as it does omniscience. If one contemplates the entire sequence of events proposed in the V/A scenario, one comes to the realization that a single application of power to redirect an inert, massive 'traveler' in space some five thousand years ago, could have been the impetus for (a) the creation of a new planet, proto-Venus, to be used for habitation in the distant future, (b) the replenishment of the Earth's oceans and atmosphere, and (c) for the introduction of mankind into this beautiful, newly finished world. There can be no better example of omnipotence and omniscience that this.

Conventionally 'religious' Christians will be aghast at the 'blasphemy' of the plurality of Gods kingdom and their human-like appearance. But the Bible is full of statements and promises of the Kingdom of God in which the chosen ones will share, and in which the 'first fruits' are already assembled. And what human being can really understand his own resurrection in something other than a human body, situated in a beautiful place like the Earth. A deep spiritual makeup does not preclude a physical body. A spiritual being just has a better body and soul, one that remains young for thousands of years, perhaps needs no food because the body can literally absorb energy from the variations in the electromagnetic fields that surround us, and has great powers that stem from his spiritual aspect.

Jesus was always vague about the great powers that become ours in the Kingdom of God, but there are small clues around us if we are but conscious of them. When in a trance, Edgar Casey could travel in his mind to an address in another town and 'find' a patient who had written to him at from that address, look inside his body and diagnose his or her illness. In one instance he was able to locate a bottle of an old tonic that no one remembered how to make, on a shelf in a drugstore, far from home. Isolated powers are sometimes revealed in humans as a result of genetic flaws or head trauma. For example, so-called idiot savants can perform feats far beyond the capabilities of normal human beings, such as the instantaneous multiplying two twelve digit numbers in their heads, or, looking at a randomly dropped box of toothpicks and instantaneously counting them. We believe that these and many more powers have merely been 'turned off' in mortal man and lie inaccessible to us, behind the mortal veil.

We see no heresy in these notions of God. This book is part of the process of coming closer to the truth - revealing a better understanding of what the Bible was trying to tell us all along. We see the example of Jesus Christ as no less important an example of how we are implored to live our lives and His sacrifice no less the means by which we are made worthy of the great promise of resurrection. We see the Lord and his Kingdom of Saints as so much more real and no less omniscient or omnipotent.

God remains the greatest mystery. One that by definition can never really be understood by mortal men. He reveals himself in the Bible to the degree that man is capable of understanding. But no knowledge is imparted to one who reads with a 'sophisticated' or cynical perspective. Only when read with the open mind of a child can one absorb its truth. It is up to each person to read it and try to understand the meaning for themselves, but not to interpret, or worse yet, to impugn Gods motives or actions. The search for ones truth is too important to be left to some 'committee.' Listen to the rabbis, the theologians, but then humbly meditate and pray to find the answer yourself.

Some people in our culture are moving away from religion and worshiping the sacred cow of science. Although often cited as a substitute for religion, any truly intelligent scientist cannot seriously believe that it can provide a valid alternative. It is a puny endeavor of man. Every theory is imperfect, as is best known by its originator, and every mathematical model is biased to show what the author desires it to show. The very idea that this imperfect body of knowledge could somehow disprove the existence of God or act as a substitute is ludicrous. The more small details that scientists discover only increase the profound nature of the great mysteries. Perhaps the most arrogant scientists are the relativists who, using a few equations on the back of an envelope claim to comprehend the creation of the universe. If the greatest scientific minds of our century have not come to even a superficial understanding of the recent history of the earth and the planets in their own solar system, how can they imagine in their remotest dreams to plumb eternity. We would recommend the ***Bible*** over ***A Brief History of Time***.

Endnotes

1. *Original Sanskrit Texts*, Muir, John, Trubner & Co., 60 Paternoster Row, London, 1868, Part Third, pp.22-24.

2. *Ibid.*, p.24.

3. *Myth and Reality*, Eliade, Mircea, Harper and Row, New York,1963, p.6.

4. *Ibid.*, p.145.

5. *Ibid.*, p. 8

6. *The Holy Bible*, Isaiah 31:8-9 and 34:4-5

7. *Ibid.*, Daniel 2:21

8. "Could Mars Have Been an Inner Planet?",. Rose L. E., in *Velikovsky Reconsidered*, Warner Books, New York, N.Y., 1966, p. 137.

9. *The Religion of the Veda*, p 105.

10. *Rig Veda*, 1.164.46

11. 9, p.357.

12. 9,p. 248.

13. *Original Sanskrit Texts*, Muir, John, Trubner & Co., 60 Paternoster Row, London, 1868, Part Second, p. xiii.

14. *Rig Veda*, X-34-8,9

15. *Hindu Mythology*, Wilkins, W.D., Rupa & Co., New Delhi, India 1988, pp. 18 & 17.

16. *Ibid.*, p.20

17. *Original Sanskrit Texts*, Muir, John, Trubner & Co., 60 Paternoster Row, London, 1868,, Part Fourth, p.18.

18. *Hindu Mythology*, Wilkins, W.D., Rupa & Co., New Delhi, India 1988, p.21

19. *The Religion of the Veda*, p.48

20. *Hindu Mythology*, p. 27

21. *Hymns of the Rig Veda*, p.20
22. *The Religion of the Veda*, p.128.
23. *Hindu Mythology*, Wilkins, W. J., Rupa & Co., New Delhi, India, 1988, p.40
24. *The Religion of the Veda*, p.121
25. *The Heart of the Rig Veda*, p 70.
26. *Hymns from the Rig Veda*, p. 5
27. *Indian Mythology*, p. 15
28. *The Religion of the Veda*, p. 123
29. *Indian Mythology*, pp. 14,15
30. *Hinduism*, pp. 30-32.
31. *The Heart of the Rig Veda*, p. 70
32. *Hinduism*, p. 31
33. *The Heart of the Rig Veda*, p. 294
34. *The Religion of the Veda*, p.91
35. *Heart of the Rig Veda*, p. 350
36. *Ibid.*, p. 266
37. *Hindu Mythology*, p. 30
38. *The Religion of the Veda*, p. 112
39. *Hindu Mythology*, p. 32,33
40. *The Religion of the Veda*, p. 74
41. *Rig Veda* 1.92.6
42. *Rig Veda* 1.123.1,2
43. *Indian Mythology*, p. 22
44. *The Heart of the Rig Veda*, p. 392

45. *Indian Mythology*, p. 22

46. *The Heart of the Rig Veda*, p. 68

47. *Hinduism*, p. 26-28

48. *Ibid.*, p. 23,24

49. *Indian Mythology*, p.15

50. *Metrical Translations*, John Muir

51. *Hindu Mythology*, Wilkins, W. J., Rupa & Co., New Delhi, India, 1988, p. 57

52. *Ibid.*, p. 57

53. *Ibid.*, p. 64

54. *Ibid.*, pp. 65,66

55. *Ibid.*, p. 67

56. *Hinduism*, p.25

57. *The Heart of the Rigveda*, p. 183

58. *Ibid.*, p. 182

59. *Hindu Myths*, ed. Doniger, Wendy, Penguin Classics, New York, N.Y., 1975, p. 58.

60. *The Heart of the Rigveda*, p.213

61. *Hinduism*, p. 33

62. *Hindu Mythology*, Wilkins, W. J., Rupa & Co., New Delhi, India, 1988, p. 125

63. *Ibid.*, p. 126

64. *Ibid.*, p. 125

65. Ibid., p. 125

66. *Original Sanskrit Texts*, Muir, John, Trubner & Co., 60 Paternoster Row, London, Part First, p. .130

67. *Hymns from the Rigveda*, p. 34

68. Ibid., p. 18

69. *Hindu Mythology*, Wilkins, W. J., Rupa & Co. New Delhi, India, 1988, p. 69, footnote

70. *Hinduism*, p. 21

71. *The Heart of the Rig Veda*, p. 138

72. Ibid., p. 331

73. Ibid., p. 329

74. Ibid., p. 327

75. Ibid., p. 389

76. Ibid., p.387

77. *Hindu Mythology*, Wilkins, Gupa & Co., Calcutta, 1988, pp. 93-96

78. Ibid, p. 100

79. Ibid, p. 99-101

80. Ibid, pp. 104-105

81. *Original Sanskrit Texts*, Muir, John, Trubner & Co., 60 Paternoster Row, London, Part Third, p. 304

82. Ibid., Part First, p. 80

83. Ibid., p.41

84. Ibid., p.51

85. Ibid., p.52

86. Ibid., p.52, (footnote)

87. Ibid., pp. 9,10

88. Atharva-Veda, X.8,7.

89. *Original Sanskrit Texts*, Muir, John, Trubner & Co., 60 Paternoster Row, London, Part First, p.28-30

90. Ibid. Part Third, p.12

91. Ibid., Part First, p. 43-44

92. Unknown

93. *Original Sanskrit Texts*, Muir, John, Trubner & Co., 60 Paternoster Row, London, Part Third, p. 14

94. Ibid., Part First, p.49

95. *The Rigveda*, Kaegi, Adolf, p. 45-46

96. *Hindu Mythology*, Wilkins, W. J., Rupa & Co., Calcutta, India, 1988, p. 263.

97. Ibid, pp. 265,266.

98. Ibid, p.271.

99. *Hindu Mythology*, Wilkins, W. J., Rupa & Co., New Delhi, India, 1988, p.90

100. *Ibid.*, p. 93

101. Ibid., p. 277

102. Ibid, p. 121

103. Ibid., p. 161

104. Ibid., p.118

105. Ibid., p.118

106. Ibid., p.120

107. Ibid., p.99 (footnote)

108. Ibid., p.121

109. Ibid., p.134

110. Ibid., p.138-141

111. Ibid., pp. 129,130

112. Ibid., p. 144

113. Ibid., p. 142

114. Baker, V. et al., University of Arizona

115. *Original Sandskrit Texts*, Muir, John, Trubner & Co., 60 Paternoster Row, Part First, 1868, p. 495-504

116. Ibid., p. 489

117. Ibid., p. 494

118. Budge, E.A. Wallis, The Egyptian Book of the Dead, (The Papyrus of Ani), Dover Publications, Inc. New York, 1967, p. cxv

119. Osiris and the Egyptian Resurrection, Budge, E.A. Wallace, Dover Publications, Vol 1, Page 384.

120. Egyptian Myths, George Hart, University of Texas, 1992, p.48,49.

121. Ancient Art, The Metropolitan Museum of Art Bulletin, Spring 1992.

122. *Worlds in Collision* , Velikovsky, Immanuel, Dell, New York, N. Y., 1950, Page 111-112.

123. Ibid., pp. 247-250.

124. *The Flood from Heaven*, Zangger, E., William Morrow and Company, Inc., New York, p. 114.

125. *Timeaus and Critias*, by Plato, Penguin Classics, 1977, p..56.

126. *Divine Encounters*, Sitchen, Zecharia, Avon Books, New York, 1995.

127. Ibid

128. *Worlds in Collision*, Velikovsky, Immanuel, Dell, N.Y., 1965.

129. "Sternkunde und Sterndienst in Babel", Schaumberger, in Kugler, 3rd. supp., 307.

130. "Early Chinese Astronomy: A Sidetracked Science", Fleming, Stuart, *Archaeology*, Nov./Dec. 1987, p. 68.

131. "Early Chinese Observations and Modern Astronomy", Stevenson, Richard, *Sky and Telescope*, Dec. 1999, pp. 48-55.

132. Ibid, p. 19

133. *Himalayas*, Olshak, Gansser, Gruschke, and Buhrer, Facts on File, N.Y., 1987, p. 11.

134. *Ibid.*, p. 193.

135. *Dragons and Dynasties*, Yuan Ke, Penguin Books, 1993, p.15-21.

136. *Oriental Mythology*, Campbell, Joseph, Arkana, 1991, p.15.

137 *Fasti*, Ovid, (transl Frazer, 1931), II 11. 489 ff.

138. *Worlds in Collision,* Velikovsky, Immanuel, Dell, 1950, Chapter 8.

139. Ibid, p.344.

140. "Observations of the North Polar Region of priori-Mars from the priori-Mars Orbital Laser Altimeter", Zuber, Maria, et al., *Science*, **282**, 11 Dec. 1998., pp. 2053-2060.

141. "Earth Without a Moon", *Velikovsky Reconsidered,* Warner Books, 1976, p. 124.

142. Velikovsky Reconsidered, page

143. *Himalayas*, Olschak, Gansser, Gruschke, and Buhrer, Facts on File, N.Y., 1987, p.17.

144. "Partially Molten Middle Crust Beneath Southern Tibet: Synthesis of Project INDEPTH Results", Nelson, K.D. et al, Science, **274**, 6 Dec 1996.

145. "Vegetation and soil feedbacks on the response of the African monsoon to orbital forcing in the early to middle Holocene", Kutzbach, et al, Nature, **384**, p.623.

146. Arnold, Dorothea, as quoted in "Animals in Egyptian Art", by Angela M. H. Schuster, Archaeology, July/August 1995, p.67.

147. *Archaeology of the Land of the Bible*, Mazar, Amihai, Doubleday, 1992, p. 88.

148. *The Rotation of the Earth, A Geophysical Discussion*, Munk, Walter, H., and MacDonald, Gordon J. F., Cambridge University Press, 1960, p. 206, 248.

149. *Geology*, Anderson, William T., February 1997.

150. "Tectonics of Quaternary Time in Middle America", King, P.B., Fig. 4, page 836, *The Quaternary of the United States,* Wright & Frey eds., 1965, Princeton University Press.

151. *Original Sanskrit Texts*, Muir, J. Rupa & Co., New Delhi, India, 1868

152. *The Outline of History*, Wells, H.G., p.77.

153. "Indus Valley Inc.", Shanti Menon, *Discover*, December 1998, p.67.

154. *Ancient Civilizations, The Near East and Mesoamerica*, Lamberg-Karlovsky & Sabloff, Benjamin/Cummings, Reading Mass., p. 206.

155. "Partially Molten Middle Crust Beneath Southern Tibet: Synthesis of Project INDEPTH Results", Nelson, K. D. et al, Science, **274**, 6 dec 1996.

156. *Himalayas*, Olschak, Gansser, Gruschke & Buhrer, Facts on File, NY 1987, p. 28

157. *Ibid.*

158. *Ibid.*, p. 19

159. *Nature,* April 1989.

160. *Nature*, September, 1989.

161. *The Holy Bible*, Joshua 24:2

162. *Geology of Egypt* , Hume, F. W., Egyptian Survey of Egypt,1925-1948, pp. 3, 27.

163. "Excavating an Ancient Merchantman", Linder, Elisha, Biblical Archaeology Review, V.18, n.6, Nov/Dec 1992, p. 25-35.

164. "The Persistence of Phoenician Culture", Stern, Ephraim, Biblical Archaeology Review, v.19, 17.3, May/June, 1993, P. 39-49.

165. *Oceans*, Karl K. Turekian, Prentice-Hall Inc., Englewood Cliffs, New Jersey, USA, Figure 5-3, p. 79.

166. "Salty Brines on the Mediterranean Sea Floor", Wallman, K., E. Suess, G. H. Westbrook, G. Winckler, M. B. Cita,, *Nature*, **387**, 32,33.

167. *Continents in Collision*, ed. Miller, Russell, Time-Life Books, p.154.

168. *Submarine Geology*, Francis P. Shepard, Harper and Row, N.Y. 1973, p.425.

169. *Archaeology of the Land of the Bible*, Mazar, Amihai, Doubleday, 1992, p.198.

170. "The Hieratic City State", Joseph Campbell, *Parabola* (magazine), v. XVIII, n. 4, winter 1993, p. 40.

171. *Serpent in the Sky*, John Anthony West, The Theosophical Publishing House, Wheaton, Il, 1993.

172. *People of the Sea*, Dothan, Trude and Moshe, Macmillan, N.Y.,1992.

173. *The Athenian Agora - Excavations in the Heart of Classical Athens*, Camp, John M., Thames and Hudson Ltd., London, 1986.,p. 25.

174. *The Influence of Sea Power on Ancient History*, Starr, C.G., Oxford University Press, 1989. p. 7-8.

175. "Transitions in the Mid-Holocene", Sandweiss, D. H., Maasch, K. A., and Anderson, D. G.,, *Science*, **283**, p. 499.

176. Personal communication from Dan Sandweiss, Dept. of Anthropology and Institute for Quaternary Studies, University of Maine, Orono, ME, 04469.

177. , "An ~15,000-Year Record of El Nino-Driven Alluviation in Southwestern Ecuador", Rodbell, D. T. et al, *Science*, **283**, pp. 516-520.

178 *Worlds in Collision*, Immanuel Velikovsky,.Macmillan, N.Y., Part Second, Chapters 1 and 2.

179. *The Holy Bible*, II Kings 19:6; Isaiah 37:7

180. *The Holy Bible*, Isaiah 38:6-8 II Kings 20:9

181. *The History and Geography of Human Genes*, L. L. Cavalli- Sforza, P. Menozzi, and A. Piazza, Princeton University Press, Princeton N.J. 1994., Figure 5.11.3, p. 293.

182. *Mummies of the Tarim Basin*, Victor M. Mair, Archaeology, Mar/Apr 1995, p.28.

183. *The Man in the Ice*, Spindler, Konrad, Random House, 1994.

184. *Original Sanskrit Texts*, John Muir, v.2, p.211-213.

185. Collerson, K.D., and B. S. Kamber, "Evolution of the Continents and the Atmosphere Inferred from Th-U-Nb Systematics of the Depleted Mantle", *Science* **283**, p. 1519.

186. *The Holy Bible*, Genesis, Chapter1, Verses 26-28.

187. *The Star of Redemption*, Rozenzweig, Franz, trans. William W. Hallo, Holt, Reinhardt & Winston, New York, pp. 154,311.

188. *There Is a River*, pub. A. R. E., Virginia Beach, VA

Made in the USA
Monee, IL
08 November 2026

68020217R00164